ALLELOPATHY

CHEMISTRY AND
MODE OF ACTION OF
ALLELOCHEMICALS

ALLELOPATHY

CHEMISTRY AND MODE OF ACTION OF ALLELOCHEMICALS

EDITED BY
Francisco A. Macías
Juan C. G. Galindo
José M. G. Molinillo
Horace G. Cutler

CRC PRESS

Boca Raton   London   New York   Washington, D.C.

Library of Congress Cataloging-in-Publication Data

Allelopathy : chemistry and mode of action of allelochemicals / edited by Francisco A. Macías, Juan C.G. Galindo, José M.G. Molinillo, and Horace G. Cutler.
    p. cm.
    Includes bibliographical references and index.
    ISBN 0-8493-1964-1 (alk. paper)
        1. Allelochemicals. 2. Allelopathic agents. 3. Allelopathy. I. Macías, Francisco A., Galindo, Juan C.G., Molinillo, Jose M.G., and Cutler, Horace G.

QK898.A43A456 2003
571.9'2—dc21   2003055404

This book contains information obtained from authentic and highly regarded sources. Reprinted material is quoted with permission, and sources are indicated. A wide variety of references are listed. Reasonable efforts have been made to publish reliable data and information, but the author and the publisher cannot assume responsibility for the validity of all materials or for the consequences of their use.

Neither this book nor any part may be reproduced or transmitted in any form or by any means, electronic or mechanical, including photocopying, microfilming, and recording, or by any information storage or retrieval system, without prior permission in writing from the publisher.

All rights reserved. Authorization to photocopy items for internal or personal use, or the personal or internal use of specific clients, may be granted by CRC Press LLC, provided that $1.50 per page photocopied is paid directly to Copyright Clearance Center, 222 Rosewood Drive, Danvers, MA 01923 USA. The fee code for users of the Transactional Reporting Service is ISBN 0-8493-1964-1/04/$0.00+$1.50. The fee is subject to change without notice. For organizations that have been granted a photocopy license by the CCC, a separate system of payment has been arranged.

The consent of CRC Press LLC does not extend to copying for general distribution, for promotion, for creating new works, or for resale. Specific permission must be obtained in writing from CRC Press LLC for such copying.

Direct all inquiries to CRC Press LLC, 2000 N.W. Corporate Blvd., Boca Raton, Florida 33431.

Trademark Notice: Product or corporate names may be trademarks or registered trademarks, and are used only for identification and explanation, without intent to infringe.

Visit the CRC Press Web site at www.crcpress.com

© 2004 by CRC Press LLC

No claim to original U.S. Government works
International Standard Book Number 0-8493-1964-1
Library of Congress Card Number 2003055404
Printed in the United States of America 1 2 3 4 5 6 7 8 9 0
Printed on acid-free paper

To our beloved families

Preface

The development of the science of allelopathy may be likened to the genesis of a painting. The initial few sketches may be highly criticized as clumsy and amateurish, but as the work builds, the skeletal structure slowly becomes animated and the content more substantial. And, like art, the science is never truly finished but continues to grow. The case of the painter Degas gives an analogous example. Often, purchasers of his work would return to their homes, after an evening out, to discover their "Degas" missing. After some panic and considerable search, they found that Degas had visited their home during their absence because he had noted, on an earlier visit, that there was an unfinished element on the canvas. He had then "borrowed" the painting, added the missing information and, later, returned the work to its owners. However, we are not told how many times this happened to a singular painting.

In allelopathy, the canvas is handed down to each generation for further development. In its entirety, the progressive art should encompass observation, chemistry, and mode of action, culminating in practical application, the latter, of course, solving practical problems to the benefit of the general population.

Initially, most of the work in allelopathy was observational, and the science was chided by purists as being clumsy and somewhat lacking in hard content and proof. But in recent years, some of the chemical causes and effects for the allelopathic phenomenon have begun to take form. Essentially, this was the substance of *Recent Advances in Allelopathy*. Volume 1. *A Science for the Future*. (Eds. F.A. Macias, J.C.G. Galindo, J.M.G. Molinillo and H.G. Cutler. University of Cadiz Press. 1999). Indeed, that publication was a mix of both observational and chemical allelopathy, and it emanated from the First Symposium of the International Allelopathy Society (IAS), held in Cadiz, Spain, in September 1996. Essentially, the present work, *Allelopathy: Chemistry and Mode of Action of Allelochemicals* is Volume II in the continuing saga of allelopathy and the title is self explanatory.

Perhaps, in the future, a further volume will cover those discoveries that have made significant contribution in the application of allelochemicals and practices of importance, not only financially, but also aesthetically.

The Editors

Contributors

G. Aliotta. Dipartimento di Scienze della Vita, II Università di Napoli, via Vivaldi, 43-81100 Caserta, Italy.
e-mail: aliotta@unina.it

A. L. Anaya. Laboratorio de Alelopatía. Instituto de Ecología, UNAM. AP. 70-275. Ciudad Universitaria, 04510, México, D.F. México.
e-mail: alanaya@miranda.ecologia.unam.mx

G. Ayala-Cordero. Laboratorio de Alelopatía. Instituto de Ecología, UNAM. AP. 70-275. Ciudad Universitaria, 04510, México, D.F. México.

U. Blum. Department of Botany, North Carolina State University. Raleigh, NC 27695-7612. USA.
e-mail: udo_blum@ncsu.edu

D. Chinchilla. Departamento de Química Orgánica, Facultad de Ciencias. Universidad de Cádiz. Avda. República Saharaui s/n, Apdo. 40. 11510-Puerto Real, Cádiz, Spain.
e-mail: david.chinchilla@uca.es

C. Ciniglia. Dipartimento di Biologia Vegetale, Università degli Studi di Napoli Federico II, Via Foria 223-80139 Napoli, Italy.

T. Coba de la Peña. Departamento Fisiología y Bioquímica Vegetal. Centro de Ciencias Medioambientales. Consejo Superior de Investigaciones Científicas. 28006-Madrid, Spain.

R. Cruz-Ortega. Laboratorio de Alelopatía. Instituto de Ecología, UNAM. AP. 70-275. Ciudad Universitaria, 04510, México, D.F. México.
e-mail: rcruz@miranda.ecologia.unam.mx

H. G. Cutler. Southern School of Pharmacy, Mercer University. 3001 Mercer University Drive, Atlanta, GA 30341-4155. USA.
e-mail: Hcutler876@aol.com

S. J. Cutler. Southern School of Pharmacy, Mercer University. 3001 Mercer University Drive, Atlanta, GA 30341-4155. USA.
e-mail: Cutler_s@mercer.edu

F. E. Dayan. Natural Products Utilization Research Unit, Agricultural Research Service, United States Department of Agriculture. P. O. Box 8048, University, MS 38677. USA.
e-mail: fdayan@olemiss.edu

M. DellaGreca. Dipartimento di Chimica Organica e Biochimica, Università Federico II, Via Cynthia 4, I-80126 Napoli, Italy.
e-mail: dellagre@unina.it

S. O. Duke. Natural Products Utilization Research Unit, Agricultural Research Service, United States Department of Agriculture. P. O. Box 8048, University, MS 38677. USA.
e-mail: sduke@ars.usda.gov

F. A. Einhellig. Graduate College, Southwest Missouri State University. Springfield, MO 65804. USA.
e-mail: FrankEinhellig@smsu.edu

A. Fiorentino. Dipartimento di Scienze della Vita, Seconda Università di Napoli, Via Vivaldi 43, I-81100, Caserta, Italy.

J. C. G. Galindo. Departamento de Química Orgánica, Facultad de Ciencias. Universidad de Cádiz. Avda. República Saharaui s/n, Apdo. 40. 11510-Puerto Real, Cádiz, Spain.
e-mail: juancarlos.galindo@uca.es

M. D. García-Díaz. Departamento de Química Orgánica, Facultad de Ciencias. Universidad de Cádiz. Avda. República Saharaui s/n, Apdo. 40. 11510-Puerto Real, Cádiz, Spain.
e-mail: lola.garciadiaz@alum.uca.es

L. González. Laboratorio de Ecofisioloxía Vexetal, Facultade de Ciencias, Universidade de Vigo. Campus Lagoas-Marcosende s/n. Vigo, Spain.

T. Haig. School of Science and Technology, and Farrer Centre for Conservation Farming. Charles Sturt University, Wagga Wagga, NSW, 2678, Australia.
e-mail: thaig@csu.edu.au

H. Hao. Chinese Academy of Science, Shanghai Institute of Organic Chemistry, 354 Fenglin Road, 25#, Shanghai 200032, China, VR

R. E. Hoagland. Southern Weed Science Research Unit, Agricultural Research Service, United States Department of Agriculture. P. O. Box 350, Stoneville, MS 38776. USA.
e-mail: rhoagland@ars.usda.gov

M. Isidori. Dipartimento di Scienze della Vita, II Università di Napoli, Via Vivaldi 43, I-81100, Caserta, Italy.

J. Jorrín. Departamento de Bioquímica y Biología Molecular, ETSIAM, Universidad de Córdoba, Apdo. 3048. 14080 - Córdoba, Spain.
e-mail: bf1jonoj@uco.es

R. Ligrone. Dipartimento di Biologia Vegetale, Università degli Studi di Napoli Federico II, Via Foria 223-80139 Napoli, Italy.

F. A. Macías. Departamento de Química Orgánica, Facultad de Ciencias. Universidad de Cádiz. Avda. República Saharaui s/n, Apdo. 40. 11510-Puerto Real, Cádiz, Spain.
e-mail: famacias@uca.es

A. Martínez. Laboratorio de Ecofisioloxía Vexetal, Facultade de Ciencias, Universidade de Vigo. Campus Lagoas-Marcosende s/n. Vigo, Spain.

D. Matesic. Southern School of Pharmacy, Mercer University. 3001 Mercer University Drive, Atlanta, GA 30341-4155. USA.

J. M. G. Molinillo. Departamento de Química Orgánica, Facultad de Ciencias. Universidad de Cádiz. Avda. República Saharaui s/n, Apdo. 40. 11510-Puerto Real, Cádiz, Spain.
e-mail: chema.gonzalez@uca.es

N. P. D. Nanayakkara. National Center for Natural Products Research, Research Institute of Pharmaceutical Sciences, School of Pharmacy, University of Mississippi. University, MS 38677. USA.

A. Oliva. Department of Molecular Genetics and Microbiology, Life Science, 130. State University of New York. Stony Brook, NY 11794-5222. USA.

e-mail: anoliva@ms.cc.sunysb.edu

G. Pinto. Dipartimento di Biologia Vegetale, Università degli Studi di Napoli Federico II, Via Foria 223-80139 Napoli, Italy.

A. Pollio. Dipartimento di Biologia Vegetale, Università degli Studi di Napoli Federico II, Via Foria 223-80139 Napoli, Italy.

F. Pellisier. Laboratoire de Dynamique des Ecosystèmes d'Altitude, Université de Savoie. Cedex 73 376 Le Bourget-du-Lac, France.

e-mail: pellissier@univ-savoie.fr

M. J. Reigosa. Laboratorio de Ecofisioloxía Vexetal, Facultade de Ciencias, Universidade de Vigo. Campus Lagoas-Marcosende s/n. Vigo, Spain.

e-mail: mreigosa@uvigo.es

J. G. Romagni. Department of Biology. University of St. Thomas, Houston, TX 77006. USA.

e-mail: romagnj@stthom.edu

T. Romero-Romero. Laboratorio de Alelopatía. Instituto de Ecología, UNAM. AP. 70-275. Ciudad Universitaria, 04510, México, D.F. México.

R. C. Rosell. Department of Biology. University of St. Thomas, Houston, TX 77006. USA.

e-mail: rrosell@stthom.edu

A. M. Sánchez-Moreiras. Laboratorio de Ecofisioloxía Vexetal, Facultade de Ciencias, Universidade de Vigo. Campus Lagoas-Marcosende s/n. Vigo, Spain.

e-mail: adela@uvigo.es

M. Schulz. Institut für Landwirtschaftliche Botanik, Universität Bonn, Karlrobert Kreiten-Str. 13, 53115 Bonn, Germany.
e-mail: ulp509@uni-bonn.de

D. Sicker. Institut für Organische Chemie, Universität Leipzig, Johannisallee 29, 04103 Leipzig, Germany.
e-mail: sicker@organik.chemie.uni-leipzig.de

M. Stanzione. Dipartimento di Biologia Vegetale, Università degli Studi di Napoli Federico II, Via Foria 223-80139 Napoli, Italy.

G. R. Waller. Past-President, International Allelopathy Society. Department of Biochemistry and Molecular Biology, Oklahoma Agricultural Experiment Station, Oklahoma State University Still water, OK 74078-3035. USA.
e-mail: george.waller@cox.net

R. D. Williams. Agricultural Research Service, United States Department of Agriculture. Langston University. P. O. Box 730, Langston, OK 73050. USA.

M. Wink. Universität Heidelberg. Institut für Pharmazeutische Biologie. Im Neuenheimer Feld 364. D-69120 Heidelberg, Germany.
e-mail: wink@uni-hd.de

Contents

Introduction
    Reality and Future of Allelopathy 1
    G. R. Waller

Chapter 1
    Ecophysiology and Potential Modes of Action for Selected Lichen Secondary Metabolites 13
    J. G. Romagni, R. C. Rosell, N. P. D. Nanayakkara, and F. E. Dayan

Chapter 2
    Bioactive Compounds from Potamogetonaceae on Aquatic Organisms .. 35
    M. DellaGreca, A. Fiorentino, and M. Isidori

Chapter 3
    Fate of Phenolic Allelochemicals in Soils – the Role of Soil and Rhizosphere Microorganisms 57
    U. Blum

Chapter 4
    Benzoxazolin-2(3*H*)-ones – Generation, Effects and Detoxification in the Competition among Plants 77
    D. Sicker, H. Hao, and M. Schulz

Chapter 5
    Heliannanes– a Structure-Activity Relationship (SAR) Study 103
    F. A. Macías, J. M. G. Molinillo, D. Chinchilla and J. C. G. Galindo

Chapter 6
    Chemistry of Host-Parasite Interactions 125
    J. C. G. Galindo, F. A. Macías, M. D. García-Díaz, and J. Jorrín

Chapter 7
    Application of Analytical Techniques to the Determination of Allelopathic Agents in Wheat Root Exudates – Practical Case Study 149
    T. Haig

Chapter 8
　　The Importance of Alkaloidal Functions 163
　　M. S. Blum
Chapter 9
　　Allelochemical Properties of Quinolizidine Alkaloids 183
　　M. Wink
Chapter 10
　　Mode of Action of Phytotoxic Terpenoids 201
　　S. O. Duke and A. Oliva
Chapter 11
　　Mode of Allelochemical Action of Phenolic Compounds 217
　　F. A. Einhellig
Chapter 12
　　Mode of Action of the Hydroxamic Acid BOA and other Related
　　Compounds 239
　　A. M. Sánchez-Moreiras, T. Coba de la Peña, A. Martínez, L. González,
　　F. Pellisier, and M. J. Reigosa
Chapter 13
　　Mode of Action of Phytotoxic Fungal Metabolites 253
　　H. G. Cutler, S. J. Cutler, and D. Matesic
Chapter 14
　　Proteomic Techniques for the Study of Allelopathic Stress Produced by
　　Some Mexican Plants on Protein Patterns of Bean and Tomato Roots ...271
　　R. Cruz-Ortega, T. Romero-Romero, G. Ayala-Cordero, and A. L. Anaya
Chapter 15
　　Application of Microscopic Techniques to the Study of Seeds and
　　Microalgae under Olive Oil Wastewater Stress 289
　　G. Aliotta, R. Ligrone, C. Ciniglia, A. Pollio, M. Stanzione, and G. Pinto
Chapter 16
　　Bioassays – Useful Tools for the Study of Allelopathy 315
　　R. E. Hoagland and R. D. Williams
Index 353

Introduction–Reality and Future of Allelopathy

G. R. Waller

CONTENT

Abstract 1
Introduction 2
Reality 3
Food Production on Limited Resources 4
World Food Consumption 7
Future 8
Conclusions 10
References 11

ABSTRACT

The world's need for research and development in allelopathy in agriculture, forestry, and ecology will be outlined. The world's agricultural and forestry production, as well as the ecological dimensions in relation to population, calls for global changes to be brought about by allelopathy. It is important, I think, for us to emphasize the evolutionary nature of these changes in priorities. The judicial use of allelopathy reflects the new priorities and new values which are evolving within our society. Allelopathic interactions are based primarily on the production of secondary chemicals by higher plants that produce a wide array of biochemical compounds that create biological changes, many of which we are still trying to understand. Allelopathy can be a challenge to all disciplines. A team approach to solve these complicated problems is both important and necessary, since seldom can all of the research, development or production be accomplished by one group. We must work together to achieve our new goals in improving the quality of life through allelopathy.

INTRODUCTION

The world's need for research and development in allelopathy in agriculture, forestry, and ecology is of extreme urgency.[1-18] The world's agricultural and forestry production, as well as the ecological dimensions in relation to population, calls for global changes to be brought about by allelopathy in connection with the other disciplines that have been involved in successful changes. We wish to call attention to the myriad workers who have been using allelopathic principles in their production and preservation of natural resources, for without them the world's population could not have increased to 5 or 6 billion.

Allelopathy interactions are based primarily on the production of secondary chemicals by higher plants that produce a wide array of biochemical compounds that create biological changes, many of which we are still trying to understand. Allelopathy can be and is a challenge to all disciplines. A team approach to solve these complicated problems is both important and necessary, since seldom can all of the research, development, or production be accomplished by one group. We must work together to achieve our new goals in improving the quality of life through allelopathy.

Comparing apples with oranges is always chancy, even when they are in the same basket. But when one tries to compare one with the other and the baskets are continents apart, it seems necessary to make a few rationalizations to obtain a reliable comparison. When I was asked to talk on the reality and future of allelopathy at the First World Congress on Allelopathy: A Science for the Future, it seemed that reality and future were somewhat like the bushel of oranges and apples. After listening to so many diverse, but outstanding presentations during the week, I was again brought to the comparison of apples and oranges. Incidentally apples and oranges contain allelopathic compounds which (based on their concentration) exert favorable or unfavorable biological effects on the trees that produce them.

But I am finally getting smart enough to work out some of the problems of the International Allelopathy Society (IAS) so that we are able to leave this meeting with a new hope and a set of ideals that can lead to a more productive group of scientists. We have answered some of the questions about the need for an IAS. The big problem that we face is how to translate our allelopathy findings to more beneficial solutions that affect mankind in a positive manner. How do we do this today, in tomorrow's world? I have thought and wondered for the past two years wether if I have been "on the right track." When I look back, I can see failures but also a lot of satisfactory things have happened. The founding members of IAS worldwide have helped immensely in bringing together some of the scientists involved in allelopathy or those who want to be involved in allelopathy to establish a framework for IAS. This new group of scientists--- YOU --- hopes to demonstrate to our supporters (the individual administrations and governments involved) that we can make statements about allelopathy that we think prudent, important, and beneficial to mankind.

REALITY

In the 1930's, crop yields in the United States, England, India, and Argentina were essentially the same. Since that time, researchers, scientists, and a host of federal policies in each country have helped farmers dramatically increase yields of corn, wheat, soybeans, cotton, and most other major commodities. Today, fewer farmers feed more people than ever before. This success, however, has not come without costs.

The environmental protection agencies of most countries have identified agriculture as the largest nonpoint source of surface water pollution. This is a major problem in each country. Pesticides and nitrates from fertilizers are detected in the groundwater in many agricultural regions. Soil erosion is a concern in many countries. Pest resistance to pesticides continues to grow, and the problem of pesticide residues in food has yet to be resolved. All nations are more competitive in international markets than a few years ago.

Because of these concerns, some farmers have begun to adopt sustainable farming practices with the goals of reducing input costs, preserving the resource base, and protecting human health. These changes are occurring all over the world. The concern of the IAS is that the allelopathy component be recognized and made an integral part of the program of each country. We recognize that it is a problem, but the time is now to realize the importance of allelopathy in the world's agricultural and forestry supplies.

Many components of sustainable agriculture are derived from conventional agronomic practices; however, they do not include allelopathy for the most part. The hallmark of a sustainable farming approach is not the conventional practices it rejects but the innovative practices it includes. In contrast to conventional farming, however, sustainable systems more deliberately integrate and take advantage of naturally occurring beneficial interactions between organisms, which means they recognize allelopathy but under different names. Sustainable agriculture systems emphasize management of biological relationships, such as those between the pest and predator, and natural processes, such as natural nitrogen fixation instead of chemically intensive methods. The objective is to sustain and enhance rather than reduce and simplify the biological interactions on which productive agriculture depends, thereby reducing the harmful off-farm effects of production practices.

Sustainable agriculture is *not* a single system of farming practices. It includes a spectrum of practical farming methods, ranging from organic systems that attempt to use no purchased synthetic chemical inputs to those involving the prudent use of pesticides or antibiotics to control specific pests or diseases. Alternative farming encompasses but is not limited to farming systems known as biological, low-input, organic, regenerative, or sustainable. It includes a range of practices such as integrated pest management; low-intensity animal production systems; crop rotations designed to reduce pest damage, improve crop health, decrease soil erosion, and in the case of legumes, fix nitrogen in the soil; and tillage and planting practices that reduce soil erosion and help control weeds. Successful farmers incorporate these and other practices into their farming operations. Farmers that practice sustainable agriculture do what all good managers do: they apply management skills and information to reduce costs, improve efficiency, and maintain production levels worldwide.

The evolutionary process is slow, and likewise the development and incorporation of allelopathy into our understanding of sustainable agriculture (which includes forestry) will proceed at rates that will be slower than we would like them to be. It is important, I think, for us to emphasize the evolutionary nature of these changes in priorities. The judicial use of allelopathy reflects the new priorities, as well as the new values which are evolving within our society.

FOOD PRODUCTION ON LIMITED RESOURCES

We have 32.5 billion acres of land in the world. Only 24% or 8 billion acres is potentially suitable for cultivation. The important groups of world food crops (cereal, food legumes, and oilseeds) utilize over 2 billion acres, producing over 1 billion metric tons of food per year (Table I.1).

About 70% of the land (23 billion acres) cannot be used for food production. This land is located where it is either too cold, too dry, or too steep, or the soil is too thin (Table 2). About 10%, or 3.2 billion acres, of our best agricultural land is developed for food production. There is another 20% or 6.5 billion acres in pasture and meadow which has the potential for cultivation but at greater costs. You can see that if the 6.5 billion acres in pasture and meadows are put in cultivation, that will bring us up to 9.7 billion acres. This is an important factor if we haven't developed control of the world population by 2025-2050 AD.

Table I.1

World food production (estimated).

Crop Group	Acres (Millions)	Metric Tons Produced (Millions)
Cereals	1734	1138
Food Legumes	156	40
Oil Seeds	279	98
Total	2169	1276

There are certain restraints to the production of food and other agricultural products. These are the effects of fertilizer, weather, pestilence, water (including irrigation), soil, energy, variety of new crops, and temperature (for example, compare Tibet (cold) and Sahara (hottest and driest, 1800 miles north of the equator), which are at the same latitude (30° N). However, Tibet has a polar climate). We might ask ourselves how much allelopathy influences the world's soil resources?

Insects, weeds, disease, and rodents destroy 30% of the world's food supply. In developing countries, the crop losses may be even higher. The World Health Organization estimates that about 12,000 people starve *daily*; that is 4.4 million per year. Not only is the waste of food inexcusable, but it represents a waste of the energy used for production.

The regions where all factors of climate and soil are favorable are generally where food will have to be produced (Table I.2). There are about 8 billion acres of potentially arable land in the world, but we are cultivating less than 4 billion acres. Most of these areas are already in production, so in most places there is little room for further land development, and the world must depend on reducing other barriers.

At present there are over 5-6 billion people in the world, so this means the food for each person is produced on less than one acre.

Water from precipitation is shielded from regions by mountain ranges. In some instances, regions are dry because prevailing winds move from continent to ocean and do not bring moisture into the region. Agricultural production is found in regions where water is available, either by precipitation or irrigation, and good temperatures prevail. Since the climate and vegetation are component parts of the soil formation process, the best soils have evolved under favorable climatic conditions. Was allelopathy involved? Furthermore, agriculture requires lands suitable for cultivation. Since climate and vegetation are not the only component parts of soil formation, not all regions with favourable climate and water are arable. These unknown factors will have an effect based on allelopathy. We do not know what the allelopathy effect is with respect to qualitative and quantitative measurement.

Table I.2

World soil resources.

% Total	Billions Acres	Situation
20	6.5	Too Cold
20	6.5	Too Dry
20	6.5	Too Steep
10	3.2	Soil Layer Too Thin
10	3.2	Used for Crops
20	6.5	Pasture and Meadows

It may even be that water in surplus (storms and floods) may cause yield reduction and put a limit on production of a large region. The Mississippi River flooding in the United States that occurred in 1996 cost several billions of dollars in industrial, domestic, and farm losses. Could allelopathy have prevented the flooding? I doubt it, but improved knowledge would have helped alleviate some of these human problems.

There are reports of the relation between annual rainfall and production of crops within regions. While these data are interesting, they are not very useful in providing estimates of worldwide production, since averages tend to prevail anyway. When one region has good rainfall, another will be deficient. Allelopathy can and does have an important and perhaps dominant role in this situation.

Water supply may be enhanced in some regions by reducing non-productive evaporation through mulching, micro-windbreaks (for example, one row of sorghum to ten rows of peanuts), and using natural reflectivity of crops, narrow-row-spacing, and wide-furrow-spacing irrigation. Some of these do not cost money and are available for water conservation to producers in impoverished areas. Allelopathy certainly plays an important role in crop production in these agricultural situations.

WORLD FOOD CONSUMPTION

The world uses about one-half of the land area potentially available for crop production, but most of the additional land lies outside densely populated countries. This could mean that increased food production will come from continuing and strengthening research, development, and extension programs to provide increasing yields. Table I.3 shows the approximate world food consumption broken down into developing and developed countries. Farmers produce food no matter where they are located.

Table I.3

World food consumption.

	Calories/per Person/Day
Developed countries	3043
Developing countries	2097
Average of the world	2386

It is obvious to you and me that allelopathy has an enormous impact on the composition of the world food consumption. Can we make the case for allelopathy in each of our countries? I hope that we can!

Food is harvested year after year without exhausting the means for renewal. Some fields, such as some in Spain, have been farmed for thousands of years and are still productive. Any process that destroys the essential productivity of the soil must ultimately destroy the civilization that depends upon that soil; hence, we must have proper environmental protection devices such as the realization of the many uses of allelopathy. We prefer, based on research, extension, and experience, to use the land in such a way that it can be expected to produce indefinitely at a maximum level. That can be done only after we recognize that improving the quality of life through understanding allelopathy will enable us to focus on achieving our new goals.

FUTURE

In the final minutes remaining, I suggest that we look into the future — not too far, since my crystal ball remains cloudy — just to the year 2100 AD. This might be called "World Changes During the New Century that Affect Allelopathy." My suggestion is based on world harmony.

The period that we are living in is characterized by anguish over population, energy, food, and agricultural concerns, environmental matters, and economic conditions. I predict that the global problems will finally force the nations of the world, developed and developing, east and west, north and south, to recognize the importance of global cooperation. This must be brought about so that it forcefully changes the people of the world. The Asiatic, Australian, Arabian, and African people are teaching Europeans and the American people. The result will be an intensive effort in international cooperation by increasing the output of food production and agriculture, forestry, energy, industry, medicine, trade, and raw materials production; all are subjected to a tight control minimizing environmental problems while maximizing the quality of life. Does allelopathy have a role in the 21st Century? It most certainly does!

Our agricultural productivity will have increased sixfold. We will have twice the land in cultivation that we now have and the land will be three times as productive as it is now. There will still be meat and fish consumers and vegetable consumers, much as we have today. According to a former director of the National Science Foundation, we will have giant international agri-industrial centers based on solar power, fusion power, and deuterium from sea water, and built primarily in the arid areas of the world. I would add aquaculture, such as Singapore has, which relies on importing foods from Malaysia and other Asiatic countries that have been grown utilizing an aqueous medium rather than soil. You cannot build these industrial plants without recognizing that a key component is allelopathy! These visionary processes will come about when the right combination of technology, capital, and international cooperation is available to put together these centers.

A world agricultural system will be made possible through close and careful international cooperation. This system will have solved the world food problems on an immense cooperative scale.

A. A worldwide system of agricultural experimental research stations with affiliated agricultural extension services. These stations, working closely with weather and climate scientists, botanists, entomologists, experts in plant pathology, agronomists, agricultural engineers, horticulturists, foresters, animal scientists, biochemists, chemists, nutrition experts, and other individuals, will continually develop and improve new genetic strains of plants and animals to counteract natural changes. This means that allelochemicals are part of that integrated system. This system will have an enormous effect in reducing the huge losses of food previously destroyed to bad weather, plant diseases, insects, and rodents. Reductions in these losses, combined with higher yields made possible by better application of water and fertilizer, would allow the world to more than triple global agricultural production in less than 150 years. We would be able to compare old and new pieces of agricultural equipment, much of which was created especially for the developing nations, to be low-cost, labor-intensive, and designed for use on small but high-yielding, multiple-cropping farms. We would also see and perhaps be able to sample a variety of new foods.

B. A similar research and extension arrangement would apply to marine and fresh-water food production. Through international cooperation we will have thoroughly researched and charted the characteristics of the oceans that control their productivity. Although allelopathy is only in its infancy with respect to marine and fresh-water, this is an important area in which we will see more research, development, and extension. Fishing in open international water will be carefully regulated. In some seas we will have experimented with anchored and ocean-bottom power stations to create regulated upwellings to stimulate fish productivity. In many areas of the world we have highly productive inland agriculture systems that give a relatively high yield of protein per acre.

A and B are themes that permeate the Constituion and By-laws of the International Allelopathy Society. I hope that we can see these themes brought together as they relate to allelopathy.

C. We see displays of a system of international agricultural economic centers that would serve as the world's food banks. They are responsible for the regulation and exchange of food and agricultural commodities between nations, making certain that all countries are able to receive substantial nutrition in exchange for the nonfood agricultural commodities they could produce most efficiently on their type of land.

D. We will still be using fossil fuels, oil, gas and coal, but their usage will be curtailed because there will have been a dramatic increase of harnessing of solar energy, wind energy, fusion and fission energy, and other sources. We propose that fusion reactors may become the usable energy source of choice, because of minimum problems of disposal and because of uses of the fissionable products (tritium). These are less of a security risk than fission products (which are plutonium and uranium).

E. In health care we see a move over the last century and a quarter toward preventive and diagnostic medicine and comprehensive health care. With the help of extremely sensitive and accurate medical practice, pharmacology, biochemistry, and electronic and computer systems every individual's health is analyzed periodically from birth. Every family and person is counseled as to the best health regimen to follow based on tests and background. We have settled the problems of national values and controls of genetic matters and other new scientific and medical procedures related to human life. This required years of scientific investigation, ethical deliberation, and new legislation which affects all countries.

CONCLUSIONS

In conclusion, it seems to me that answers to the searching questions about the exploratory role of allelopathy in how it affects what is happening will unravel for use only if we put in our energy and time and hard work.

We have to stop finding reasons why they must be done. We must do this before we wipe ourselves out or wipe out what remaining faith we have in one another. The question remains whether we can civilize and humanize our international relations, not simply by improving our traditional way of doing things but also by devising and using new techniques and developing new attitudes within our capacity to meet our needs.

I propose the following, which was suggested to me by Russell Peterson,[18] former Chairman of the President's Council of Environmental Quality:

A Declaration of Interdependence

"We, the people of planet earth, with respect for the dignity of each human life, with concern for future generations, with growing appreciation of our relation to our environment, with recognition of limits to our resources, and with need for adequate food, air, water, shelter, health, protection, justice, and self-fulfillment, hereby declare our interdependence and resolve to work together in brotherhood and in harmony with our environment to enhance the quality of life everywhere."

If these broad concepts have a chance for growth, then perhaps also they will bring with them a new environmental quality which is all encompassing: growth, population, food and agriculture, energy, space, *allelopathy*, and quality of life.

We must remember that *change* is inevitable; *progress* is not! All of us believe in *change* through *progress*!

REFERENCES

(1) Brown, A. W. A., Byerly, T. C., Gibbs, M., San Pietro, A., **1975**. *Crop Productivity-Research Imperatives.* Michigan Agricultural Experiment Station, East Lansing, MI and Charles F. Kettering Foundation, Yellow Springs, OH, 399p.

(2) Brown, L. R., **1974**. In: *Bread Alone*, Praeger Publishers, Inc., New York, NY, 272p.

(3) Chou, C.-H. and Waller, G. R. **1983**. In: *Allelochemicals and Pheromones.* Institute of Botany, Academia Sinica, Taipei, Taiwan, 314p.

(4) Chou, C.-H. and Waller, G. R. **1989**. In: *Phytochemical Ecology: Allelochemicals, Mycotoxins, and Insect Pheromones and Allomones.* Institute of Botany, Academia Sinica, Taipei, Taiwan, 504p.

(5) Chou, C.-H., Waller, G. R., and Reinhardt, C. **1999**. In: *Biodiversity and Allelopathy: From Organisms to Ecosystems in the Pacific.* Institute of Botany, Academia Sinica, Taipei, Taiwan, 358p.

(6) Kohli, R. K., Singh, H. P., and Batish, D. R. **2001**. In: *Allelopathy in Agroecosystems.* Food Products Press, The Haworth Press, Inc., New York, NY, 447p.

(7) Inderjit, Dakshani, K. M. M., Einhellig, F. A. **1995**. In: *Allelopathy: Organisms, Processes, and Applications.* ACS Symposium Series 582, American Chemical Society, Washington, D. C., 382p.

(8) Inderjit, Dakshini, K. M. M., Foy, C. L. **1999**. In: *Principles and Practices in Plant Ecology: Allelochemical Interactions.* CRC Press, New York, NY, 589p.

(9) Macias, F. A., Galindo, J. C. G., Molinillo, J. M. G., and Cutler, Horace, G. **1999**. In: *Recent Advances in Allelopathy: A Science for the Future.* Servicio de Publicaciones de la Universidad de Cádiz, Cadiz, Spain, 515p.

(10) National Research Council. *Alternative Agriculture.* National Academy Press, Washington, D. C., 448p.

(11) Putnam, A.R. and Tang, C.-S. **1986**. In: *The Science of Allelopathy.* John Wiley and Sons, New York, NY, 317p.

(12) Reigosa, M. and Pedrol, N. **2002**. In: *Allelopathy: from Molecules to Ecosystems.* Science Publishers Inc., Enfield, NH, 316p.

(13) Rice, E.L. **1995**. In: *Biological Control of Weeds and Plant Diseases: Advances in Applied Allelopathy.* University of Oklahoma Press, Norman, OK, 439p.

(14) Rice, E.L. **1984**. In: *Allelopathy.* Second Edition. Academic Press, New York, NY, 442p.

(15) Thompson, A.C. **1985**. In: *The Chemistry of Allelopathy: Biochemical Interactions Among Plants.* ACS Symposium Series 268, American Chemical Society, Washington, D. C., 470p.

(16) Rizvi, S.J.H. and Rizvi, V. **1992**. In: *Allelopathy: Basic and Applied Aspects.* Chapman and Hall, New York, NY, 480p.

(17) Waller, G.R. **1987**. In: *Allelochemicals: Role in Agriculture and Forestry.* ACS Symposium Series, 330, American Chemical Society, Washington, D.C., 606p.

(18) Waller, G.R. and Edison, Jr., L.F. **1974**. *Growth with Environmental Quality?* McCormack-Armstrong Co. Inc., Wichita, KS, 524p.

1 Ecophysiology and Potential Modes of Action for Selected Lichen Secondary Metabolites

J. G. Romagni, R. C. Rosell, N. P. D. Nanayakkara, and F. E. Dayan

CONTENT

Abstract 13
Introduction 14
Results and Discussion 16
    Usnic Acid 16
    Anthraquinones 20
    Whitefly Bioassays 25
Methodology 26
References 30

ABSTRACT

Lichens, a symbiosis between a fungal and algal partner, produce secondary compounds that are unique to the symbiosis. Due to the high energy investment in these compounds, which can comprise up to 25% of the dry thallus weight, they must have an important role in lichen ecology. Our group is beginning to elucidate specific allelopathic roles and modes of action for these compounds. One lichen compound, (-)-usnic acid, was found to inhibit 4-hydroxyphenylpyruvate dioxygenase, a key enzyme in carotenoid biosynthesis. A series of lichen emodin analogues have been found to cause bleaching in grasses. Continued research suggested a decrease in photosystem II (PSII) activity, but the putative mode of action for these compounds remains to be determined. Another group of anthraquinone analogues has been found to inhibit germination and primary root formation. The preliminary data suggest that each lichen secondary compound has several ecological roles. Many inhibit pathways crucial for seedling development. This may decrease interspecific competition, especially in the canopy. Finally, we determined several compounds to be effective against phloem-feeding insects, particularly whiteflies (*Bemisia tabaci*). Both (-)-usnic acid and vulpinic acid caused highly significant mortality in whiteflies. Other functions of the same compounds, such as (-)-usnic, may include antiherbivory mechanisms.

INTRODUCTION

Lichens are a classic example of symbiosis. This partnership may contain up to three kingdoms, including a fungal (mycobiont) and algal and/or cyanobacterial (photobiont) partners. These organisms produce a variety of secondary compounds, most of which arise from the secondary metabolism of the fungal component and are deposited on the surface of the hyphae rather than compartmentalized in the cells. Many of these compounds are unique to lichens, with a small minority (*ca.* 60) occurring in other fungi or higher plants.[12] Due to a long history of chemotaxonomic study, the secondary chemistry of lichen compounds is better documented than in any other phylogenetic group; however, the bioactivity associated with these compounds has been generally ignored.

Of the more than 20,000 known species of lichens, only a few have been analyzed and identified as containing biologically active secondary compounds. Most of the unique secondary metabolites that are present in lichens are derived from the polyketide pathway, with a few originating from the shikimic acid and mevalonic acid pathways (Table 1.1). Previous studies have suggested that the *para*-depsides are precursors to *meta*-depsides, depsones, diphenyl ethers, depsidones and dibenzofurans.[9,12]

Lichen secondary products may comprise up to 20% of thallus dry weight,[15] although 5-10% is more common. Due to the high cost of carbon allocation, it is probable that these compounds have important ecological roles, either as protection against biotic factors such as herbivory[37] and competition or abiotic factors such as UV light.[15] Of those species tested, over 50% of them synthesize substances with some degree of antimicrobial activity. This may play some role in general lichen ecology and/or ecosystem dynamics. The antimicrobial activity, however, appears to be unrelated to other ecological roles, such as herbivory.[28]

Several anthraquinones with high antimicrobial activity have been isolated and characterized from some species in the lichen genus Xanthoria.[30]

Table 1.1

Major classes of secondary lichen metabolites.

Biosynthetic Origin	Chemical Class	Examples
Polyketide	Depsides	lecanoric acid
	Depsone	picrolichenic acid
	Depsidone	physodic acid
	Dibenzofurans	pannaric acid
	Usnic acids	usnic acid
	Chromones	sordinone, eugenitin
	Xanthones	lichexanthone
	Anthraquinone	emodin
Mevalonate	Diterpenes	16α-hydroxykaurane
	Triterpenes	zeorin
	Steroids	ergosterol
Shikimate	Terphenylquinones	polyporic acid
	Pulvinic acid	pulvinic acid

Source: From Elix, J. A. **1996**. *Lichen Biology*. Cambridge University Press, Cambridge, U.K. pp. 154-180. With permission.

Antiherbivory roles of metabolites have been well documented.[9,25,26] Proksch[31] reported that lichens produced secondary metabolites that acted as feeding deterrents which protected them from animal consumption. Several insects appeared to selectively avoid the medullary region, which contained most of the lichen metabolites, grazing primarily on the algal layer.[25]

Several lichen metabolites are known to inhibit the growth and development of fungal species. For example, crude aqueous extracts of lichens inhibit wood-decaying fungi, and other lichen products inhibit certain pathogenic fungi. Crude lichen extracts inhibit spore germination and may also cause decreased mycorrhyzal growth.[18,27,28]

The potential role of lichen metabolites in allelopathic interactions has recently been reviewed.[9,27] The phytotoxic effect of certain lichen metabolites may play a role in the establishment of lichen populations. The depsides, barbatic acid and lecanorin, and the tridepside, gyrophoric acid, have been shown to inhibit photosynthetic electron transport in isolated chloroplasts.[13,34] Another aspect of the allelopathic potential of lichens is related to the ability of (-)-usnic, one of the two enantiomers known to exist in nature, to inhibit carotenoid biosynthesis through the enzyme 4-hydroxyphenyl pyruvate dioxygenases.[33] The *in vitro* activity of usnic acid is superior to that of other synthetic inhibitors of this herbicide target site.

Despite these experimental results, the ecological impact of these lichen secondary metabolites is not well understood. Primary lichen successional species do not have fewer secondary compounds than do subsequent successional species. There are also those species that thrive although they do not have high levels of secondary metabolites. Some theories attempting to explain why certain species produce more secondary products than others include the possibility that those producing high levels of compounds are able to grow in more severe environments, such as those with limited nutrient supplies or those with high nitrogen and phosphorus content.[15]

The objectives of this paper are broad. Our first objective is to describe the primary mechanism of action of usnic acid on plants as ascertained by our laboratory.[33] A second objective is to describe the phytotoxic activity of selected lichen anthraquinone analogues. In addition to the phytotoxic activity, we describe the effects of these secondary metabolites on phloem-feeding insects. Finally, we provide a hypothesis to explain the functional roles of these metabolites in the ecosystem.

RESULTS AND DISCUSSION

USNIC ACID

(-)-Usnic acid [2,6-diacetyl-7,9-dihydroxy-8,9b-dimethyl-1,3(2H,9βH)-dibenzofurandione] is one of two naturally occurring biologically active enantiomers (Fig. 1.1) that are found in most yellow-green lichens. This compound is biosynthesized via the polyketide pathway and is categorized as either a dibenzofuran or triketone. The enantiomers, which differ in the orientation of the methyl group at 9*b* on the otherwise rigid molecule, have been identified as showing different biological activities and mechanisms of action. Usnic acid has been documented to have antihistamine, spasmolytic, antiviral, and antibacterial activities.[12] Proska et al.[32] reported that (-)-usnic acid inhibited urease and arginase activity. There are several reports[24] that the (+)-enantiomer is a more effective antimicrobial agent, although no specific mode of action was determined.

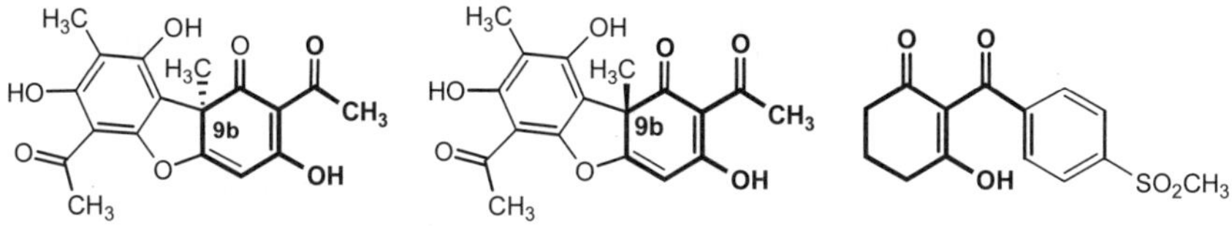

Figure 1.1

Structure illustrating triketone moiety of A. (-)-usnic, B. (+)-usnic, and C. sulcotrione.

Limited studies have documented phytotoxic effects of usnic acid including inhibition of transpiration and oxygen evolving processes in maize and sunflower seedlings.[23] Studies of mouse mitochondria have suggested that (+)-usnic acid uncouples oxidative phosphorylation at levels of 1 μM.[1] However, a definitive explanation of the phytotoxicity of usnic acid had, to our knowledge, never been reported. Thus, we have attempted to determine the phytotoxic mode of action for (-)-usnic acid.[33]

(-)-Usnic caused a dose-dependent bleaching of the cotyledonary tissues (Fig. 1.2) that ultimately led to the death of the seedlings, whereas (+)-usnic did not cause any significant changes in chlorophyll content. Loss of chlorophylls in response to phytotoxins can be associated with light-dependent destabilization of cellular and subcellular membranes, but usnic acid apparently acts differently since both enantiomers caused membrane leakage in the absence of light (Fig. 1.3).

Many photobleaching herbicides act by inhibiting the enzyme protoporphyrinogen oxidase (Protox), which catalyzes the last step in common between chlorophyll and heme biosynthesis. Usnic acid shares some structural features in common with these herbicides, such as the diphenyl ether scaffolding. The inhibitory activity of (-)-usnic acid on Protox was similar to that of the herbicide, acifluorfen, (I_{50} ca. 3 μM). However, these compounds did not displace acifluorfen from its binding site on Protox (data not shown), indicating that this natural product interacts with Protox differently than other photobleaching inhibitors.

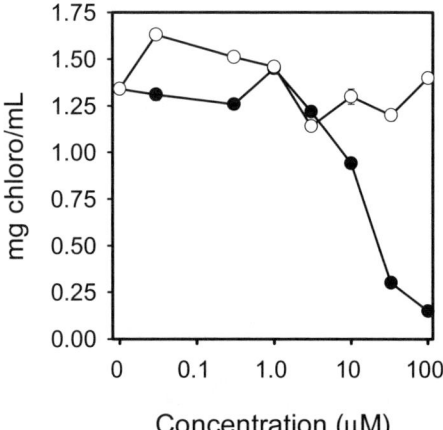

Figure 1.2

Effect of (-)-usnic acid (-●-) and (+)-usnic acid (-○-) (0.03 – 100 µM; no data for 0.1µM) on chlorophyll concentration in lettuce cotyledons after 6 days of growth.

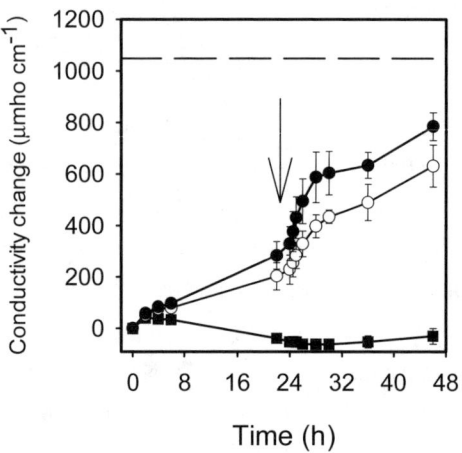

Figure 1.3

Effect of (-)-usnic acid (-●-) and (+)-usnic acid (-○-) on electrolyte leakage from cucumber cotyledons. The arrow represents the time when the samples were exposed to 325 µmol m^{-2} s^{-1} light. Dashed line represents maximum leakage obtained from boiled samples and leakage from untreated samples are shown (-■-).

Inhibitors of carotenoid synthesis also lead to chlorophyll destruction by destabilizing the photosynthetic apparatus. Total carotenoid content decreased with increased (-)-usnic concentration (Fig. 1.4). Carotenoid biosynthesis can be interrupted by inhibiting the enzyme phytoene desaturase that converts phytoene to carotenes or by inhibiting the enzyme HPPD responsible for plastoquinone (required for phytoene desaturase activity) synthesis.[14] Usnic acid possesses some of the structural features of the triketone HPPD inhibitors, such as sulcotrione (Fig. 1.1C).[8] (-)-Usnic acid had a strong inhibitory activity on HPPD, with an apparent IC_{50} of 70 nM, surpassing the activity obtained with the commercial herbicide sulcotrione (Fig. 1.5).

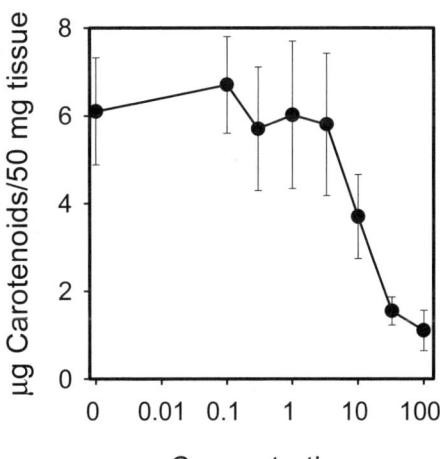

Figure 1.4
Effect of (-)-usnic acid (-●-) (0.1-100µM) on carotenoid concentration in lettuce cotyledons after 6 days of growth.

Carotenoids play an important role in quenching excess excitation energy in the photosynthetic apparatus. Under normal conditions, chlorophyll reaches the singlet excitation state following absorption of a photon. Chlorophyll transfers an electron to plastoquinone and returns to ground state rapidly by receiving an electron from the splitting of water. Under excessive light intensity, the photosynthetic transfer of electrons may become saturated, and chlorophyll can reach the more stable triplet state. Normally, this excess energy is transferred to carotenoids through intersystem crossing and is harmlessly released in a non-radiative way. In the absence of carotenoids, the photosynthetic apparatus is destabilized.[41] The excess energy from the chlorophyll in their triplet state is transferred to oxygen, causing formation of singlet oxygen. Singlet oxygen is highly reactive and causes bleaching of pigments and lipid peroxidation of membranes. As stated above, these symptoms (*e.g.* chlorophyll degradation and electrolyte leakage) were observed *in vivo* in seedlings treated with (-)-usnic acid, suggesting that the primary mechanism of action of this natural product is associated with inhibition of HPPD.

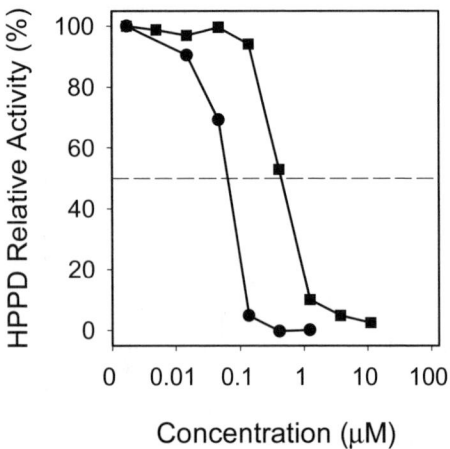

Figure 1.5

Effect of (-)-usnic acid (-●-) (0.01-100μM) on activity of HPPD. The activity of the herbicide sulcotrione was added for comparison (-■-).

Lichens containing usnic acid may exhibit phytotoxic activity. In nature, (-)-usnic acid may decrease interspecific competition by inhibiting growth of seedlings in the canopy. Since it is primarily an HPPD inhibitor and secondarily inhibits Protox, there may be a decreased chance for resistance in those competing species.

ANTHRAQUINONES

Emodin analogues

Emodin is a naturally occurring anthraquinone produced in many species of lichen, fungi, and higher plants (e.g., the genus, *Rhamnus*). Extracts containing emodin have been used in traditional medicine as herbal laxatives. It has also been identified as having antioxidant activity.[38] Emodin has been determined to be a nucleotide-binding site-directed inhibitor.[3] It has been reported to inhibit the enzyme CK2, a protein kinase originally misnamed casein kinase 2, with an IC_{50} value of 1 μM.[38]

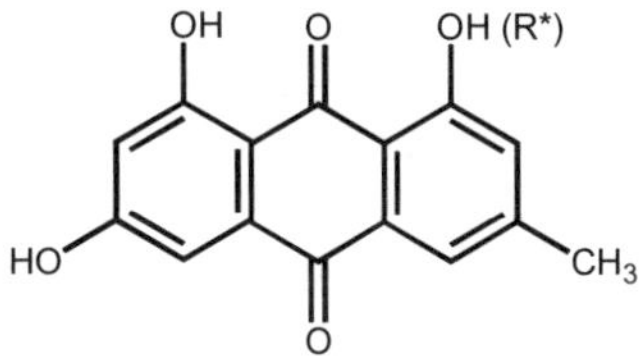

Figure 1.6
Structure of emodin with the substituted group noted (R*).

Emodin has been found to penetrate the active site of the α subunit of CK2, partially overlapping the ATP binding site, thereby preventing binding of the natural substrate.[39] It has also been identified as an inhibitor of tyrosine protein kinases, especially the receptor kinase HER-2 neu;[40] however, the IC_{50} value (21 μM) is much higher, suggesting that this may be a secondary effect.

Currently, we are testing two sets of analogues from emodin (Fig. 1.6). Series 1 consists of a group of compounds with aliphatic R-groups ending in a terminal hydroxyl. Series 2 has a terminal methyl.

All of the emodin analogues with a terminal hydroxyl caused selective dose-dependent bleaching in monocots (Fig. 1.7) that eventually led to the death of the seedlings, whereas those compounds containing a terminal methyl did not cause any significant changes in chlorophyll content. There was no effect on dicot tissues for either group. As previously mentioned, loss of chlorophylls in response to phytotoxins can be associated with light-dependent destabilization of cellular and subcellular membranes, but the anthraquinones apparently act differently since there was no significant leakage in the light (data not shown).

Previously, we determined that the only known mode of action that is selective for grasses/monocots is acetyl CoA carboxylase inhibition.[6] This enzyme is the first of two enzymes involved in *de novo* fatty acid biosynthesis. This mode of action prevents the synthesis of many essential wax compounds. In order to screen our compounds for this mode of action, we used resistant oat (*Avena* sp.) seeds. The emodin analogues caused dose-dependent bleaching (Fig. 1.7) and a severe decrease in germination for both resistant and nonresistant grasses.

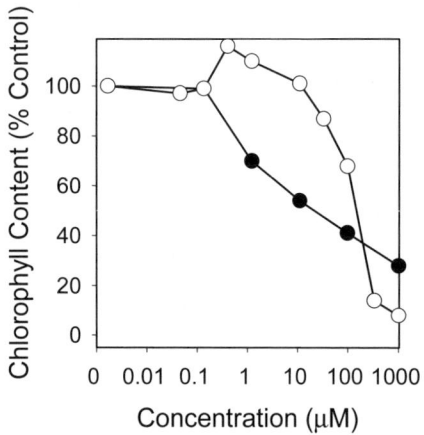

Figure 1.7

Average effect of emodin analogue (series 1) on (0.03 – 1000 µM) on chlorophyll concentration in ACCase resistant *Avena* sp. (-●-) and *Agrostis* sp. (-○-) after 7 days of growth.

Because bleaching can also be caused by inhibition of carotenoid biosynthesis, we tested our compounds for effects on carotenoid content. There was a dose-dependent increase in carotenoid content using the series 1 emodin analogues (Fig. 1.8). This was seen in both monocots and dicots and was not observed with the series 2 analogues.

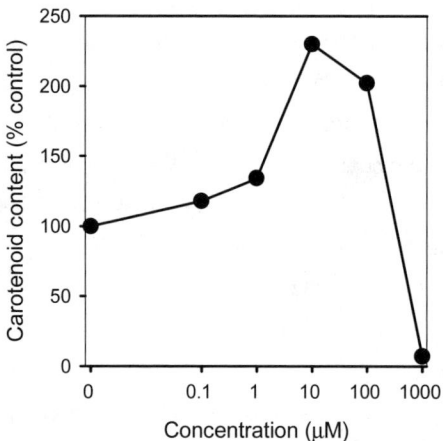

Figure 1.8

Effect of series 1 emodin analogues (0.1-1000µM) on carotenoid concentration in lettuce cotyledons after 6 days of growth.

Due to the anthraquinone moiety, we tested all compounds for photosystem II (PS II) inhibition using both spinach and corn thylakoids. By using both monocot and dicot thylakoids, we accounted for any differences in activity. There was no effect on PS II activity for series 2 analogues (-CH$_3$) (data not shown). However, for series 1 analogues (-OH), there was a 50% decrease in PS II activity at 0.1 µM in thylakoids isolated from spinach (Fig. 1.9A). This was similar for thylakoids isolated from corn (Fig. 1.9B).

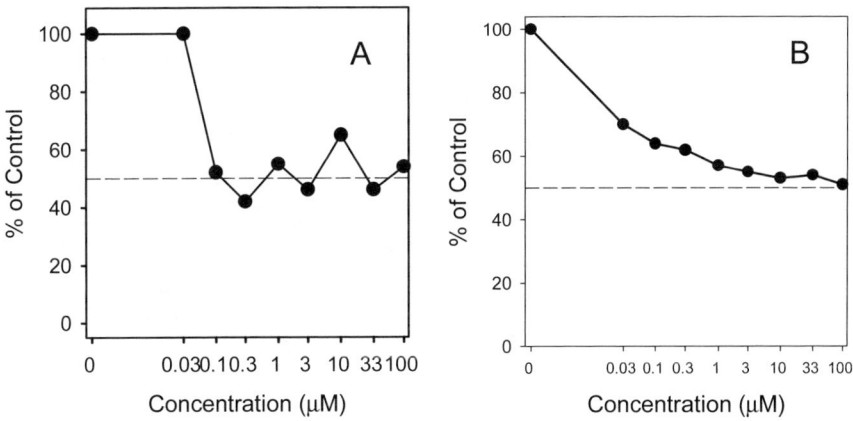

Figure 1.9

Effects of emodin analogues, series 1, on PS II activity. (A) Effects on PS II using spinach thylakoids. (B) Effects on PS II using corn thylakoids. Dashed line indicates 50% inhibition.

Rhodocladonic acid analogues

Rhodocladonic acid is an anthraquinone that occurs in several lichen species, especially in the family Roccellaceae).[20] Little research has been done documenting bioactivity, particularly phytotoxic activity. Similar to emodin, we tested two sets of analogues (Fig. 1.10). Series 1 consisted of a group of compounds with aliphatic R-groups ending in a terminal hydroxyl. Series 2 had a terminal methyl. The R-group substitutions were identical to those of emodin.

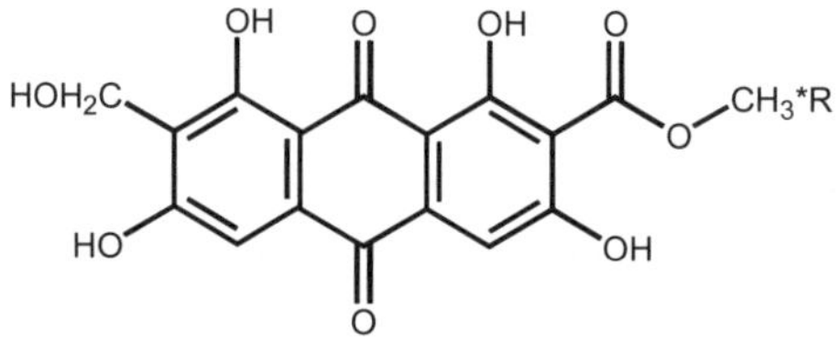

Figure 1.10
Structure of rhodocladonic acid with the substituted group noted (*R).

Unlike the emodin analogues, rhodocladonic acid analogues exhibited different phytotoxic effects for similar substitutions. There was no dose-dependent bleaching associated with any of these compounds. In addition, there was no selectivity in phytotoxic activity between monocots and dicots. Despite the lack of bleaching, structure-activity relationships would suggest the potential for PS II inhibition. Both series of Rhodocladonic acid analogues, those with terminal $-CH_3$ and those with a terminal $-$ OH (Fig. 1.11 A&B, respectively), caused an increase in PSII activity with regard to controls.

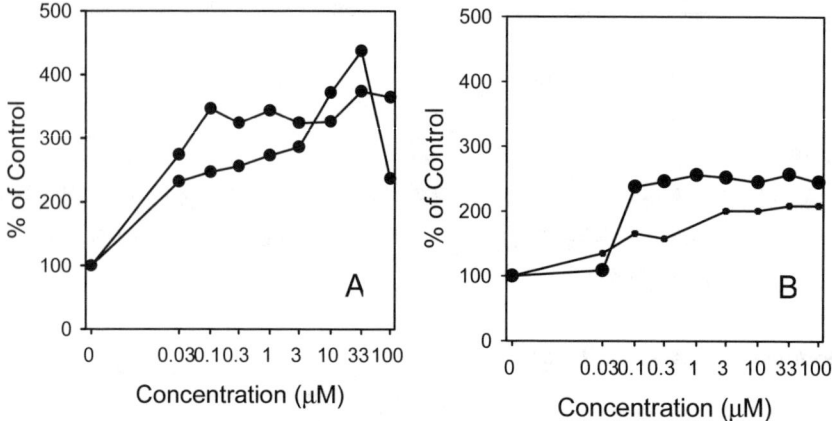

Figure 1.11
Effect of rhodocladonic acid analogues on PS II activity. Average activity of (A) series 2 analogues (terminal $-CH_3$); and (B) series 1 analogues (terminal $-OH$).

Although we have not yet determined the actual mode of action for these compounds, the roles these anthraquinone analogues play in nature is extremely diverse. Anthraquinones are known to inhibit cyt *b6f* in photosynthesis and cyt *b* in respiration. Depending upon minor substitutions, compounds with the same backbone can have different effects upon other organisms in the ecosystem. Some may inhibit respiration, an important mechanism against herbivores. Some anthraquinones may have allelopathic activity. Studies have determined that emodin influences the availability of soil ions, specifically by decreasing Mn^{2+} and increasing Na^{2+} and K^+.[21] We have determined that these anthraquinones were highly phytotoxic. Emodin analogues with a terminal –OH inhibited monocot growth and caused dose-dependent bleaching. Rhodocladonic acid analogues containing a terminal –CH_3 inhibited seedling germination and caused severe root malformation (data not shown).

WHITEFLY BIOASSAYS

Although not directly allelopathic, lichen metabolites have also been found to be effective deterrents for phloem-feeding insects, specifically *Bemisia tabaci* whiteflies, which are agricultural pests of temperate and subtropical environments.[5] They can reduce agricultural yield and crop loss through feeding damage, and they serve as vectors for plant viruses, in the genus Begomoviridae.[4] Without integrated pest management techniques that incorporate parasitoids, predators, diseases, and cultural manipulation of crop and chemicals to suppress whitefly populations, they can reach high density, causing leaf withering, premature dehiscence, and plant death.[7,17] Due to its global importance to agriculture, it is important to use novel strategies for control of *B. tabaci* because of increased resistance to many commercial insecticides after extensive applications in the field.[11] Thus, new, environmentally safer sources are needed to compensate for pesticidal demands, and plant secondary compounds are being tested which may prove useful as pesticides.

We used a leaf disk bioassay adapted from Rowland et al.[35] to test the insecticidal activity of lichen secondary products on *B. tabaci*. Two lichen secondary compounds, (-)-usnic acid and vulpinic acid, showed significant results when compared to the controls. Vulpinic acid had an average mortality of 18%, and (-)-usnic acid had an average mortality of 14%. From the dose response of (–)-usnic acid, LD50 was not reached at 1000µM, but a positive correlation was established with increasing concentration and whitefly population response (data not shown).

While the lichen compounds were not as effective as Isotox®, a known insecticide used as a positive control, in causing high mortality rates, the insecticidal activity could be further exploited synthetically with other active compounds or manipulated to form more active derivatives. By showing that lichen compounds do have insecticidal activity, a new area of natural compounds can be explored for effective control of insects that threaten crop yield.

METHODOLOGY

Dose-response assays

Twenty-five lettuce seeds (*Lactuca sativa* cv. iceberg), onion (*Allium cepa* L. cv. Evergreen Bunching), or a small scoop of bentgrass (*Agrostis tenuis* Sibth) were placed on sterile 55-mm dia. filter paper fitted to 60 x 15-mm petri dishes. Filter paper was premoistened with 3 mL of the test solutions. Stock solutions of test compounds were prepared in acetone to obtain final assay concentrations of 100, 33, 10, 3, 1, 0.3, 0.1, and 0.03 µM with volume of carrier solvent being <1% in the assay. Controls received similar amounts of acetone. The plates were maintained in a growth chamber (25°C, 200 µmol m^{-2} s^{-1}, 16/8 h light cycle) for six days. Growth was monitored by measuring root length in mm. Each treatment was triplicated and the experiment was repeated.

Determination of chlorophyll and carotenoid content

Chlorophyll was extracted from 40 mg of leaf tissue per treatment in 3 mL of dimethyl sulfoxide.[19] Total chlorophyll concentration was determined spectrophotometrically according to Arnon[2] measuring optical absorbance at 645 and 663 nm.

Chlorophyll (µg/mL) = 20.2(A_{645}) + 8.02(A_{663}).

Carotenoid analysis was determined spectrophotometrically using methods from Sandmann.[36] Under dim light, 20 mL of methanol containing 6% KOH was added to 40 mg of leaf material and heated to 60°C for 20 min. The methanol extract was poured into a separatory funnel with 10 mL of 10% diethyl ether (petroleum ether). 10 mL of saturated NaCl solution was added. The upper layer was collected, and the lower one was re-extracted with another 10 mL of diethyl ether/petrol. Total carotenoid content was determined in the combined upper layers by determining volume and measuring optical absorbance at 445 and 550 nm. The total amount of extracted carotenoids was calculated by the following equation:

$$\text{Carotenoids } (\mu g) = 3.92(A_{445} - A_{550}) * \text{volume [diethyl ether (mL)]}$$

Determination of chlorophyll / carotenoids spectra by HPLC

Treated and control cucumber cotyledons were collected in dim green light and homogenized in 2 mL of basic methanol (methanol: 0.1N NH_4OH; 9:1 v:v). The samples were centrifuged at 9000 x g for 20 min and the supernatant was collected. The pellet was resuspended in an additional 1 mL basic methanol and spun at 9000 x g for an additional 20 min. The resultant supernatant was collected and added to the original 2 mL, and filtered through a 0.2-µm nylon syringe filter before analysis by HPLC.

HPLC conditions

HPLC conditions were modified from Dayan et al.[10] The HPLC system was composed of Waters Associates (Milford, MA 01757, USA) components, which included a Model 510, pump, a Model 712 autosampler, a Millenium 2010 controller, and Models 470 fluorescence and 990 photodiode spectrophotometric detectors. The column was a 4.6 x 250 mm (ID) Spherisorb 5 µm ODS-1 reversed phase preceded by a Bio-Rad ODS-5S guard column. The solvent system consisted of a gradient beginning at 60% HPLC-grade methanol and 40% ddH_2O. At 10 mins, the gradient was 100% methanol, and at 30 mins, it returned to the original settings. The injection volume was 100 µL. Pigment detection was performed with fluorescence detector with excitation and emission wavelength settings at 440 and 630 nm, respectively, and the peaks were confirmed by scanning them from 300 to 700 nm with the photodiode array detector.

Effect of compounds on membrane integrity (leakage studies)

Cucumber seedlings (*Cucumis sativus* 'Long Green Improved') were grown in a growth chamber maintained at a temperature of 25°C and a photon flux density (PFPD) of 200 µmol s^{-1} m^{-2} continuous illumination. Electrolyte leakage induced by compounds was determined using 4 mm cotyledon discs as described by Kenyon et al.[22] The Petri dishes were placed in the dark at 25°C for 22 hours and then exposed to 325 µmol m^{-2} s^{-1} of light for the remainder of the experiment.

Protoporphyrinogen oxidase (protox) assay

Crude etioplast preparations were obtained from 10-d-old dark-grown barley seedlings according to the method of Dayan et al.[10] Barley seedlings (25 g) were homogenized on ice in 125 mL of grinding buffer [330mM sorbitol, 10 mM HEPES (pH 7.7 at 4°C), 1 mM EDTA, 1 mM MgCl$_2$, 5mM cysteine]. The homogenate was strained through cheesecloth and centrifuged for 20 min at 9000 x g. Each pellet was resuspended in 0.5 mL resuspension buffer [330 mM sorbitol, 10 mM HEPES (pH 7.7 at 4°C), 1 mM EDTA, 1mM MgCl$_2$, 1 mM DTT]. The suspension was diluted to ca. 4 mg protein/mL. Preparation of the enzyme substrate protoporphyrinogen and the Protox assay were performed as described by Dayan et al.[10] The substrate must be prepared fresh and kept cold.

To assay, incubate etioplast suspension with compound for at least 15 min on ice. Combine in a cuvette, 100 µL assay buffer [50 mM HEPES (pH 7.5), 25 mM EDTA], 2 µL 1M DTT, 20 µL substrate (protogen), 678 µL DIH$_2$O, and 100 µL etioplasts. Invert to mix and begin recording on a spectrofluorometer (Fixed λ; excitation = 395 nm; emit. = 622 nm) for 60 s.

Determination of p-hydroxyphenylpyruvate dioxygenase (HPPD) activity in vitro

Recombinant HPPD from A. thaliana was overexpressed in E. coli JM105 with pTrc 99A-AT4-HPPD plasmid grown as described by Maniatis et al.[29] Expression of the vector was induced by IPTG (1mM) when bacterial growth was equivalent to an A$_{600}$ of 0.6. The cells were incubated for another 17 h at 30°C and harvested by centrifugation (6,000 x g). The pellet was resuspended in buffer (20 mM potassium phosphate, pH 6.8, 1 mM EDTA, 1 mM DTT, 1 mM 6-aminohexanoic acid, 1 mM benzamidine), lysed by sonication (Branson Sonifier 450, Danbury, CT). A cell-free supernatant was obtained by centrifugation at 35,000 x g for 30 min.

The reaction mixture, consisting of 185 µL of assay buffer (50 mM sodium ascorbate in 100 mM Tris-HCl, pH 7.5) and 50 µg protein, was incubated for 15 min on ice with various concentrations of inhibitors. Compounds were tested at final concentrations ranging from 0.01 to 100 µM in semi-log increments. Controls received the same volume (4 µL) of solvent used to deliver the inhibitors. The reaction was initiated by adding 5 µL of 4-hydroxyphenylpyruvate (10 mM in methanol) for a total volume of 200 µL.

The reaction was stopped after 15 min incubation at 30°C by addition of 70 µL of 20% perchloric acid (v/v). The supernatant obtained after centrifugation (20,000 x g for 5 min) was subjected to HPLC analysis for the determination of homogentisic acid produced. The HPLC system was identical to the one described above, except chromatographic separation was obtained with a 3.9 mm x 15 cm Pico Tag reversed phase column preceded by a Bio-Rad ODS-5S guard column. The solvent system consisted of a linear gradient beginning at 0% (100% A) to 70% B from 0 to 17 min, 70% to 100% B from 17 to 20 min, 100% B from 20 to 24 min, 100% to 0% B from 24 to 28 min. and 0% B from 28 to 35 min. The flow rate was 1 mL/min and the injection volume was 100 µL. Solvent A was 0.1% (v/v) trifluoroacetic acid in ddH$_2$O and solvent B was 0.07% (v/v) trifluoroacetic acid in 80% (v/v) HPLC-grade acetonitrile / ddH$_2$O.

Whitefly bioassay

The leaf disk bioassay, adapted from Rowland et al.,[35] employed scintillation vials containing disks of pumpkin leaves that were dipped into test solutions. Twenty-five mating pairs of whiteflies were collected from pumpkin leaves, cooled at 4°C for 1 min to arrest movement, and then tapped into each vial. The vials were capped with dialysis membrane secured with a rubber band. After 48 hours, whitefly mortality was assessed in each chamber by observing insects under a dissecting microscope. Negative and positive controls were established with distilled water and Isotox©, a known pesticide of whiteflies. Lichen compounds were isolated according to Huneck and Yoshimura.[20] The final lichen compound solution concentration was 100µM. For the dose response, we varied the concentration of the secondary compounds from 10µM to 100µM to 1000µM.

ACKNOWLEDGMENTS

The authors would like to thank the students who provided excellent technical support, in particular, Debbie Aguilar, Christin Rivera, Chris Dauterive, Mai Le, Graciela Sanabria, Simy Parambil, and Jennifer Sutherland. We would also acknowledge partial support from the USDA-ARS Specific Cooperative Agreement #58-6408-1-002 and the University of St. Thomas Undergraduate Research Program.

REFERENCES

(1) Abo-Khatwa, A. N., Al-Robai, A. A., and Al-Jawhari, D. A., **1996**. Lichen acids as uncouplers of oxidative phosphorylation of mouse-liver mitochondria. *Nat. Toxins* **4**, 96-102.

(2) Arnon, D. I., **1949**. Copper Enzymes in isolated chloroplasts. Polyphenoloxidase in *Beta vulgaris*. *Plant Physiol.* **24**, 1-15.

(3) Battistutta, R., Sarno, S., De Moliner, E., Papinutto, E., Zanotti, G., and Pinna, L. A., **2000**. The replacement of ATP by the competitive inhibitor emodin induces conformational modifications in the catalytic site of protein kinase CK2. *J. Biol. Chem.* **275**, 29618-29622.

(4) Bedford, I. D., Markham, P. G., Brown, J. K., and Rosell, R. C., **1994**. Geminivirus transmission and biological characterization of whitefly (*Bemisia tabaci*) biotypes from different geographic regions. *Ann. Appl. Biol.* **125**, 311-325.

(5) Brown, J. K., Frohlich, D. R., and Rosell, R. C., **1995**. The sweetpotato or silverleaf whiteflies: biotypes of *Bemisia tabaci* or a species complex? *Annu. Rev. Entomol.* **40**, 511-534.

(6) Burton, J. D., Gronwald, J. W., Somers, D. A., Connelly, J. A., Gengenbach, B. G., and Wyse, D. L., **1987**. Inhibition of plant acetyl-coenzyme A carboxylase by the herbicides sethoxydim and haloxyfop. *Biochem. Biophys. Res. Comm.* **148**, 1039-1044.

(7) Byrne, D. N. and Bellows, T. S., **1991**. Whitefly biology. *Annu. Rev. Entomol.* **36**, 431-57.

(8) Dayan, F. E. and Allen, S. N., **2000**. Predicting the activity of the natural phytotoxic diphenyl ether cyperine using comparative molecular field analysis. *Pest Manag. Sci.* **56**, 717-722.

(9) Dayan, F. E. and Romagni, J. G., **2001**. Structural diversity of lichen metabolites and their potential use in pest management. In: *Advances in Microbial Toxin Research and Its Biotechnological Exploitation* (in press).

(10) Dayan F. E., Duke S. O., Reddy K. N., Hamper B. C., and Leschinsky, K. L., **1997**. Effects of isoxazole herbicides on protoporphyrinogen oxidase and porphyrin physiology. *J. Agric. Food Chem.* **45**, 967-975.

(11) Devine, G. J. and Denholm, I., **1998**. An unconventional use of piperonyl butoxide for managing the cotton whitefly, *Bemisia tabaci* (Hemiptera: Aleyrodidae). *Bull. Ent. Res.* **88**, 601-610.

(12) Elix, J. A., **1996**. Biochemistry and secondary metabolites. In: Nash III, T.H. (Ed), *Lichen Biology*. Cambridge University Press, Cambridge, U.K. 154-180.

(13) Ellis, M. K., Whitfield, A. C., Gowans, L. A., Auton, T. R., Provan, W. M., Lock, E. A., and Smith, L. L., **1995**. Inhibition of 4-hydroxyphenylpyruvate dioxygenase by 2-(2-nitro-4-trifluoromethylbenzoyl)-cyclohexane-1,3-dione and 2-(2-chloro-4-methanesulfonyl-benzoyl)-cyclohexane-1,3-dione. *Toxicol. Appl. Pharmacol.* **133**, 12-19.

(14) Endo, Y., Hayashi, H., Sato, T., Maruno, M., Ohta, T., and Nozoe, S., **1994**. Confluentic acid and 2'-O-methylperlatolic acid, monoamine oxidase B inhibitors in a Brazilian plant, *Himatanthus sucuuba*. *Chem. Pharm. Bull.* **42**, 1198-1201.

(15) Fahselt, D., **1996**. Individuals, populations and population ecology. In: Nash III, TH. (Ed), *Lichen Biology*. Cambridge University Press, Cambridge, U.K. 181-198.

(16) Fernandez, E., Reyes, A., Hidalgo, M. E., and Quilhot, W., **1998**. Photoprotector capacity of lichen metabolites assessed through the inhibition of the 8-methoxypsoralen photobinding to protein. *J. Photochem. Photobiol.* **42**, 195-201.

(17) Gerling, D., **1996**. Status of *Bemisia tabaci* in the Mediterranean countries: opportunities for biological control. *Biological Control* **6**,11-22.

(18) Goldner, W. R., Hoffman, F. M., and Medve, R. J., **1986**. Allelopathic effects of *Cladonia cristatella* on ectomycorrhyzal fungi common to bituminous strip-mine spoils. *Can. J. Bot.* **64**, 1586-1590.

(19) Hiscox, J. D. and Isrealstam, G. F., **1979**. A method for the extraction of chlorophyll from leaf tissue without maceration. *Can. J. Bot.* **57**, 1332-1334.

(20) Huneck, S. and Yoshimura, I., **1996**. In: *Identification of Lichen Substances*, Springer-Verlag, Berlin, 228-229.

(21) Inderjit and Nishimura, O., **1999**. Effect of the anthraquinones emodin and physcion on availability of selected soil inorganic ions. *Ann. Appl. Biol.* **135**,

425-429.

(22) Kenyon W. H., Duke, S. O., and Vaughn, K. C., **1985**. Sequence of effects of acifluorfen on physiological and ultrastructural parameters in cucumber cotyledon discs. *Pestic. Biochem. Physiol.* **24**, 240-250.

(23) Lasceve, G. and Gaugain, F., **1990**. Effects of usnic acid on sunflower and maize plantlets. *J. Plant Phys.* **136**, 723-727.

(24) Lauterwein, M., Oethinger, M., Belsner, K., Peters, T., and Marre, R., **1995**. *In vitro* activities of the lichen secondary metabolites vulpinic acid, (+)-usnic acid and (-)-usnic acid against aerobic and anaerobic microorganisms. *Antimicrob. Agents & Chemo.* **39**, 2541-2543.

(25) Lawrey, J. D., **1983**. Lichen herbivory preference: A test of two hypotheses. *Amer. J. Bot.* **70**, 1188-1194.

(26) Lawrey, J. D., **1986**. Biological role of lichen substances. *Bryologist* **9**, 111-122.

(27) Lawrey, J. D., **1993**. Lichen allelopathy. *Amer. J. Bot.* (S) **80,** 103.

(28) Lawrey, J. D., **1995**. Lichen allelopathy: A review. *Am. Chem. Soc. Symp. Ser.* **582**, 26-38.

(29) Maniatis, T., Fritsch E. F., and Sambrook J., **1982**. In: *Molecular Cloning: Laboratory Manual*. Cold Spring Harbor Laboratory Press, Cold Spring Harbor, NY, 545 pp.

(30) Manojilovic, N. T., Solujic, S., Sukdolak, S., and Krstic, L. J., **2000**. Isolation and antimicrobial activity of anthraquinones from some species of the lichen genus *Xanthoria*. *J. Serb. Chem. Soc.* **65**, 555-560.

(31) Proksch, P., **1995**. The protective system of lichens against being consumed by animals. *Dtsch. Apoth. Ztg.* **135**, 21-24.

(32) Proska, B., Sturdikova, M., Pronayova, N., and Liptaj, T., **1996**. (-)-Usnic acid and its derivatives. Their inhibition of fungal growth and enzyme activity. *Pharmazie* **51**, 195-196.

(33) Romagni, J. G., Meazza, G., Nanayakkara, and D., Dayan, F. E., **2000**. The Phytotoxic Lichen Metabolite, Usnic Acid, is a Potent Inhibitor of Plant *p*-Hydroxyphenylpyruvate Dioxygenase. *FEBS Letters* **480** (2-3), 301-305.

(34) Rojas, I. S., Lotina-Hennsen, B., and Mata, R., **2000**. Effect of lichen metabolites on thylakoid electron transport and photophosphorylation in isolated spinach chloroplasts. *J. Nat. Prod* **63**, 1396-1399.

(35) Rowland, M., Hackett, B., and Stribley, M., **1991**. Evaluation of insecticides in field-control simulators and standard laboratory bioassays against

resistant and susceptible *Bemisia tabaci* (Homoptera:Aleyrodidae) from Sudan. *Bull. Entomol. Res.* **81**,189-199.

(36) Sandmann, G., **1993**. Detection of bleaching by benzoyl cyclohexanediones in cress seedlings by HPLC separation and determination of carotenoids. In: *Target Assays for Modern Herbicides and Related Phytotoxic Compounds*. P. Böger & G. Sandmann (Eds), Lewis Publishers, Boca Raton, USA, 9-13.

(37) Slansky, F., **1979**. Effect of lichen chemicals atranorin and vulpinic acid upon feeding and growth of larvae of the yellow-striped armyworm, *Spodoptera ornithogalli*. *Env. Entomol.* **8**, 865-868.

(38) Yen G.-C., Chen, H.-W., and Duh, P.-D., **1998**. Extraction and identification of an antioxidative component from Jue Ming Zi (*Cassia tora* L.). *J. Agric. Food Chem.* **46**, 820-824.

(39) Yim, H., Lee, Y. H., Lee, C. H., and Lee, S. K., **1999**. Emodin, an anthraquinone derivative isolated from the rhizomes of *Rheum palmatum*, selectively inhibits the activity of casein kinase II as a competitive inhibitor. *Planta Med.* **65**, 9-13.

(40) Zhang, L., Lau, Y.-K., Xia, W., Hortobagyi, G. N., and Hung, M.-C., **1999**. Tyrosine kinase inhibitor emodin suppresses growth of HER-2/neu-overexpressing breast cancer cells in athymic mice and sensitizes these cells to the inhibitory effect of paclitaxel. *Clin. Cancer Res.* **5**, 343-353.

(41) Zubay, G., **1993**. In: *Biochemistry*, 3rd Edition. W. C. Brown, Publishers, DuBuque, IA, 420-422.

2 Bioactive Compounds from Potamogetonaceae on Aquatic Organisms

M. DellaGreca, A. Fiorentino, and M. Isidori

CONTENT

Abstract 35
Introduction 35
Results and Discussion 36
    Ruppia maritima 36
    Potamogeton natans 39
    Antialgal Assays 43
    Aquatic Invertebrate Assays 45
Methodology 48
References 54

ABSTRACT

Twenty *ent*-labdane diterpenes, isolated from the aquatic plants *Ruppia maritima* and *Potamogeton natans*, were tested to detect their effects on aquatic organisms from different trophic levels. Toxicity tests were performed on aquatic producers (the alga *Selenastrum capricornutum*) and consumers, including a rotifer (*Brachionus calyciflorus*), a cladoceran crustacean (*Daphnia magna*), and two anostracan crustaceans (*Thamnocephalus platyurus* and *Artemia salina*). Furano-*ent*-labdanes exhibited high toxicity toward all of these organisms. 15,16-Epoxy-12(*S*)-hydroxy-8(17),13(16),14-*ent*-labdatrien-20,19-olide had a high toxicity only toward the algae and the rotifers, while it was inactive for the crustaceans.

INTRODUCTION

Aquatic macrophytes living in lakes and rivers face strong competition with other primary producers for light and nutrients. Strategies developed by submersed macrophytes to overcome shading by phytoplankton and epiphytes include fast growth, canopy formation and the production of growth inhibitors for algae and cyanobacteria. Several allelopathic active compounds have been isolated so far, and the ability of some natural products to inhibit the *in vitro* development of microalgae has been reported by our research group in recent years.[16,13] We also have shown that bioactive products isolated from *Pistia stratiotes* are released into the environment.[17]

In pursuing our chemical investigation of aquatic plants distributed in Italy, as well as the assessment of antialgal properties of their components, we have focused on Potamogetonaceae, which grow in Volturno, the largest river of Southern Italy for its length. The first, *Potamogeton natans,* commonly known as water tongue, is a fresh water species, while *Ruppia maritima*, commonly known as sea hay, lives at the mouth of the river in brackish waters.

The plants were air dried and extracted with solvents of increasing polarity. Chromatographic processing of the extracts led to the isolation of twenty *ent*-labdane diterpenes, identified on the basis of their spectroscopic properties and by chemical correlation.

The biological properties of labdane diterpenes as antimicrobials,[32] insect antifeedants[2] and their cytotoxic activity[33] have been extensively reported, but little data have been given for their phytotoxicity.[26] In this study, we determined the toxic potential of these metabolites on algae and even on aquatic species from various phylogenetic groups to provide a wider range of ecotoxicity information.

RESULTS AND DISCUSSION

Plants of *R. maritima* and *P. natans* were air-dried and extracted with light petroleum, ethyl acetate and methanol.

RUPPIA MARITIMA

Chromatographic separation (CC, TLC and HPLC) of the light petroleum extract led to the isolation of seven *ent*-labdanes (Fig. 2.1), identified on the basis of their spectroscopic data (Tables 2.1 and 2.2).[15] Five labdanes (1-5) have a furano ring in the skeleton and one is a *nor-bis*-diterpene (7). Compounds 1 and 4 had already been isolated from *Psiadia altissima*[4] and *Daniella oliveri*[20] respectively, while all the other compounds were isolated here for the first time from

R. maritima.[15] The diterpene 3 gave 1 by reduction with NaBH$_4$ and compound 1 gave 2 by acetylation. These chemical correlations (Scheme 2.1) confirmed the structure assigned on the basis of spectral data.

Scheme 2.1

Chemical interconvertion of some diterpenes.

$$3 \xrightarrow{\text{NaBH}_4} 1$$

$$1 \xrightarrow{\text{Ac}_2\text{O}/\text{Py}} 2$$

$$11 \xrightarrow{\text{LiAlH}_4} 10 \xleftarrow{\substack{1.\ \text{NaBH}_4 \\ 2.\ \text{KOH/MeOH}}} 8 \xrightarrow{\text{oxidation}} 9$$

$$12 \xrightarrow{\text{NaBH}_4} 13 + 13a$$

$$16 \xrightarrow{\text{Ac}_2\text{O}/\text{Py}} 17$$

$$20 \xrightarrow{\text{KOH/MeOH}} 19$$

Table 2.1

Selected ^{1}H-NMR spectral data of compounds **1 – 7**.

H	1	2	3	4	5	6	7
14	6.25 dd (0.9, 1.9)	6.28 dd (0.9, 1.9)	6.24 dd (1.0, 1.9)	6.26 dd (0.9, 2.0)	6.78 dd (0.9, 1.8)	5.41 t (7.4)	2.11 s
15	7.35 dd (1.3, 1.9)	7.35 dd (1.4, 1.9)	7.20 dd (1.2, 1.9)	7.34 dd (1.3, 2.0)	7.43 dd (1.4, 1.8)	4.17 d (7.4)	-
16	7.20 dd (0.9, 1.3)	7.20 dd (0.9, 1.4)	7.38 dd (0.9, 1.3)	7.18 dd (0.9, 1.3)	8.13 dd (0.9, 1.4)	1.68 s	-
17	4.56 s / 4.86 s	4.58 s / 4.48 s	4.60 s / 4.95 s	4.60 s / 4.88 s	4.38 s / 4.78 s	4.46 s / 4.82 s	4.44 s / 4.83 s
18	0.97 s	1.17 s	1.01 s	0.96 s	1.20 s	0.87 s	0.88 s
19	3.43 d (11.5) / 3.74 d (11.5)	-	9.78 s	3.85 d (11.2) / 4.22 d (11.2)	-	0.80 s	0.81 s
20	0.66 s	0.51 s	0.60 s	0.70 s	0.59 s	0.68 s	0.70 s
OAc	-	-	-	2.04 s	-	-	-
OMe	-	3.60 s	-	-	3.63 s	-	-

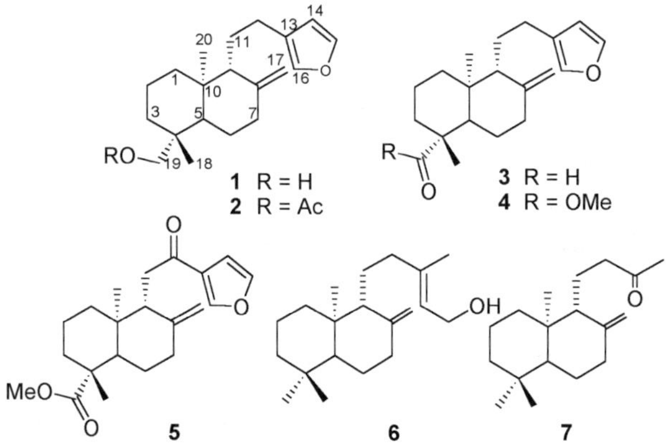

Figure 2.1

Ent-Labdane-diterpenes isolated from *Ruppia maritima*.

Table 2.2

^{13}C-NMR spectral data of compounds **1 – 7**.

C	1	2	3	4	5	6	7
1	38.7	39.1	38.3	38.8	39.4	39.1	39.0
2	18.9	19.9	19.2	18.9	19.9	19.4	19.3
3	35.5	38.2	38.3	36.2	38.0	42.2	42.1
4	38.9	44.0	48.6	39.4	44.0	33.6	33.6
5	56.1	56.3	55.9	56.1	56.0	56.3	56.3
6	24.1	26.3	24.2	24.1	25.8	24.4	24.4
7	38.5	38.7	34.3	38.5	38.2	38.3	38.2
8	147.9	147.9	147.2	147.8	149.0	148.6	148.3
9	56.0	55.2	54.5	56.0	50.4	55.5	55.9
10	38.9	40.2	39.9	39.4	39.6	39.7	39.7
11	23.5	23.6	23.5	23.5	36.5	21.7	17.5
12	24.3	24.2	24.0	24.4	194.0	33.6	42.9
13	125.4	125.0	125.3	125.5	125.0	140.7	209.5
14	110.9	110.9	110.8	110.9	108.8	122.9	30.0
15	142.6	142.6	142.7	142.7	144.1	59.4	-
16	139.6	138.7	138.7	138.7	146.6	16.4	-
17	106.5	106.3	107.2	106.7	106.4	106.3	106.2
18	27.0	28.8	24.3	27.5	28.8	38.3	33.6
19	65.0	177.7	205.7	66.8	181.1	21.7	21.7
20	15.3	12.6	13.5	15.3	13.1	14.5	14.3
OMe	-	51.5	-	-	51.5	-	-
Ac1	-	-	-	171.0	-	-	-
Ac2	-	-	-	21.0	-	-	-

POTAMOGETON NATANS

Chromatographic separation of the extracts of the water tongue afforded six furano-*ent*-labdanes (**8, 10-14**), six γ-lactone-*ent*-labdanes (**15-20**), two of them as glucosides (**19** and **20**), and one *nor*-diterpene (**9**) (Fig. 2.2).[3,14] The ^{1}H-NMR (Table 2.3) and ^{13}C-NMR (Table 2.4) were assigned by combination of COSY, NOESY, DEPT, HMQC and HMBC experiments. All the furano-diterpenes **8-14** were isolated for the first time from *P. natans*, except potomagetonin (**11**), already isolated by Smith et al.[31] Some chemical correlations, illustrated in Scheme 2.1, confirmed the structure of diterpenes. To establish the configuration at C-12 of the compound **13**, 12-oxo-potomagetonin (**12**) was reduced with NaBH$_4$ to give the natural **13** and its epimer **13a** at C-12. Mosher's method[11] was applied to both the products. The R-MPTA and S-MTPA esters showed differences in the chemical shifts in agreement with the *S* configuration for **13** and, subsequently, an *R* configuration for its epimer (**13a**). The lactone-diterpenes **15-20** were isolated for the first time from *P. natans*. Andrograpanine (**16**)[18] and neoandrographolide (**19**)[5] have been previously isolated from *Andrographys paniculata*. Finally, the acetylation of neoandrographolide (**19**) gave **20**.

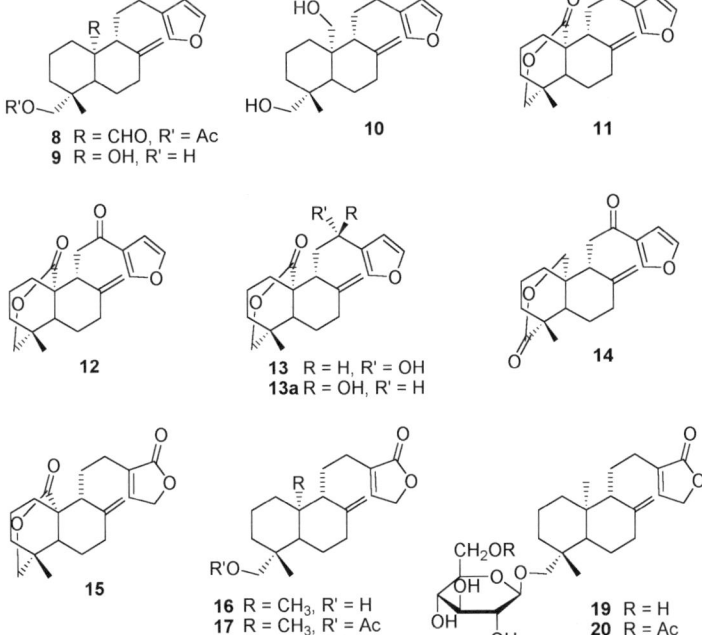

Figure 2.2

    ent-Labdane-diterpenes isolated from *Potamogeton natans*.

Table 2.3

Select ^{1}H-NMR spectral data of compounds **8 – 20**

H	8	9	10	11	12	13	13a	14
14	6.23 dd (0.9, 1.6)	6.26 dd (0.8, 1.8)	6.25 dd (0.9, 1.8)	6.27 dd (0.9, 1.6)	6.80 dd (0.8, 1.7)	6.43 dd (0.7, 1.9)	6.43 dd (0.7, 1.8)	6.78 dd (1.0, 1.9)
15	7.34 dd (1.3, 1.6)	7.36 dd (1.4, 1.8)	7.36 dd (1.3, 1.8)	7.34 dd (1.3, 1.6)	7.43 dd (1.3, 1.7)	7.40 dd (1.2, 1.9)	7.40 dd (1.2, 1.8)	7.46 dd (1.5, 1.9)
16	7.17 dd (0.9, 1.3)	7.20 dd (0.8, 1.4)	7.20 dd (0.9, 1.3)	7.20 dd (0.9, 1.3)	8.20 dd (0.8, 1.3)	7.39 dd (0.7, 1.2)	7.34 dd (0.7, 1.2)	8.09 dd (1.0, 1.5)
17	4.64 brs 4.96 brs	4.81 s 5.13 s	4.83 s 5.95 s	4.84 brs 4.96 s	4.61 s 4.69 s	4.76 s 4.96 s	4.97 s 4.97 s	4.46 s 4.88 s
18	0.98 s	0.80 s	1.02 s	0.91 s	0.93 s	0.92 s	0.89 s	1.22 s
19	3.75 d (11.5 Hz) 3.85 dd (1.6, 11.5)	3.20 d (11.5) 3.54 dd (1.5, 11.5)	3.58 d (11.6) 3.77 d (11.6)	3.98 d (11.9) 4.17 dd (2.1, 11.9)	4.18 d (11.8) 4.23 dd (2.0, 11.8)	4.03 d (11.8) 4.19 dd (2.0, 11.8)	4.00 d (11.5) 4.16 dd (1.5, 11.5)	4.19 d (11.6) 4.43 dd (1.9, 11.6)
20	9.82 s	-	3.80 m	-	-	-	-	-
OAc	-	-	-	-	-	-	-	-

H	15	16	17	18	19	20
14	7.18 m	7.10 m	7.10 m	7.09 m	6.10 m	5.98 m
15	4.78 m	4.75 m	4.77 m	4.76 m	4.72 m	4.61 m
17	4.75 s 4.96 s	4.55 s 4.83 s	4.59 s 4.87 s	4.63 s 4.96 s	4.63 s 4.83 s	4.64 s 4.85 s
18	0.92 s	0.95 s	0.97 s	0.99 s	0.94 s	0.98 s
19	4.03 dd (11.9, 0.4) 4.19 dd (11.9, 2.1)	3.36 d (11.0) 3.72 d (11.0)	3.85 d (11.0) 4.22 d (11.0)	3.75 d (10.8) 3.89 d (10.8)	3.63 d (10.4) 3.86 d (10.4)	3.84 d (10.0) 4.28 d (10.0)
20	-	0.63 s	0.70 s	9.81 s	0.66 s	0.70 s
1'	-	-	-	-	4.30 d (8.0)	4.30 d (7.8)
OAc	-	-	2.04 s	2.01 s	-	2.01 s

Table 2.4

^{13}C NMR chemical shift assignments for labdanes **8 – 20**

C	8	9	10	11	12	13	13a
1	31.9	37.7	36.3	41.8	36.4	36.0	35.9
2	19.2	19.8	18.8	20.9	20.9	20.9	20.8
3	36.6	38.2	38.5	36.0	41.1	41.1	41.0
4	37.4	38.1	38.8	33.4	33.5	33.5	33.7
5	54.5	53.6	55.2	49.4	49.2	49.5	50.9
6	24.4	23.8	23.6	28.1	27.6	28.0	27.8
7	38.2	35.9	34.5	36.9	36.4	36.9	36.8
8	145.6	148.4	151.8	145.2	146.6	145.9	145.8
9	52.7	53.3	55.5	51.6	46.2	48.1	49.0
10	53.7	74.1	44.6	51.2	50.4	50.8	50.9
11	24.1	23.0	26.2	23.7	36.4	34.3	34.1
12	22.8	25.5	24.8	25.7	194.8	65.1	66.1
13	124.7	125.0	125.0	125.2	128.0	130.0	128.3
14	110.8	110.8	110.9	111.0	108.7	108.4	108.2
15	142.9	142.8	142.8	142.7	144.2	143.3	143.5
16	138.9	138.8	138.8	138.8	147.4	138.6	139.7
17	108.0	109.6	106.1	108.2	107.9	108.7	108.6
18	26.7	27.8	27.9	23.7	23.6	23.7	23.6
19	66.5	68.6	67.4	76.4	76.7	76.5	76.4
20	206.8	-	65.8	172.9	173.6	173.2	178.0
Ac 1	173.0	-	-	-	-	-	-
Ac 2	20.9	-	-	-	-	-	-
glc-1	-	-	-	-	-	-	-
glc-2	-	-	-	-	-	-	-
glc-3	-	-	-	-	-	-	-
glc-4	-	-	-	-	-	-	-
glc-5	-	-	-	-	-	-	-
glc-6	-	-	-	-	-	-	-

Table 2.4 (cont.)

^{13}C NMR chemical shifts assignments for labdanes **8 – 20**.

C	14	15	16	17	18	19	20
1	38.1	36.9	39.0	38.9	31.9	40.1	40.8
2	21.1	20.8	18.9	18.9	19.1	20.3	20.8
3	39.7	41.1	35.3	36.2	38.0	34.4	35.1
4	43.6	33.4	39.5	39.5	37.3	40.8	41.4
5	49.0	49.5	56.2	56.2	53.5	57.9	58.4
6	26.5	28.0	24.5	24.5	24.0	24.7	25.2
7	35.8	36.1	38.5	38.4	36.5	36.8	38.1
8	146.5	145.0	147.5	147.2	145.1	149.6	149.9
9	45.2	52.6	56.4	56.4	54.3	57.9	58.3
10	37.9	51.2	38.8	37.3	53.3	40.3	39.3
11	36.0	23.7	21.7	21.7	22.2	20.3	21.6
12	192.7	24.9	24.4	24.4	23.9	25.8	26.5
13	127.6	134.7	134.8	134.9	134.2	135.2	138.6
14	108.6	143.9	143.9	143.8	144.0	140.8	148.1
15	144.4	70.2	70.1	70.1	70.2	69.4	69.9
16	146.8	174.3	174.4	174.3	174.2	171.0	174.1
17	109.1	108.4	107.0	107.0	108.1	107.6	108.5
18	23.1	23.8	27.1	27.5	26.6	28.1	28.9
19	176.0	76.4	65.0	66.7	66.4	65.1	68.7
20	73.1	173.0	15.2	15.2	206.6	16.2	16.6
Ac 1	-	-	-	171.3	170.9	-	174.1
Ac 2	-	-	-	20.9	20.9	-	20.8
glc-1	-	-	-	-	-	104.2	104.7
glc-2	-	-	-	-	-	75.3	75.9
glc-3	-	-	-	-	-	78.2	79.0
glc-4	-	-	-	-	-	71.8	72.4
glc-5	-	-	-	-	-	78.3	78.8
glc-6	-	-	-	-	-	62.9	63.6

ANTIALGAL ASSAYS

All the natural products (**1-20**), and the synthetic **13a**, were assayed against the green alga *Selenastrum capricornutum*. In standard algal toxicity tests, a rapidly growing algal population in a nutrient-enriched medium is exposed to a chemical for three days. *S. capricornutum* is one of the most frequently used and recommended to measure the potentially adverse effect of chemicals on the aquatic environment.[1,27] The toxicity was reported as IC_{50}, the inhibiting concentration that reduces algal growth by 50% (Table 2.5).

Table 2.5

Effects of diterpenes **1-20** on *S. capricornutum*, reported as IC_{50} (µM) with $p<0.05$.

Compound	IC_{50}	Compound	IC_{50}
1	0.8 (0.43-1.41)	11	28.58 (23.85-36.63)
2	13.62 (10.95-16.96)	12	60.89 (57.8-62.8)
3	7.57 (6.90-8.33)	13	4.40 (3.11-6.21)
4	ND[a]	13a	ND[a]
5	1.45 (1.07-2.0)	14	ND[a]
6	ND[a]	15	44.16 (25.11-61.9)
7	ND[a]	16	217.13 (200.4-221.8)
8	58.27 (37.94-89.49)	17	696.25 (680.7-703.4)
9	2.84 (1.11-3.02)	18	NR[b]
10	18.45 (13.73-24.81)	19	80.54 (68.52-94.70)
		20	53.70 (34.27-61.91)

[a] ND = not determinable at the highest concentration assayed
[b] NR = no relation dose-response

Results revealed that the toxicity was highly dependent not only on the skeleton but even on the substituent. The presence of the furano group is important for antialgal activity. Among all the furano-diterpenes, the most toxic was compound **1**, showing an IC_{50} of 0.8 µmol/L. The free hydroxyl group at C-19 seems to be important for such activity. In fact, the acetylation of the C-19 position (compare **1** and **2**) increases the IC_{50} value (13.62 µmol/L) retaining a dose-response relationship in compound **2**, as reported in Figure 2.3, but diminishing the activity.

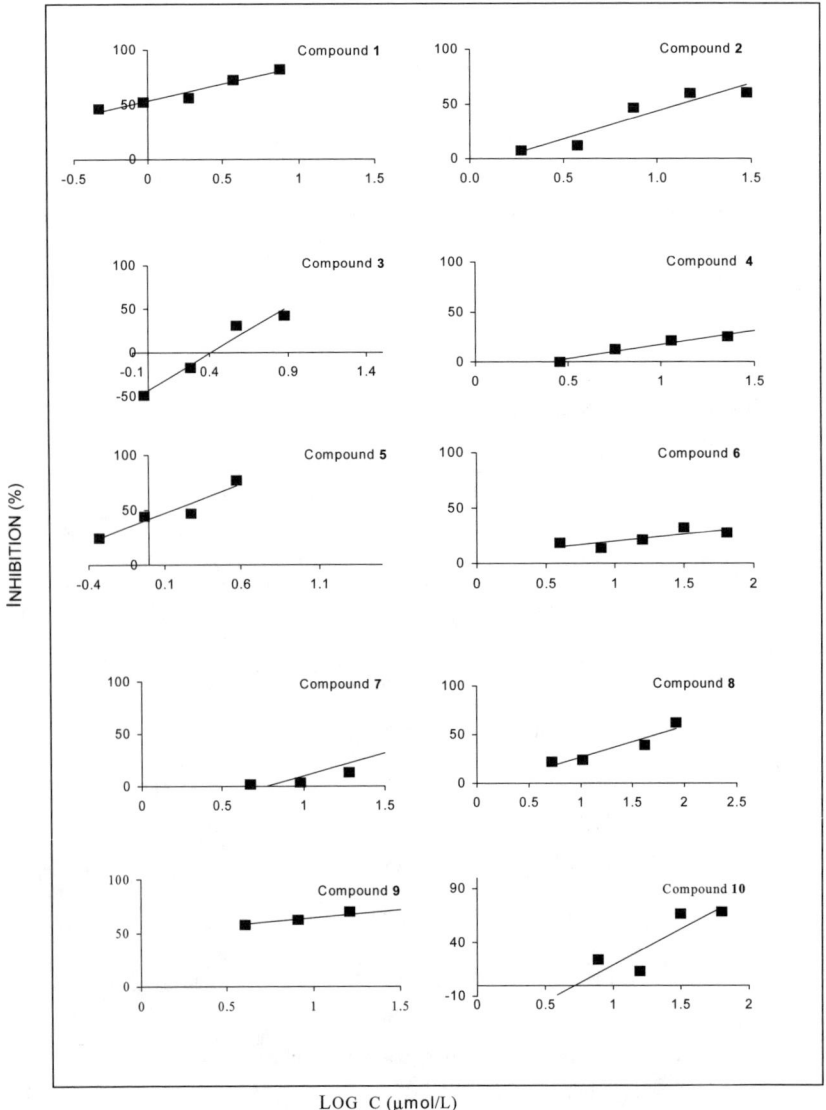

Figure 2.3

Algal growth inhibition (%) of diterpenes **1-10** from Potamogetonaceae.

Compound **3** is a growth stimulant at low concentrations, and the toxic effect was revealed only by increasing the concentrations (IC_{50} = 7.57 µmol/L). The toxicity of **4** was significant for concentrations higher than 9.5 µmol/L, and its IC_{50} could not be found at the highest tested concentration (45 µmol/L). The carbonyl group at C-19 decreases activity unless another carbonyl group is present at C-12 (compound **5**).

Also, the lactone ring between the C-19 and C-20 decreases the inhibitory effect. In fact, the IC_{50} of potamogetonin **11** was 28.58 µmol/L, while the 12-oxo derivative **12** was bio-stimulating up to the concentration of 25.85 µmol/L and became slightly inhibiting at the highest concentration (60.89 µmol/L), as is well observable in Figure 2.4. The activity becomes significant again because of the presence of a hydroxyl group at C-12, as shown by compound **13**, which has an IC_{50} of 4.40 µmol/L. It is important to note that epimer **13a** acts in a completely different way, being only slightly active (20% inhibition) at the highest tested concentration (136.4 µmol/L) and thus indicating that the configuration of carbon 12 is fundamental to the biological activity.

The absence of the furano group (compounds **6** and **7**) reduces the activity of the chemicals considerably. Also, oxidation of the heterocyclic ring to γ-lactone decreases the toxicity, as is evident from the comparison of compounds **11**, **16**, **17** and **18** with the analogues **15**, **1**, **2** and **8**. Furthermore, a statistical analysis performed using Student's *t* test (comparison between averages) shows that the furano compounds are significantly more toxic than lactones ($P < 0.01$).

Finally, the glucosylation of the lactone-diterpenes (**19** and **20**) causes a bio-stimulation at the low concentrations. These results can be explained by the fact that detoxification often occurs through glucosylation.[21]

AQUATIC INVERTEBRATE ASSAYS

To evaluate the impact of these diterpenes on aquatic invertebrates, present in the same ecosystems as the investigated Potamogetonaceae, we tested the most active and abundant chemicals on consumers including a rotifer (*Brachionus calyciflorus*), a crustacean cladoceran (*Daphnia magna*), and two anostracans (*Thamnocephalus platyurus* and *Artemia salina*). Acute toxicity of labdane diterpenes was detected by exposing the organisms to solutions of varying concentrations and counting the number of dead (*T. platyurus*, *B. calyciflorus*, and *A. salina*) or immobile (*D. magna*) organisms. We report the toxicity as LC_{50} or EC_{50}, the lethal or immobilizing concentration that affects 50% of the exposed organisms (Table 2.6).

The activity of the diterpenes on *D. magna* is lower in respect to that on *S. capricornutum*. In fact, compounds **2**, **3**, **13** and **13a** are completely inactive at the highest concentrations tested. The carbonyl group at carbon 20 and the absence of an oxygenated function at the carbon 12 (compounds **8** and **11**) seem to enhance toxicity.

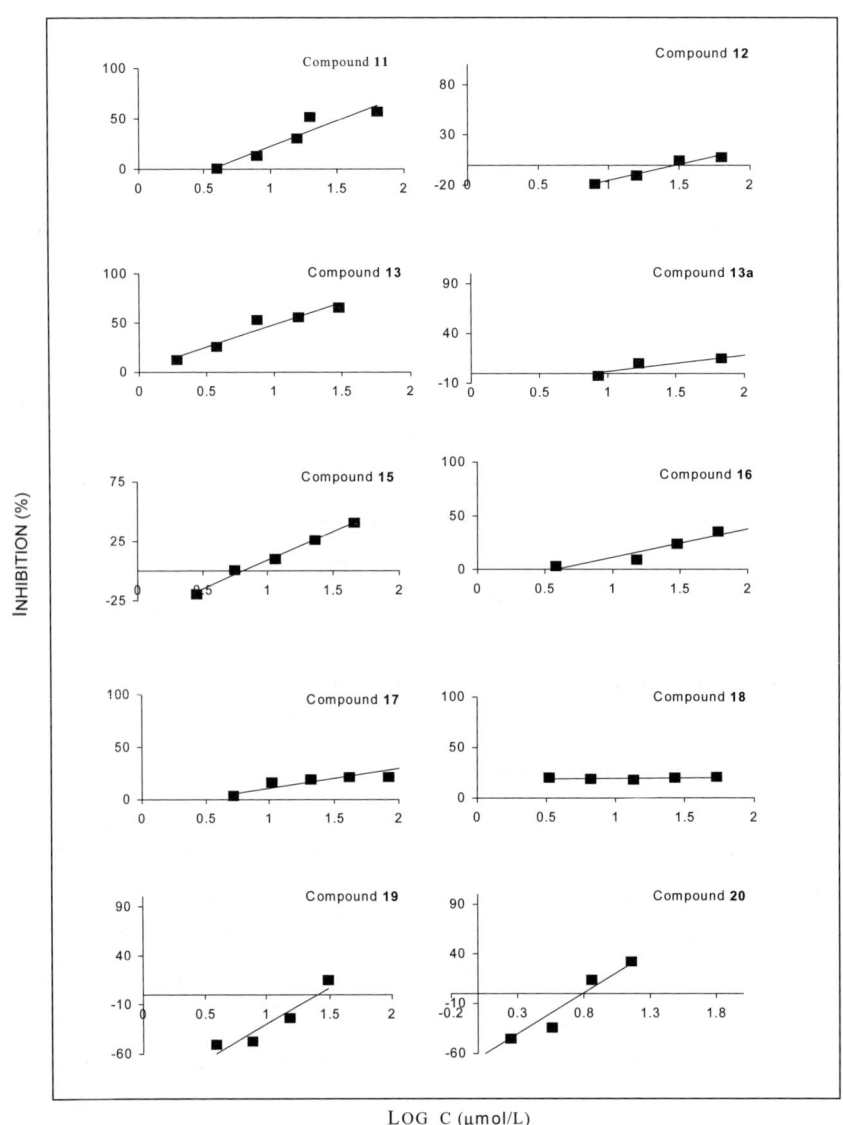

Figure 2.4

Algal growth inhibition (%) of diterpenes **11-20** from Potamogetonaceae.

Table 2.6

Effects of the furano-diterpenes on aquatic organism reported as L(E)C$_{50}$ (µM) with the 95% confidence limits.

Compound	D. magna (EC$_{50}$)	T. platyurus (LC$_{50}$)	B. calyciflorus (LC$_{50}$)	A. salina (LC$_{50}$)
1	157.8 (113.0-248.0)	0.84 (0.76-0.94)	66.59 (59.54-74.47)	ND^a
2	ND^a	126.1 (87.85-143.1)	6.60 (5.23-7.85)	ND^a
3	ND^a	ND^a	4.54 (1.64-7.73)	ND^a
8	19.8 (9.78-30.11)	9.41 (5.84-15.17)	56.11 (44.63-74.86)	ND^a
10	47.17 (40.41-55.06)	ND^a	74.87 (65.38-85.75)	ND^a
11	14.59 (11.27-18.88)	23.56 (14.68-37.86)	36.97 (30.32-45.22)	ND^a
12	35.0 (23.04-53.20)	74.30 (56.19-98.23)	ND^a	ND^a
13	ND^a	ND^a	1.48 (1.21-1.82)	ND^a
13a	ND^a	ND^a	ND^a	ND^a

^{a} ND = not determined at the highest concentration assayed

Unlike in the previously mentioned species, compounds **1** and **8** were very active on *T. platyurus*. The LC$_{50}$ value of the first compound (0.84 µmol/L) was comparable with its IC$_{50}$ value found for *S. capricornutum* (0.80 µmol/L). Compounds **3**, **10**, **13** and **13a** were inactive at the tested concentrations, while *ent*-labdanes **2** and **12** showed high values of LC$_{50}$.

B. calyciflorus was sensitive to most of the tested compounds. In fact, only compounds **12** and **13a** were inactive for this invertebrate. The most toxicant for rotifers was **13**, showing an LC$_{50}$ of 1.48 µmol/L. Also, diterpenes **2** and **3** showed a high toxicity, with LC$_{50}$ values lower than 7 µmol/L.

A. salina, already known as a resistant organism,[29] was not affected by exposure to these chemicals, clearly indicating the difference in sensitivity between this marine crustacean and the other freshwater crustacean species used.

It should be noted that the synthetic **13a** was active on all the tested species, while its natural epimer **13** showed high toxicity values only for algae and rotifers. Treatment of **12** with NaBH$_4$ afforded the alcohol **13a** in two-fold yield with respect to the epimer **13**. The absence of **13a** in the plant allows one to presume the involvement of an appropriate enzymatic system in the biosynthesis of **13** and a plausible specific role of this metabolite.

Except for compound **13a**, no one compound was active or inactive toward all the organisms. In fact, compound **3** was inactive for *D. magna* and *T. platyurus*, while it had low LC$_{50}$ and IC$_{50}$ values for *B. calyciflorus* and *S. capricornutum*. Moreover, compound **1** was highly toxic toward *T. platyurus* and algae, while showing high LC$_{50}$ and EC$_{50}$ values for *B. calyciflorus* and *D. magna*, respectively.

The biological activity of diterpenes is copiously reported in the literature. Singh et al. reviewed the biological activity of labdane diterpenes from 1987 to 1997.[30] The antibacterial, antifungal, anti-inflammatory, cytotoxic and other biological activities, detected in the labdane diterpenes, have been widely discussed in relation to chemical structures. In recent years, the role of diterpenes in allelopathy has been proven,[24,25] but so far the antialgal activity of these compounds has never been reported. The only exception is geniculol, an irregular diterpenoid isolated from an ecdophytic fungus of the genus *Geniculosporium*, which inhibits the growth of the green alga *Chlorella fusca*.[23] Also, the toxicity of the labdanes on other aquatic organisms is but slightly known, and only the brine shrimp (*Artemia salina*) has been extensively employed in bioassays.[19]

One of the factors supporting the use of algae and invertebrates in the bioassays is their niche in aquatic ecosystems. If these organisms are adversely affected by a toxicant, then the surrounding ecosystem may also feel the effects, either directly or indirectly, from the lack of a food source. Algae and invertebrates occupy low trophic levels, and their disturbance may exhibit a chain effect throughout the ecosystem.

The results obtained from the phytochemical study of *P. natans* and *R. maritima* suggest that the high concentrations of the bioactive diterpenes found in these plants could interfere with other aquatic organisms, thus playing a role in the equilibrium of the aquatic systems.

METHODOLOGY

General experimental procedures

NMR spectra were recorded at 400 MHz for ^{1}H and 100 MHz for ^{13}C on a

Bruker AC 400 spectrometer in $CDCl_3$ or CD_3OD solutions, at 27°C. Proton-detected heteronuclear correlations were measured using HMQC (optimized for $^1J_{HC}$ = 145 Hz) and HMBC (optimised for $^1J_{HC}$ = 7 Hz). Optical rotations were measured on a Perkin-Elmer 343 polarimeter. IR spectra were determined in $CHCl_3$ solutions on a FT-IR Perkin-Elmer 1740 spectrometer. UV spectra were obtained on a Perkin-Elmer Lambda 7 spectrophotometer in EtOH solutions. MS spectra were obtained with a HP 6890 apparatus equipped with a MS 5973 N detector. FAB mass were obtained with a Kratos VG ZAB 2SE. The HPLC apparatus consisted of a pump (Shimadzu LC-10AD), a refractive index detector (Shimadzu RID-10A) and a Shimadzu Chromatopac C-R6A recorder. Preparative HPLC was performed using RP-8 (Luna 10 µm, 250 x 10 mm i.d., Phenomenex), SiO_2 (Maxsil 10 µm, 250 x 10 mm i.d., Phenomenex) or RP-18 (Kromasil 10 µm, 250 x 10 mm i.d., Phenomenex) columns. Analytical TLC was performed on Merck Kieselgel 60 F_{254} or RP-18 F_{254} plates with 0.2 mm layer thickness. Spots were visualized by UV light or by spraying with H_2SO_4 – AcOH – H_2O (1:20:4). The plates were then heated for 5 min at 110 °C. Preparative TLC was performed on Merck Kieselgel 60 F_{254} plates, with 0.5 or 1 mm film thickness. Flash column chromatography (FCC) was performed on Merck Kieselgel 60 (230-400 mesh) at medium pressure. Column chromatography (CC) was performed on Merck Kieselgel 60 (70-240 mesh) or on Sephadex LH-20® (Pharmacia).

Extraction and isolation of diterpenes from R. maritima

Air-dried plants (10 kg) were sequentially extracted with light petroleum, EtOAc and MeOH. The light petroleum extract (36 g) was chromatographed on neutral Al_2O_3 (grade III), and the elution with hexane – Et_2O (19:1) gave fractions A-D. Fraction A was purified on Sephadex LH-20 [hexane – $CHCl_3$ – MeOH (4:1:1)] to give compound **1** (23 mg). Fraction B was chromatographed on silica gel eluting with hexane – $CHCl_3$ (4:1) to give pure **3** (31 mg) and **4** (37 mg). Fraction C was chromatographed on silica gel eluting with hexane - benzene mixtures. Hexane - benzene (49:1) gave crude **6**, which was purified by preparative TLC [hexane - benzene (9:1)]. The fraction eluted with hexane - benzene (9:1) was chromatographed by RP-18 HPLC [MeOH – MeCN (19:1)] to give compounds **2** (8 mg) and **7** (35 mg). Fraction D was chromatographed on Sephadex LH-20 with hexane – $CHCl_3$ – MeOH (4:1:1) to give compound **5** (8 mg), which was purified by RP-18 HPLC [MeOH – MeCN (19:1)].

Allelopathy: Chemistry and Mode of Action of Allelochemicals

Extraction and isolation of diterpenes from P. natans

The air-dried and powdered plant material (13 kg) was sequentially and exhaustively extracted with light petroleum, EtOAc and MeOH. The light petroleum extract (66 g) was chromatographed on silica gel eluting with hexane and EtOAc solutions to give fraction A-B. Fraction A eluted with hexane – EtOAc (19:1), was chromatographed using hexane – EtOAc (17:3) as eluent to give pure compounds **8** (57 mg) and **11** (125 mg). Fraction B, eluted with hexane – EtOAc (17:3), was filtered on Sephadex LH-20 using hexane – $CHCl_3$ – MeOH (3:1:1) as eluent; fractions 12-38 were purified by HPLC, using a preparative silica gel column and a mixture of hexane – iPr_2O (3:1) as eluent, and gave pure **17** (15 mg) and **18** (11 mg); fractions 56-89 were purified by preparative TLC, using a mixture of benzene – EtOAc (19:1) as eluent, to give pure **12** (85 mg) and **14** (4 mg). The EtOAc extract (79 g) was chromatographed on silica gel eluting with hexane and EtOAc solutions to give fraction C-E. Fraction C, eluted with hexane – EtOAc (4:1), was chromatographed on Sephadex LH-20 with hexane – $CHCl_3$ – MeOH (3:1:1). The eluate obtained was purified by HPLC using a preparative RP-8 column and a mixture of MeOH – H_2O (4:1) as eluent, to give pure **9** (11 mg) and **13** (35 mg). Fraction D, eluted with hexane – EtOAc (2:1), was chromatographed on preparative TLC [$CHCl_3$–EtOAc (3:1)] to give compound **10** (31 mg). Fraction E, eluted with hexane – EtOAc (3:1), was chromatographed on preparative TLC [benzene – EtOAc (3:2)] to give compounds **15** (13 mg) and **16** (51 mg). The methanolic extract (120 g) was treated with H_2O and EtOAc. The aqueous layer, concentrated to 500 mL, was chromatographed on Amberlite XAD-4. The fraction eluted with MeOH was chromatographed on Sephadex LH-20 [MeOH – H_2O (1:1)] to give a mixture of glycosides, which were separated on preparative RP-18 eluting with MeOH –H_2O (4:1) to give pure **19** (21 mg) and **20** (8 mg).

Preparation of compound 1. $NaBH_4$ (8 mg) was added to a solution of **3** (25 mg) in MeOH (2.5 mL), and the reaction mixture was stirred for 5 h at room temperature. The hydride in excess was broken with AcOH (2 drops). The mixture was then directly purified by preparative TLC with hexane – $CHCl_3$ (4:1) to give compound **1** (20 mg).

Preparation of compound **9**. Pure aldehyde **8** (30 mg) was dissolved in C_6H_6 and the solution evaporated in a large flask to obtain a thin layer. After 2 days at room temperature, the autoxidation mixture was dissolved in peroxides-free Et_2O

and the ether solution extracted with 10% Na$_2$CO$_3$ aq. The residue was purified by prep. TLC with hexane–Me$_2$CO (4:1) to give compound **9** (4 mg).

*Preparation of compound **10***. LiAlH$_4$ (15 mg) was added to a solution of **11** (30 mg) in dry Et$_2$O (5 ml), and the reaction was stirred for 6 hr at room temperature. The mixture, after the hydride in excess had been destroyed, was purified by prep. TLC [CHCl$_3$–EtOAc (3:1)] to give compound **10** (23 mg). The analogue reduction of aldehyde **8** with NaBH$_4$, after alkaline hydrolysis, gave the same compound **10**.

*Preparation of **13** and **13a***. 30 mg of **12** were reduced as described for **1**. The mixture obtained was purified by preparative TLC with CHCl$_3$ – EtOAc (17:3) to give the unreacted **12** (11 mg), compound **13** (5 mg) and 12(S)-hydroxy-15,16-epoxy-8(17),13(16),14-*ent*-labdatrien-20,19-olide (**13a**, 11 mg).

*Preparation of (R)- and (S)-MTPA esters of **13***. (S)-(+)-MTPA-Cl (5 µL) was added to a solution of pure **13** (1.5 mg) in dry pyridine (50 µL). After 4 h at room temperature under magnetic stirring, ethyl ether (5 mL) and water (5 mL) were added to the reaction mixture. The organic layer gave crude ester, which was purified by preparative TLC eluting with hexane – EtOAc (5:1). The ester (R)-MTPA had ^{1}H-NMR (CDCl$_3$, 400 MHz) δ: 6.45 (1H, s, H-14), 5.95 (1H, d, J = 10.5 Hz, H-12), 5.01 (1H, s, H-17), 4.69 (1H, s, H-17), 4.18 (1H, dd, J = 1.5, 11.6 Hz, H-19), 4.01 (1H, d, J = 11.5 Hz, H-19), 2.65 (1H, m, H-11), 2.11 (1H, m, H-11), 0.90 (3H, s, H-18).

(S)-MTPA of **13** was prepared by the same procedure. ^{1}H-NMR (CDCl$_3$, 400 MHz) δ:6.38 (1H, s, H-14), 5.91 (1H, dd, J = 4.7, 10.5Hz, H-12), 5.02 (1H, s, H-17), 4.76 (1H, s, H-17), 4.14 (1H, dd, J = 1.5, 11.5Hz, H-19), 3.98 (1H, d, J = 11.5 Hz, H-19), 2.75 (1H, m, H-11), 2.26 (1H, m, H-11), 0.89 (3H, s, H-18).

*Preparation of (R)- and (S)-MTPA ester of **13a***. The procedure used for the MTPA esters of **13** was also used for the preparation of (R)-MTPA and (S)- MTPA of **13a**. The (R)-MTPA ester had ^{1}H-NMR (CDCl$_3$, 400 MHz) δ: 6.29 (1H, s, H-14), 6.03 (1H, dd, J = 4.7, 10.5Hz, H-12), 4.99 (2H, s, H-17), 4.16 (1H, dd, J = 1.5, 11.5Hz, H-19), 4.00 (1H, d, J = 11.5 Hz, H-19), 2.95 (1H, m, H-11), 2.23 (1H, m, H-11), 0.89 (3H, s, H-18).

The ester (S)-MTPA had ^{1}H-NMR (CDCl$_3$, 400 MHz) δ: 6.41 (1H, s, H-14), 6.02 (1H, dd, J = 4.7, 10.5Hz, H-12), 5.01 (1H, s, H-17), 4.99 (1H, s, H-17), 4.16 (1H, dd, J = 11.5, 1.5 Hz, H-19), 4.00 (1H, d, J = 11.5 Hz, H-19), 2.85 (1H, m, H-11), 2.20 (1H, m, H-11), 0.88 (3H, s, H-18).

Toxicity tests

Various organisms were selected to address the effects of diterpenes on different organizational structures and sensitivities. Alternative, small-scale aquatic toxicity tests known as microbiotests were used. These tests are independent of the culturing of live organisms and based on immobilized or dormant (cryptobiotic) stages of aquatic species set free or hatched when needed. The following bioassays for freshwater supplied by Creasel, Deinze, Belgium were applied:

- the Algaltoxkit F™, a 0-72 h algal growth inhibition assay with *Selenastrum capricornutum*,[9]
- the Rotoxkit F™, a 24 h rotifer mortality test with the test species *Brachionus calyciflorus*,[7]
- the Thamnotoxkit F™, a 24 h crustacean mortality test with the anostracan test species *Thamnocephalus platyurus*,[8]
- the Daphtoxkit F™, a 24 h crustacean immobilization test with cladoceran test species *Daphnia magna*.[10]

All the microbiotests were performed following the methodology prescribed in the Standard Operational Procedures of the respective microbiotests and in accordance with testing conditions and culturing media of International Organizations.[1,22,27]

The toxicity tests were conducted by exposing the organisms to solutions of at least five concentrations with a number of variable replicates from test to test. Single chemicals of high purity were initially dissolved in dimethylsulfoxide (DMSO) not exceeding the concentration of 0.01% (v/v) in the test solution.[15,12]

The algal test was performed after deimmobilization of algae from beads of alginate. The inoculum of *S. capricornutum* was 1×10^4 cells/mL in flasks already containing 25 mL of test solutions prepared in five toxicant concentrations for each compound with OECD dilution medium.[27] The flasks, in three replications for each concentration and control, were placed in a growth chamber at 25°C under continuous illumination (8,000 lux). The determination of the effect on algal growth was measured at time zero and every 24 h for 3 days by an electronic particle dual threshold counter (Coulter Counter Z2, 100 µm capillary).

The rotifer toxicity assay was performed with juveniles of *B. calyciflorus* hatched from cysts after 16-18 h of incubation at 25°C in synthetic freshwater (moderately hard U.S. Environmental Protection Agency medium)[28] under continuous illumination (light source 3,000-4,000 lux). Test solutions were prepared in five toxicant concentrations (two-fold dilutions) for each compound, each with six replicates of five animals in disposable multiwell test plates containing 0.3 mL of test solution per well. After 24 h in a 25°C incubator in the dark, the dead rotifers were counted.

The crustacean toxicity test to assess the mortality of *T. platyurus* was conducted by hatching these anostraca from cysts after 20-22 h of incubation at 25°C in the same synthetic freshwater used for rotifers and at the same illumination conditions. The test solutions in five concentrations (two-fold dilutions) with three replicates of 10 animals were poured in disposable multiwell test plates (1 mL per well). After 24 h in a 25 °C incubator in the dark, the number of dead crustaceans was recorded.

The test to assess the immobilization of *D. magna* Straus was performed by hatching neonates from ephippia after 3 to 4 days of incubation at 20°C in synthetic freshwater, according to the formula recommended by International Standardization Organization[22] under continuous illumination (light source 10,000 lux). Test solutions were prepared in five toxicant two-fold concentrations with four replicates of five animals in glass beakers with 10 mL test solution. After 24 h the number of immobile daphnids for each concentration was recorded. In this study to the test battery we added a further bioassay with the anostracan *Artemia salina*, a marine species of crustaceans, to detect the impact of metabolites released by *R. maritima* and *P. natans* in proximity of the mouth of the river where salinity is substantially higher. The crustacean toxicity test to assess the mortality of *A. salina* was conducted by hatching these anostraca from cysts incubated for 2 days at 25 °C in synthetic seawater in the dark. The solutions were prepared in five toxicant concentrations (two-fold concentration increments) for each compound, each with 3 replicates of 10 animals in disposable multiwell plates (1 mL per well). After 24 h in a 25 °C incubator in the dark, the number of dead crustaceans was counted.[6]

Data analysis

Raw data for all bioassays, except algal test, were analyzed using the Toxcalc Toxicity Data Analysis and Database Software, vers. 5.0 (Tidepool Scientific Software). Point estimations were calculated by concentration/response regression using Probit Analysis as first choice and Trimmed Spearman-Karber as second, if Probit was not possible. If the raw data did not allow the respective $L(E)C_{50}$ to be calculated at the highest tested concentrations, then these values were determined by direct graphical method. Raw test data from Algaltoxkit, were analyzed by a Microsoft Excel 5.0 program tailored for this test. The algal median growth inhibition was estimated by integrating the mean values from time 0 to 72 h (area under the curve). Inhibition (%) values were reported against log-transformed data of concentrations (µmol/L) to evaluate the slope, the trend of regression, and the IC_{50} value.

REFERENCES

(1) ASTM, **1990**. *Standard Guide for Conducting Static 96h Toxicity Tests with Microalgae*, E1218-90. American Society for Testing and Materials, Philadelphia, USA

(2) Bohlmann, F., Ahmed, M., Borthakur, N., Wallmeyer, M., Jakupovic, J., King, R. M., and Robinson, H., **1985**. Diterpenes related to grindelic acid and further constituents from *Grindelia* species. *Phytochemistry* **21**, 167-17

(3) Cangiano, T., DellaGreca, M., Fiorentino, A., Isidori, M., Monaco, P., and Zarrelli, A. **2001**. Lactone diterpenes from the aquatic plant *Potamogeton natans*. *Phytochemistry* **56**, 469-473

(4) Canonica, L., Rindone, B., Scolastico, C., Ferrari, G., and Casagrande, C., **1969**. Costituenti estrattivi della *Psiadia altissima* Benth e Hook. *Gazz. Chim. Ital.* **99**, 260-275

(5) Chan, W. R., Taylor, D. R., Willis, C. R., and Bodden, R. L., **1971**. The structure and stereochemistry of neoandrographolide, a new diterpene glucoside from *Andrographys paniculata* Ness. *Tetrahedron.* **27**, 5081-5091

(6) Creasel, **1990**. Artoxkit M™. *Artemia Toxicity Screening Test for Estuarine and Marine Waters. Standard Operational Procedure*. Creasel, Deinze, Belgium

(7) Creasel, **1992**. Rotoxkit F™. *Rotifer Toxicity Screening Test for Freshwater. Standard Operational Procedure*. Creasel, Deinze, Belgium

(8) Creasel, **1995**. Thamnotoxkit F™. *Crustacean Toxicity Screening Test for Freshwater. Standard Operational Procedure*. Creasel, Deinze, Belgium

(9) Creasel, **1996**. Algaltoxkit F™. *Freshwater Toxicity Test with Microalgae. Standard Operational Procedure*. Creasel, Deinze, Belgium

(10) Creasel, **1996**. Daphtoxkit F™ magna. *Crustacean Toxicity Screening Test for Freshwater. Standard Operational Procedure*. Creasel, Deinze, Belgium

(11) Dale, J. A. and Mosher, H. S., **1973**. Nuclear magnetic resonance enantiomer reagents. Configurational correlations via nuclear magnetic

(12) DellaGreca M., Fiorentino A., Isidori M., and Zarrelli, A., **2001**. Toxicity evaluation of natural and synthetic phenanthrenes in aquatic systems. *Environ. Toxicol. Chem.* **20(8)**, 1824-1830

resonance chemical shifts of diastereomeric mandelate, O-methylmandelate, and α-methoxy-α-trifluoromethyl-phenylacetate (MTPA) esters. *J. Am. Chem. Soc.* **95**, 512-519

(13) DellaGreca, M., Ferrara, M., Fiorentino, A., Monaco, P., and Previtera, L., **1998**. Antialgal compounds from *Zantedeschia aethiopica*. *Phytochemistry* **49**, 1299-1304

(14) DellaGreca, M., Fiorentino, A., Isidori, M., Monaco, P., Temussi, F., and Zarrelli, A., **2001**. Antialgal furano-diterpenes from *Potamogeton natans*. *Phytochemistry* **58**, 299-304

(15) DellaGreca, M., Fiorentino, A., Isidori, M., Monaco, P., and Zarrelli, A., **2000**. Antialgal *ent*-labdane diterpenes from *Ruppia maritima*. *Phytochemistry* **55**, 909-913

(16) DellaGreca, M., Fiorentino, A., Monaco, P., Pinto, G., Pollio, A., and Previtera, L., **1996**. Action of antialgal compounds from *Juncus effusus* L. on *Selenastrum capricornutum*. *J. Chem. Ecol.* **22**,587-603

(17) DellaGreca, M., Fiorentino, A., Monaco, P., Previtera, L., Pinto, G., and Pollio, A., **1999**. Release of potential allelochemicals from aquatic plants. *In*: Macias, F. A., Galindo, J. C. G., Molinillo J. M. G., and Cutler, H. G. (Eds); *Recent Advances in Allelopathy - A Science for the Future*, Vol. 1. Servicio de Publicaciones de la Universidad de Cádiz, pp 255-262

(18) Fujita, T., Fujitani, R., Takeda, Y., Takaishi, Y., Yamada, T., Kido, M., and Miura, I., **1984**. On the diterpenoids of *Andrographys paniculata*: X-ray crystallographic analysis of Andrapholide and structure determination of new minor diterpenoids. *Chem. Pharm. Bull.* **32**, 2117-2125

(19) Gray, C. A., Davies-Coleman, M. T., and McQuaid, C., **1998**. Labdane diterpenes from the South African marine pulmonate *Trimusculus costatus*. *Nat. Prod. Lett.* **12**, 47-53

(20) Haeuser, J. and Lombard, R., **1961**. Isolament et structure d'un nouveau diterpène: l'acide daniellique. *Tetrahedron* **12**, 205-214

(21) Harborne, J. B., **1977**. *Introduction to Ecological Biochemistry*. Academic Press, London, U.K.

(22) International Organization for Standardization **1996**. Water Quality – Determination of the inhibition of the mobility of *Daphnia magna* Straus (Cladocera, Crustacea) – Acute toxicity test. ISO/6341

(23) König, G. M., Wright, A. D., Aust H.-J., Draeger S., and Schulz, B., **1999**. Geniculol, a new biologically active diterpene from the endophytic fungus *Geniculosporium* sp. *J. Nat. Prod.* **62**, 155-157

(24) Macias, F. A., Molinillo, J. M. G., Galindo, J. C. G., Varela R. M., Torres, A., and Simonet, A. M., **1999**. Allelopathic studies in cultivar species. 13. Terpenoids with potential use as natural herbicides templates. *J. Biol. Act. Nat. Prod. Agrochem.* 15-31

(25) Macias, F. A., Varela, R. M., Simonet, A. M., Cutler, H. G., Cutler, S. J., Ross, S. A., Dunbar, D. C., Dugan, F. M., and Hill, R. A., **2000**. Allelochemicals from New Zealand fungi. (+)-Brevione A. The first member of a novel family of bioactive spiroditerpenoids isolated from *Penicillium brevicompactum* Dierckx. *Tetrahedron Lett.* **41**, 2683-2686

(26) Munesada, K., Siddiqui, H. L., and Suga, T., **1992**. Biologically active labdane-type diterpene glycosides from the root-stalks of *Gleichenia japonica*. *Phytochemistry* **31**, 1533-1536

(27) OECD. **1984**. Algal growth inhibition test, *OECD Guideline for Testing Chemicals*, No. 201. Organization for Economic Cooperation and Development, Geneva, Switzerland

(28) Peltier, W. H., and Weber, C. I., Eds. **1985**. *Methods for Measuring the Acute Toxicity of Effluents to Freshwater and Marine Organisms*. EPA-600/4-85-013. U.S. Environmental Protection Agency, Washington, D.C., USA

(29) Ruck, J. G., Martin, M., and Mabon, M., **2000**. Evaluation of Toxkits as method for monitoring water quality in New Zealand. *In*: Persoone, G., Janssen, C., and De Coen, W. (Eds); *New Microbiotests for Routine Toxicity Screening and Monitoring*. Kluwer Academic/Plenum Publishers, NY, USA, 103-119

(30) Singh M., Pal, M., and Sharma R. P., **1999**. Biological activity of the labdane diterpenes. *Planta Med.* **65**, 2-8

(31) Smith, C. R., Jr., Madrigal R. V., Weisleder D., Mikolajczak K. L., and Highet R.J., **1976**. Potamogetonin, a new furanoid diterpene. Structural assignment by carbon-13 and proton magnetic resonance. *J. Org. Chem.* **41**, 593-596

(32) Ulubelen, A., Miski, M., Johansson, C., Lee, E., Mabry, T. J., and Matlin, S. A., **1985**. Terpenoids from *Salvia palaestina*. *Phytochemistry* **24**, 1386-1387

(33) Zani, C. L., Alves, T. M. A., Queiroz, R., Fontes, E. S., Shin, Y. G., and Cordell, G. A., **2000**. A cytotoxic diterpene from *Alomia myriadenia*. *Phytochemistry* **53**, 877-880

3 Fate of Phenolic Allelochemicals in Soils – the Role of Soil and Rhizosphere Microorganisms

U. Blum

CONTENT

Abstract 57
Introduction 58
Results and Discussion 60
    Estimating Soil Microbial Activity 60
    Microbial Transformation and Utilization of Simple Free Phenolic
    Acids – Substrates and Products 62
    Phenolic Acid Utilizing Microbial Populations in Soil
    and Rhizosphere 65
    Effects of Free Phenolic Acids on Bulk-Soil and Rhizosphere
    Populations 67
    Influence of Bulk-Soil and Rhizosphere Microorganisms
    on Phytotoxicity 70
    One Possible Approach to Providing the Necessary Data
    to Test the Hypothesis 71
References 72

ABSTRACT

It has been hypothesized that simple phenolic acids are so readily metabolized by soil and rhizosphere microorganisms that allelopathic effects by such phenolic acids are highly unlikely. However, convincing data to support or reject this hypothesis are presently lacking. Here I discuss how microbial activity may be estimated and what data are presently available on a) rates of microbial transformation and utilization, b) phenolic acid effects on soil and rhizosphere microbial populations, and c) the influences of soil and rhizosphere microbial populations on phenolic acid phytotoxicity. The resulting insight is then used to suggest a possible approach by which this hypothesis may be tested experimentally.

INTRODUCTION

There is some confusion in the literature as to when it is appropriate to apply the term allelochemical to phenolic acids. Since phenolic acids and their derivatives are found essentially in all terrestrial soils, it should be understood that the presence of phenolic acids in soil does not automatically imply that these phenolic acids are functionally allelochemicals. In theory, phenolic acids in soils, depending on their chemical state, concentrations, and the organisms involved, can have no effect, a stimulatory effect, or an inhibitory effect on any given plant or microbial process. For phenolic acids in the soil to be classified as allelochemicals requires that a) the phenolic acids are in an active form (e.g., free and protonated), b) they are involved in chemically mediated plant, microbe, or plant/microbial interactions and c) the concentrations of the active forms in the soil solution are sufficient to modify plant or microbial behavior, either in a positive or negative manner.[8,49] However, changes in microbial behaviour associated with the utilization of phenolic acids as a carbon or energy source would not qualify as an allelopathic response.

The phenolic acids of interest here [caffeic acid (3,4-dihydroxycinnamic acid), ferulic acid (4-hydroxy-3-methoxycinnamic acid), p-coumaric acid (p-hydroxycinnamic acid), protocatechuic acid (3,4-dihydroxybenzoic acid), sinapic acid (3,5-dimethoxy-4-hydroxyxinnamic acid), p-hydroxybenzoic acid, syringic acid (4-hydroxy-3,5-methoxybenzoic acid), and vanillic acid (4-hydroxy-3-methoxybenzoic acid)] (Fig. 3.1) all have been identified as potential allelopathic agents.[8,32,34] The primary allelopathic effects of these phenolic acids on plant processes are phytotoxic (i.e., inhibitory); they reduce hydraulic conductivity and net nutrient uptake by roots.[1] Reduced rates of photosynthesis and carbon allocation to roots, increased abscisic acid levels, and reduced rates of transpiration and leaf expansion appear to be secondary effects. Most of these effects, however, are readily reversible once phenolic acids have been depleted from the rhizosphere and rhizoplane.[4,6] Finally, soil solution concentrations of phenolic acids, not root uptake, appear to be the ultimate determinant of inhibition of sensitive species.[28]

Since the actual or potential phytotoxicity of a phenolic acid is determined by its physical and chemical properties and the susceptibility of the plant process involved, the actual or potential phytotoxicity of a given phenolic acid is best determined in nutrient culture in the absence of soil processes. The phytotoxicity observed in soil systems represents a realized or observed phytotoxicity, not the actual phytotoxicity, of a given phenolic acid. For example, the actual relative phytotoxicities (or potencies) for cucumber seedling leaf expansion were 1 for ferulic acid, 0.86 for p-coumaric acid, 0.74 for vanillic acid, 0.68 for sinapic acid, 0.67 for syringic acid, 0.65 for caffeic acid, 0.5 for p-hydroxybenzoic acid and 0.35 for protocatechuic acid in a pH 5.8 nutrient culture.[5] In Portsmouth B_t-horizon soil (Typic Umbraquaalts, fine loamy, mixed, thermic; pH 5.2), they were 1, 0.67, 0.67, 0.7, 0.59, 0.38, 0.35, and 0.13, respectively.[19] The differences in phytotoxicity of the individual phenolic acids for nutrient culture and Portsmouth soil bioassays were due to various soil processes listed in the next paragraph and reduced contact (e.g., distribution and movement)[36] of phenolic acids with roots in soils.

The sources (i.e., input) of free phenolic acids in soil solutions are primarily from leachates of living leaves and litter, root exudates, release of bound forms and transformations by soil microorgansims.[8,32,34] The inactivation and losses (i.e., outputs) of free phenolic acids from soil solutions are primarily a result of ionization, oxidation, sorption onto soil particles, fixation into the recalcitrant organic matter (e.g., polymerization), seed and root uptake, and transformation and utilization by soil microorganisms.[7,9,10,15,21,22,29,31,40,46,47] Of these processes, transformation and utilization of phenolic aids by soil microorganisms are thought to have the greatest influence on phenolic acid concentrations in soil solutions. In fact, it has been hypothesized that simple phenolic acids are so readily metabolized by soil and rhizosphere microorganisms that allelopathic effects by phenolic acids are highly unlikely.[35,36] However, convincing evidence to support or reject this hypothesis is presently lacking.

In this chapter I describe how microbial activity may be estimated and what data are presently available on a) rates of microbial transformation and utilization, b) phenolic acid effects on soil and rhizosphere microbial populations, and c) the influences of soil and rhizosphere microbial populations on phenolic acid phytotoxicity. The resulting insight is then used to suggest a possible approach by which this hypothesis may be tested experimentally.

RESULTS AND DISCUSSION

Because phenolic acid concentrations in soil solutions are determined not only by input processes (e.g., leaching, exudation, release of bound forms) but also by output processes (e.g., sorption, polymerization, utilization by microorganisms), simply determining soil solution concentrations over time cannot provide information on how any one of these processes may actually influence the soil solution concentrations of phenolic acids. The effects of each process must be characterized separately. The impact of soil or rhizosphere microorganisms, for example, could be estimated by coupling changes in soil solution concentrations of phenolic acids with the activity of soil or rhizosphere microorganisms that can utilize phenolic acids as a carbon source. This approach, however, assumes that all the other output process rates remain constant.

ESTIMATING SOIL MICROBIAL ACTIVITY

Characterizing the release of bound forms associated with decomposition and/or mineralization and the transformation and the utilization of phenolic acids by soil microbial populations is an enormous challenge. Species of microbes in soil are very diverse (i.e., perhaps as many as 10^4 to 10^6 bacterial species per gram of soil),[43] have a broad range of metabolic capacities, catalyze reactions that can change soil physical and chemical properties, and are either in an active or an inactive state. Populations of microbial species range from a few per gram of soil to millions per gram of soil. The relative population density of a given microbial species in the soil is determined by its ability to reproduce and compete with other organisms (e.g., microbes, roots) for resources whenever physical and chemical requirements are above the required minimum level for that species. Most microbial species will form inactive resting states (i.e., quiescent) and/or survival structures (e.g., spores, endospores, sclerotia; i.e., dormant stages) whenever conditions fall below their minimum requirements. These quiescent and dormant stages become reactivated as soon as their minimum resource requirements are exceeded. Finally, the physical, chemical, and biological properties of soils are constantly changing; as soil properties change, so do the activities of microbial species within the soil.

Changes in microbial activity and thus, for example, phenolic acid transformation and/or utilization, can be monitored by changes in enzyme activity (e.g., peroxidases, catalases, dioxygenases), substrates (e.g., phenolic acids), products (transformed phenolic acids, breakdown products), respiration or population size (e.g., phenolic acid utilizing bacteria, actinomycetes and/or fungi). Because of the complexity and the interactions of soil processes, none of these changes, unfortunately, are precisely and consistently related to phenolic acid transformation and/or utilization by soil microbes.

Whenever energy-rich compounds are added to soil, changes in constitutive and/or induced microbial enzyme activity can be detected within just a few minutes to a few hours, as long as nothing else is limiting and enzyme activities are at sufficient levels to be detected. However, the induction of enzymes from previously unexpressed metabolic pathways (nonconstitutive enzymes) is generally much more easily recognized than changes in constitutive enzymes, which are also likely to be present in other soil organisms (e.g., roots, fauna). After depletion of energy-rich compounds, such as phenolic acids, the induced enzyme activity should return to preinduction levels as quickly as it was induced. The longevity of any enzyme activity in the soil will be determined by rates of enzyme synthesis and degradation. Thus, such changes in constitutive and induced enzymes are largely ephemeral unless inputs of energy-rich compounds at nontoxic levels are continued for extended periods and end products and toxic byproducts do not accumulate. However, even under the most ideal conditions, identifying, isolating, and quantifying constitutive and/or induced microbial enzymes in soils that relate directly to the transformation and/or utilization of specific organic molecules are extremely difficult. This is due to the fact that natural soils have a large number of free and bound enzymes not directly associated with soil microorganisms (e.g., roots, fauna), soil enzymes may act on a range of organic molecules, and the addition of organic molecules to soil frequently leads to the release of other bound or sorbed energy-rich organic molecules. At present, I am not aware of any research that specifically links soil enzyme activity with phenolic acid transformations or depletion in soil solutions.

Changes in soil respiration can also be detected within minutes to a few hours after addition of energy-rich compounds, such as phenolic acids, again as long as nothing else is limiting. After microorganisms have utilized the energy-rich compounds, their respiratory activity should return to preinduction levels fairly quickly unless the addition of the energy-rich compounds has modified the physicochemical and biotic status of the soil. I am not aware of any research to specifically link soil respiration with rates of phenolic acid depletion in soil solutions;

however, increases in soil respiration or rates of labeled $^{14}CO_2$ released after enrichment have been determined.[7,29,30,37]

Detectable changes due to cellular replication may take days, weeks, or even months depending on the soil environment, the microbes involved, and the starting population levels. Once microbial populations have developed, they may survive as quiescent or dormant structures for decades or longer, even though the conditions leading to their development no longer exist. Thus, obtaining separate estimates of active and inactive microbes in soils is essential. Unfortunately, this is extremely difficult. In most instances the best that can be done is to observe net changes of functional groups on selective media over time. I am aware of only one publication which has made an attempt to specifically link phenolic acid depletion from soil solutions with soil microbial populations.[3]

Although all of these approaches to estimate microbial activity (i.e., enzyme activity, utilization of substrates, formation of products, respiration, or changes in microbial populations) could be determined, only changes in microbial populations that can utilize phenolic acids as a sole carbon source have been related to phenolic acid depletion from soil solutions.[3]

MICROBIAL TRANSFORMATION AND UTILIZATION OF SIMPLE FREE PHENOLIC ACIDS – SUBSTRATES AND PRODUCTS

There is a substantial literature on the transformation of simple phenolic acids by microorganisms.[2,7,11,16,18,20,22,25,29,44] For example, ferulic acid is transformed by fungi to either caffeic acid or vanillic acid, and these are transformed to protocatechuic acid. Next the ring structure of protocatechuic acid is broken to produce β-carboxy-*cis,cis*-muconic acid, which is then converted to β-oxoadipic acid (Fig. 3.1), which in turn is broken down to acetic acid and succinic acid, and these ultimately are broken down to CO_2 and water.[11,18,29] However, distribution of residual ^{14}C-activity after growth of *Hendersonula toruloidea*, a fungus, in the presence of specifically ^{14}C-labeled ferulic acid ranged from 32 to 45% in CO_2, 34 to 45% in cells, 9 to 20% in humic acid and 4 to 10% in fulvic acid.[29] Thus, a considerable portion of the ferulic-acid carbon was bound/fixed over a 12-week period, and the initial ferulic acid transformation products (e.g., caffeic acid, vanillic acid and protocatechuic acid) were clearly of a transitory nature. Similar observations have also been made for other simple phenolic acids;[22,23] however, the proportions metabolized to CO_2 and fixed into cells and the soil organic matter varied considerably depending on the phenolic acid, the soil environment, and the microorganism involved.

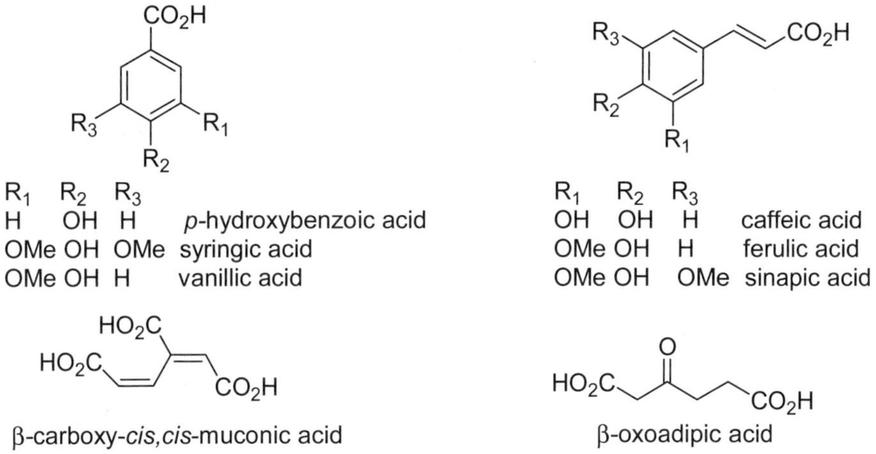
CO2H
R3
R1
R2
R1 R2 R3
H OH H p-hydroxybenzoic acid
OMe OH OMe syringic acid
OMe OH H vanillic acid
R3
CO2H
R2
R1
R1 R2 R3
OH OH H caffeic acid
OMe OH H ferulic acid
OMe OH OMe sinapic acid
HO2C
HO2C
CO2H
β-carboxy-cis,cis-muconic acid
O
HO2C
CO2H
β-oxoadipic acid

Figure 3.1

Allelopathic agents used in this study and some of their degradation products.

Rates of transformation and/or disappearance of phenolic acids in soil solutions have also been determined under a variety of circumstances and in various soils.[2,22,23,31,42,44] In general, there is a rapid initial transformation (e.g., loss of the carboxylic acid group) of phenolic acids. For example, 90% of the carboxylic acid carbon of *p*-hydroxybenzoic acid, syringic acid, and vanillic acid was lost within 1 week.[22] Losses of other side chain carbons or ring carbons, however, took considerably longer, on the order of weeks to months.[22,23,29]

In addition, there are a number of other factors besides chemical structure that determine rates;[17] for example, transformation of *p*-coumaric acid was delayed in the presence of glucose, phenylalanine, and *p*-hydroxybenzoic acid, but not methionine, in Cecil B_t - horizon soils.[31] The delays suggested preferential utilization of carbon sources by soil microorganisms. Additional evidence for differential carbon utilization in soils has also been provided by Martin and Haider[29] and Haider et al.,[23] who observed that mineralization of ^{14}C-labeled ring carbon of glucose was more rapid than from phenolic acids, and by Sugai and Schimel,[42] who observed in taiga soils from a series of successional stages that glucose, *p*-hydroxybenzoic acid and salicylic acid were processed very differently by soil microorganisms. More than twice as much glucose was converted to biomass than either of the phenolic acids, and although both phenolic acids were metabolized, only *p*-hydroxybenzoic acid was assimilated by the microbes. Finally, any environmental factors (e.g., temperature, pH, moisture, nutrition) that influence the activity of soil microorganisms or make phenolic acids less available to microbes (e.g., increased sorption by soil particles) will influence the transformation rates of free phenolic acids. To provide some insight into how a range of soil physicochemical and biotic factors may influence the rates of transformation of phenolic acids, I will compare the rates in an open and a closed system.

Steady state rates of transformation, as measured by the disappearance of *p*-coumaric acid after adjustment for soil fixation, in a continuous flow Cecil A_p - horizon soil (pH 5.0; typic Kanhapludult, clayey, kaolinitic, thermic) system (i.e., open system) supplied with a range of nutrient concentrations (0 to 50% Hoagland's solution) and 187 µg/h (53.4 µg/mL, 3.5 mL/h) of *p*-coumaric acid for 72 h at room temperature ranged from 0.035 to 0.076 picomoles/CFU (colony forming units) of *p*-coumaric acid utilizing bacteria/h, with a mean ± standard error of 0.047 ± 0.006.[3] Bacterial populations that could utilize *p*-coumaric acid ranged from 1.07 x 10^5 to 4.00 x 10^5 CFU/g soil depending on nutrient levels supplied to the system. These microbial communities, we suspect, were derived from quiescent and dormant bacteria, since laboratory stored air-dried soil was used for this study. For these calculations it was assumed that all the CFU determined after the 72 h treatment represented active *p*-coumaric acid utilizing bacteria. The initial *p*-coumaric acid utilizing bacterial populations for the air-dried soil were 0.64 x 10^5 CFU/g soil. The impact of other soil microorganisms on phenolic acid transformation were not determined.

In test tubes (i.e., closed system; unpublished data) containing 1 g air-dried autoclaved Cecil A_p - horizon soil (pH 5.0), 82 µg *p*-coumaric acid, Hoagland's solution (all solutions adjusted to pH 5.0), and soil extract for inoculum (total of 1.5 ml) the average linear transformation rates for *p*-coumaric acid over 48 hr, once microbial utilization was evident, were 3.6 x 10^{-4} ± 1.7 x 10^{-4} picomole/CFU of *p*-coumaric acid utilizing bacteria/h, about 130 times slower than what was observed for the mean utilization in the steady-state continuous flow system. The CFU of *p*-coumaric acid utilizing bacteria/g soil in the test tube system averaged 1.46 x 10^8 over the 48 h interval. Initial CFU of *p*-coumaric acid utilizing bacterial populations/g soil 24 hr after addition of inoculum were 105 ± 15. Utilization of *p*-coumaric acid by microbes in the test tubes was determined by 0.25 M EDTA (pH 7.0) extractions at 6 h intervals and HPLC analyses.[2] CFU for bacteria that utilized *p*-coumaric acid as a sole carbon source were also determined at 6 h intervals by the plate-dilution frequency technique[24] utilizing 0.5 mM *p*-coumaric acid as the sole carbon source in a basal medium.[3,7]

Differences in rates of transformation and/or utilization between the two systems are possibly due to a) constant input vs. single input of *p*-coumaric acid and nutrient solution, b) aerobic (open system) vs. more anaerobic (closed system) conditions, c) little chance for accumulation of transformation products and/or toxic microbial byproducts (constant flushing of system) vs. potential build up of transformation products and/or toxic microbial byproducts (closed system), d) different microbial communities both in terms of species (air-dried soil vs. autoclaved-inoculated soil) and numbers (10^5 vs. 10^8), and e) input of *p*-coumaric acid (53 µg/mL/h or 187 µg/h vs. 58 µg/mL one time addition) added to different amounts of soil (60 g of soil for the flow-through system vs. 1 g of soil for the test tube system).

Given the bacterial populations that utilized *p*-coumaric acid as a sole carbon source and the physicochemical (e.g., constant temperature, adequate nutrition and moisture) and biotic conditions of these two laboratory systems, utilization of *p*-coumaric acid ranged from 0.6 to 5.0 µg/g soil/h for the open systems and 8.6 µg/g soil/h for the closed system. The µg values for the open system represent steady-state rates as modified by nutrition, while the µg values for the closed system represent maximum rates. Whether such rates ever occur in field soils is not known, since the physicochemical and biotic environments of field soils are so different from those of laboratory systems. Laboratory soil systems provide potential rates of utilizations, but until field rates are determined the importance of microbial activity in phenolic acid depletion from soil solutions will not be known.

PHENOLIC ACID UTILIZING MICROBIAL POPULATIONS IN SOILS AND RHIZOSPHERE

Unfortunately there are only limited data available on soil microbial populations that can utilize phenolic acids as a carbon source. The CFU/g soil ranged from 10^4 to 10^{10} for bacteria in phenolic acid enriched air-dried soil and not enriched field soils (Table 3.1). However, these numbers of active, quiescent, and/or dormant individuals represent only a small fraction of the bacteria in these soils, since < 1% of the viable bacteria in soils can be cultured.[43] The assumption is that the CFU based on cultures are representative or consistently related to the total phenolic acid utilizing microorganisms in the soil.

Table 3.1

Observed range of soil bacterial populations that can utilize phenolic acids as a sole carbon source in Cecil A-horizon soil.

CFU/g soil	Soil	Phenolic Acid Enriched	Reference
Laboratory soils			
10^4 to 10^5	air-dried[a]	yes	Blum et al.[3]
10^7 to 10^{10}	air-dried	yes	Blum et al.[9]
Field soils (June 2001)[b]			
10^7	corn field	no	unpublished data
10^7	pine plantation	no	
10^8	wheat no till	no	
10^7	wheat conventional till	no	

[a] Collected in 1992

[b] Soil samples were collected from field plots located on North Carolina State University Field Stations and immediately processed; 0.5 mM phenolic acid mixture composed of 0.125 mM ferulic, *p*-coumaric, vanillic and *p*-hydroxybenzoic acid was used as sole carbon source in a basal medium.[7]

Table 3.2

Observed range of soil microbial populations in the rhizosphere that can utilize phenolic acids as a carbon source.

CFU/g root[a]	Soil	Enrichment	Reference
Laboratory soils - Cucumber seedling rhizosphere			
10^9	Portsmouth B	individual phenolic acids	Shafer and Blum[33]
10^9 to 10^{10}	Cecil A	mixtures of phenolic acids	Blum et al.[9]
10^{11} to 10^{13}	Cecil A	individual or mixtures of phenolic acids	Staman et al.[39]
10^{11} to 10^{12}	Cecil A	sunflower tissue	
10^{11} to 10^{12}	Cecil A	wheat tissue	
Field soils (June 2001)[b] - Wheat rhizosphere			
10^7	Cecil A	none	unpublished data

[a] Determined with basal medium[7] containing phenolic acids as the sole carbon source

[b] Soil samples were collected from field plots located on North Carolina State University Field Stations and immediately processed. 0.5 mM phenolic acid mixture composed of 0.125 mM ferulic, *p*-coumaric, vanillic and *p*-hydroxybenzoic acid was used as sole carbon source in a basal medium.[7]

As with soil microbes, the data available for microbes in the rhizosphere that can utilize phenolic acids as a carbon source are also very limited. Phenolic acid utilizing microbial populations observed within the rhizosphere of cucumber seedlings and mature wheat plants range from 10^7 to 10^{13} CFU/g root (Table 3.2).

For both the soil and the rhizosphere data (Tables 3.1 and 3.2) the bacteria that can co-metabolize phenolic acids but not utilize phenolic acids as a sole carbon source are not included in these numbers. Other soil and rhizosphere microbes (e.g., fungi and actinomycetes) that can utilize phenolic acids as a carbon source have also not been determined. The present data base for phenolic acid utilizing microorganisms in the bulk-soil and the rhizosphere is clearly inadequate.

EFFECTS OF FREE PHENOLIC ACIDS ON BULK-SOIL AND RHIZOSPHERE POPULATIONS

Phenolic acid enrichment of soils may stimulate, inhibit or have no obvious effect on microbial populations in the bulk soil and/or rhizosphere.[3,7,8,9,26,27,33,44] For example, populations of bacteria, fungi, and actinomycetes in Portsmouth A- and B-horizon soils were affected differently by repeated enrichment with ferulic acid, p-coumaric acid, p-hydroxybenzoic acid, or vanillic acid.[7] Responses varied with the type of soil material (i.e., A- vs. B-horizon), phenolic acid concentration, and inorganic nutrient status of the soil (Table 3.3). Population changes in response to phenolic acid enrichment were greater in B- than in A-horizon soils. In general, bacterial and fungal populations were stimulated by these phenolic acids at concentrations $\leq$ 0.5 µmol/g soil (Table 3.3). Concentrations $\geq$ 0.5 µmol/g soil (maximum concentration used 0.75 µmol/g soil) in some instances reduced or inhibited populations but in others continued to increase or stimulate populations. Since glucose was used as the carbon source in the selective media for this study, the CFU and the changes in CFU in Table 3.3 include all organisms that can utilize glucose as a carbon source; thus, only a fraction of these CFU actually represent microbes that can utilize phenolic acids as a carbon source.

Detectable changes in microbial populations after phenolic acid enrichment of Portsmouth soil took only days.[7] Bacterial populations in soil treated with multiple treatments of 0.5 µmol/g ferulic acid reached maximum population levels in 5 and 3 days in the A- and B- horizon soils, respectively, and then declined. Bacterial populations in soil not treated with phenolic acids increased somewhat over the first 24 to 72 h and thereafter remained fairly constant. Fungal populations in soil treated with 0.5 µmol/g ferulic acid increased most rapidly after 5 days, while fungal populations of soil not treated with phenolic acids did not change significantly.

Table 3.3

Significant responses of bacteria, fungi, and actinomycetes in Portsmouth to phenolic acid treatments.[a]

Soil	Microorganism	Phenolic Acid	log CFU/g soil 0^c	0.5^c	Change CFU/g soil	(%)
A	bacteria[b]	ferulic acid	5.30	5.81	$+4.46 \times 10^5$	(224)
		p-coumaric	5.29	5.88	$+5.64 \times 10^5$	(289)
		vanillic	5.32	5.89	$+5.67 \times 10^5$	(272)
B	bacteria	ferulic acid	5.06	6.52	$+3.20 \times 10^6$	(2784)
		p-coumaric	5.11	6.09	$+1.10 \times 10^6$	(855)
		p-hydroxybenzoic	5.08	6.25	$+1.66 \times 10^6$	(1379)
		vanillic	5.05	5.62	$+3.05 \times 10^5$	(272)
A	fungi	ferulic acid	2.18	3.06	$+9.97 \times 10^2$	(659)
B	fungi	ferulic acid	1.57	2.84	$+6.55 \times 10^2$	(1762)
		p-hydroxybenzoic	1.76	2.58	$+3.23 \times 10^2$	(561)
B	actinomycetes	p-coumaric acid	3.03	2.14	-9.33×10^2	(-87)

[a] For selective media used and details regarding soil treatments see Blum and Shafer.[7] Selective media used glucose as the carbon source.
[b] Bacteria — fast growing bacteria that produce colonies $\geq$ 1 mm diameter in 6 days
[c] µmol/g phenolic acid treatments

Blum et al.[9] isolated bacterial colonies from Cecil A-horizon soils treated with individual phenolic acids, either p-coumaric acid or vanillic acid, and then tested these isolated bacterial colonies for their ability to utilize only p-coumaric acid, only vanillic acid, or both phenolic acids. They found that the majority of isolates (>72%) could utilize both phenolic acids while a much smaller fraction (<28%) could only use the phenolic acid with which the soil had been treated. Since soils contain a variety of phenolic acids, as well as other organic molecules, Blum et al.[9] subsequently determined changes in phenolic acid utilizing bacterial populations after Cecil A-horizon soils were enriched with an equal molar mixture composed of 7 phenolic acids plus or minus glucose. Since the addition of glucose did not modify the increase of phenolic acid utilizing bacteria (approximately 1000% for the 0.25 µmol/g soil phenolic acid treatment), Blum et al.[9] concluded that the reduced microbial utilization of phenolic acids observed in the presence of glucose by Pue et al.[31] was due to preferential utilization of glucose over phenolic acids and not due to a reduction in induction and/or selection of phenolic acid utilizing bacteria in the soil.

Adequate mineral nutrition is extremely important in determining the soil populations of bacteria that can utilize phenolic acids as a carbon source. Blum et al.[3] observed that when Cecil A-horizon soil (initially nutrient limited) was supplied with 53 µg/mL (3.5 mL/h) of *p*-coumaric acid and a range of nutrient concentrations for 72 h, the populations of phenolic acid utilizing bacteria increased in a linear manner as nutrient concentration was increased.

Table 3.4

Significant response of phenolic acid utilizing cucumber seedling rhizosphere bacteria to phenolic acid, sunflower leaf tissue, or wheat plant tissue.

Soil	Treatment	per g soil	Maximum % Stimulation	Reference
Portsmouth-B	individual phenolic acids	0.5 µmol	600[a]	Shafer and Blum[33]
Cecil -A	7 phenolic acid mixture	0.5 µmol	2542	Blum et al.[9]
	4 phenolic acid mixture	0.6 µmol	1201	
	individual phenolic acids	0.6 µmol	989	
Cecil-A	4 phenolic acid mixture	0.6 µmol	655	Staman et al.[39]
	wheat shoot	5 mg	496	
autoclaved[b] Cecil - A	sunflower shoot	5 mg	4798[c] or 279[d]	

[a] Ferulic acid as the carbon source in selective medium
[b] Autoclaved only to reduce initial microbial populations in the soil; soil was not sterilized
[c] Chlorogenic acid as carbon source in selective medium
[d] A mixture of four phenolic acids as carbon source in selective medium

Phenolic acid enrichment of soils containing roots can also lead to an increase in rhizosphere microbial populations which can utilize phenolic acids as a carbon source.[9,33,39] Shafer and Blum[33] observed that the addition of up to 0.25

μmol/g of a phenolic acid to Portsmouth B-horizon soil did not detectably modify rhizosphere bacteria or fungi of cucumber seedlings, but concentrations of 0.5 or 1 μmol/g soil did, suggesting that at the lower concentrations bulk-soil microorganisms and soil fixation prevented phenolic acids from reaching the rhizosphere or that other organic molecules dominated as the carbon source for the rhizosphere microorganisms. Blum et al.[9] observed that the maximum stimulation of rhizosphere bacteria occurred when a 0.6 μmol/g of a 4-way equal molar phenolic acid mixture was supplied to Cecil A-horizon soil containing cucumber seedling roots (Table 3.4). Furthermore they observed that there was an inverse linear relationship between phenolic acid utilizing bacteria and the absolute growth rate of cucumber leaf expansion. Staman et al.[39] also found such a relationship between cucumber leaf expansion and phenolic acid utilizing rhizosphere bacteria when sunflower tissue, containing high phenolic acids concentrations, was added to Cecil A-horizon soil, but only when this soil was first autoclaved to reduce the initial microbial populations. This suggests that increases in microbial populations associated with enrichments can only be observed when energy-rich compounds, such as phenolic acid, are initially limiting.

In summary, detectable population changes of soil and rhizosphere phenolic acid utilizing microorganisms in laboratory systems to phenolic acid enrichment are a function of a variety of soil physicochemical and biotic factors, including phenolic acid enrichment concentrations, presence of other available organic molecules, nutrition, soil type, and initial microbial populations. We have no similar data for field soils systems, something that needs to be determined.

INFLUENCE OF BULK-SOIL AND RHIZOSPHERE MICROORGANISMS ON PHYTOTOXICITY

That microorganisms can reduce the observed phytotoxic effects of phenolic acids has been observed by a number of researchers.[3,7,8,33,37,38,39,41,45] I am, however, not aware of any study that has attempted to quantify how changes in bulk-soil bacteria might influence the phytotoxicity of phenolic acids. I am aware of only one study that has attempted to quantify how changes in rhizosphere microbial populations may influence the phytotoxicity of phenolic acids. Blum et al.[9] observed that a 500% increase of phenolic acid utilizing bacteria in the rhizosphere of cucumber seedlings growing in Cecil A-horizon soil enriched with an equimolar mixture of 0.6 μmol/g *p*-coumaric acid, ferulic acid, *p*-hydroxybenzoic acid, and vanillic acid resulted in a 5% reduction in the inhibition of absolute growth rates of their leaves. Assuming that this relationship were true for the soil-plant-microbe systems in Table 3.4, and ignoring the autoclaved sunflower tissue soil system, the inhibition of leaf expansion of cucumber seedlings would be reduced by a maximum of 25% for these systems. This is much less than what would be expected by some,[35,36] but not by others.[48] Much more data will be required to resolve this controversy.

What is needed at this point are quantitative data on phenolic acid utilizing microbes (e.g., bacteria, fungi, actinomycetes) in field soils ± phenolic acid enrichment during the spring, summer, and fall for various crop and forest systems. In addition, we need quantitative data describing the relationships between bulk-soil and rhizosphere phenolic acid utilizing microbes and the observed phytotoxicity of phenolic acids for sensitive species, i.e., a form of dimension analysis utilizing equations to predict useful and/or consistent relationships.

ONE POSSIBLE APPROACH TO PROVIDING THE NECESSARY DATA TO TEST THE HYPOTHESIS

A range of concentrations of individual phenolic acids, mixtures of phenolic acids, or organic mixture including phenolic acids plus or minus nutrients and water plus or minus nutrients (control) could be added in solution form to soil surfaces in the field by direct applications or drip irrigation. However, surface applications of such solutions to field soil without some confinement are not advisable. The insertion of pipes into the soil (i.e., open ended soil cores) would be an effective way to eliminate horizontal movement and at the same time provide better phenolic acid distribution throughout the soil column. Such soil columns could also be brought to the laboratory, sterilized, and then used to determine soil sorption.[13] Sufficient amounts or rates of solutions should be added for gravitational water to reach the bottom of the columns. Since soil columns will function as chromatographic columns, sufficient amounts and concentrations should also be used to obtain the desirable depth of phenolic acid penetration. The depth of penetration could be determined by subsampling of the columns or by the use of lysimeters at various depths. There is an additional benefit to the use of soil columns: seedlings could be planted into the soil cores, providing a means of regulating phenolic acid contact with roots. A laboratory version of this system has been used and described by Blum et al.[3]

Phenolic acids from lysimeters and/or extracted from the soil within the cores by water or EDTA could be analyzed by a high-performance liquid chromatography to determine transformation products and the rates of disappearance of available (free and reversibly sorbed) phenolic acids.[10,12,14] Soil and rhizosphere microbial populations that can utilize phenolic acids as a carbon source could be determined by the plate-dilution frequency technique[24] using appropriate selective media containing phenolic acids as the sole carbon source.[7,9,33,39] Observed phytotoxicity could be determined by monitoring leaf expansion of an appropriate species, for example, since that can be done nondestructively.[4] The relationships between soil and rhizosphere phenolic acid utilizing microorganisms and observed phytotoxicity could thus be determined for these core systems using the approach described by Blum et al.[9]

ACKNOWLEDGMENTS

The author wishes to thank Dr. Tom Wentworth for reviewing this manuscript and for his valuable and helpful suggestions.

REFERENCES

(1) Blum, U., **1995**. The value of model plant-microbe-soil systems for understanding processes associated with allelopathic interactions: One example. In: Inderjit, Dakshini, K. M. M., Einhellig, F. A. (Eds.), *Allelopathy: organisms, processes, and applications*. ACS Symposium Series No. 582. American Chemical Society, Washington D.C, 127-131

(2) Blum, U., **1998**. Effects of microbial utilization of phenolic acids and their phenolic acid breakdown products on allelopathic interactions. *J. Chem. Ecol.* **24**, 685-708

(3) Blum, U., Austin, M. F. and Shafer, S. R., **1999**. The Fate and effects of phenolic acids in a plant-microbial-soil model system. In: Macías, F. A., Galindo, J. C. G., Molinillo, and J. M. G., Cutler, H. G. (Eds.), *Recent Advances on Allelopathy: a Science for the Future*. Cadiz University Press, Cadiz, Spain, 159-166

(4) Blum, U. and Dalton, B. R., **1985**. Effects of ferulic acid, an allelopathic compound, on leaf expansion of cucumber seedlings grown in nutrient culture. *J. Chem. Ecol.* **11**, 279-301

(5) Blum, U., Dalton, B. R. and Shann, J. R., **1985**. Effects of various mixtures of ferulic acid and some of its microbial metabolic products on cucumber leaf expansion and dry matter in nutrient culture. *J. Chem. Ecol.* **11**, 619-641

(6) Blum, U. and Rebbeck, J., **1989**. Inhibition and recovery of cucumber roots given multiple treatments of ferulic acid in nutrient culture. *J. Chem. Ecol.* **15**, 917-928

(7) Blum, U. and Shafer, S. R. **1988**. Microbial populations and phenolic acids in soil. *Soil Biol. Biochem.* **20**, 793-800

(8) Blum, U., Shafer, S. R. and Lehman, M. E., **1999**. Evidence for inhibitory allelopathic interactions involving phenolic acids in field soils: concepts vs. an experimental model. *Crit. Rev. Plant Sci.* **18**, 673-693

(9) Blum, U., Staman, K. L., Flint, L. J. and Shafer, S. R., **2000**. Induction and/or selection of phenolic acids-utilizing bulk-soil and rhizosphere bacteria and their influence on phenolic acid phytotoxicity. *J. Chem. Ecol.* **26**, 2059-2078

(10) Blum, U., Worsham, A. D., King, L. D. and Gerig, T. M., **1994**. Use of water and EDTA extractions to estimate available (free and reversibly bound) phenolic acids in Cecil soils. *J. Chem. Ecol.* **20**, 341-359

(11) Dagley, S. **1971**. Catabolism of aromatic compounds by microorganisms. *Advan. Microbial. Physiol.* **6**, 1-42

(12) Dalton, B. R., **1999**. The occurrence and behavior of plant phenolic acids in soil environments and their potential involvement in allelochemical interference interactions: Methodological limitations in establishing conclusive proof of allelopathy. In: Inderjit, Dakshini, K. M. M., and Foy, C. L., (Eds.), *Principles and Practices in Plant Ecology: Allelochemical Interactions*, CRC Press, Boca Raton, FL, 57-74

(13) Dalton, B. R., Blum, U. and Weed, S. B., **1989**. Plant phenolic acids in soils: Sorption of ferulic acid by soil and soil components sterilized by different techniques. *Soil Biol. Biochem.* **21**, 1011-1018

(14) Dalton, B. R., Weed, S. B. and Blum, U., **1987**. Plant phenolic acids in soils: a comparison of extraction procedures. *Soil Sci. Soc. Am. J.* **51**, 1515-1521

(15) Dao, T. H., **1987**. Sorption and mineralization of phenolic acids in soil. In: Waller, G. R., (Ed.), *Allelochemicals: Role in Agriculture and Forestry*. ACS Symposium Series No. 330. American Chemical Society, Washington D.C., 358-370

(16) Dennis, D. A., Chapman, P. J., and Dagley, S., **1973**. Degradation of protocatechuate in *Pseudomonas testosteroni* by a pathway involving oxidation of the products of metafission. *J. Bact.* **113**, 521-523

(17) Elliot, L. F. and Cheng, H. H., **1987**. Assessment of allelopathy among microbes and plants. In: Waller, G. R., (Ed.), *Allelochemicals: Role in Agriculture and Forestry*. ACS Symposium Series No. 330. American Chemical Society, Washington D.C., 504-515

(18) Evans, W. C., **1963**. The microbial degradation of aromatic compounds. *J. Gen. Microbiol.* **32**, 177-185

(19) Gerig, T. M. and Blum, U. **1991**. Effects of mixtures of four phenolic acids on leaf area expansion of cucumber seedlings grown in Portsmouth B_1 soil materials. *J Chem. Ecol.* **17**, 29-40

(20) Gibson, D. T., **1968**. Microbial degradation of aromatic compounds. *Science* **161**, 1093-1097

(21) Greenland, D. J., **1965**. Interactions between clays and organic compounds in soils. Part 1. Mechanisms of interactions between clays and defined organic compounds. *Soil Fertil.* **28**, 415-425

(22) Haider, K. and Martin, J. P., **1975**. Decomposition of specifically carbon-14 labeled benzoic and cinnamic acid derivatives in soil. *Soil Sci. Soc. Amer. Proc.* **39**, 657-662

(23) Haider, K., Martin, J. P., and Rietz, E., **1977**. Decomposition in soil of ^{14}C-labeled coumaryl alcohols; free and linked into dehydropolymer and plant lignins and model humic acids. *Soil Sci. Soc. Am. J.* **41**, 556-562

(24) Harris, R. F. and Sommers, L. E. **1968**. Plate-dilution frequency technique for assay of microbial ecology. *Applied Microbiology* **16**, 330-334

(25) Hartley, R. D. and Whitehead, D. C., **1985**. Phenolic acids in soils and their influence on plant growth and soil microbial processes. In: Vaughan, D. and Malcolm, R. E., (Eds.), *Soil Organic Matter and Biological Activity*. Martinus Nijhoff. Dr. W. Junk Publishers, Dordrech, Netherlands, 109 -149

(26) Henderson, M. E. K., **1956**. A study of the metabolism of phenolic compounds by soil fungi using spore suspension. *J. General. Micro.* **14**, 684-691

(27) Kunc, F., **1971**. Decomposition of vanillin by soil microorganisms. *Folia Microbiologia* **16**, 41-50

(28) Lehman, M. E. and Blum, U., **1999**. Evaluation of ferulic acid uptake as a measurement of alleochemical dose: effective concentration. *J. Chem. Ecol.* **25**, 2585-2600

(29) Martin, J. P. and Haider, K., **1976**. Decomposition of specifically carbon-14-labeled ferulic acid: free and linked into model humic acid-type polymers. *Soil Sci. Soc. Am. J.* **40**, 377-380

(30) Martin, J. P. and Haider, K., **1979**. Effect of concentration on decomposition of some ^{14}C-labeled phenolic compounds, benzoic acid, glucose, cellulose, wheat straw, and *Chlorella* protein in soil. *Soil Sci. Soc. Amer. J.* **43**, 917-920

(31) Pue, K. J., Blum, U., Gerig, T. M., and Shafer, S. R., **1995**. Mechanism by which noninhibitory concentrations of glucose increase inhibitory activity of *p*-coumaric acid on morning-glory seedling biomass accumulation. *J. Chem. Ecol.* **21**, 833-847

(32) Rice, E. L., **1984**. *Allelopathy*. Academic Press, New York, 422

(33) Shafer, S. R. and Blum, U., **1991**. Influence of phenolic acids on microbial populations in the rhizosphere of cucumber. *J. Chem. Ecol.* **17**, 369-389

(34) Siqueira, J. O., Nair, M. G., Hammerschmidt, R. and Safir, G. R., **1991**. Significance of phenolic compounds in plant-soil-microbial systems. *Crit. Rev. Plant Sci.* **10**, 63-121

(35) Smith, S. K., **1988**. Degradation of juglone by soil bacteria. *J. Chem. Ecol.* **14**, 1561-1571

(36) Smith, S. K. and Ley, R. E., **1999**. Microbial competition and soil structure limit the expression of allelopathy. In: Inderjit, Dakshini, K. M. M., Foy, C. L., (Eds.), *Principles and Practices in Plant Ecology: Allelochemical Interactions*. CRC Press, Boca Raton, FL, 339-351

(37) Sparling, G. P., Ord, B. G. and Vaughan., D., **1981**. Changes in microbial biomass and activity in soils amended with phenolic acids. *Soil Biol. Biochem.* **13**, 455-460

(38) Sparling, G. P. and Vaughan, D., **1981**. Soil phenolic acids and microbes in relation to plant growth. *J. Sci. Food Agric.* **32**, 625-626

(39) Staman, K., Blum, U., Louws, F. and Robertson, D., **2001**. Can simultaneous inhibition of seedling growth and stimulation of rhizosphere bacterial populations provide evidence for phytotoxin transfer from plant residues in the bulk soil to the rhizosphere of sensitive species? *J. Chem. Ecol.* **27**, 807-829

(40) Stevenson, F. J., **1982**. *Humus Chemistry, Genesis, Composition, Reaction*. John Wiley & Sons, New York, 443

(41) Stowe, L. G. and Osborn, A., **1980**. The influence of nitrogen and phosphorus levels on the phytotoxicity of phenolic compounds. *Can. J. Bot.* **58**, 1149-1153

(42) Sugai, S. F. and Schimel, J. P., **1993**. Decomposition and Biomass incorporation of ^{14}C-labeled glucose and phenolics in taiga forest floor: effects of substrate quality, successional state, and season. *Soil Biol. Biochem.* **25**, 1379-1389

(43) Sylvia, D. M., Fuhrmann, J. J., Hartel, P. G. and Zuberer, D. A., **1998**. *Principles and Applications of Soil Microbiology*. Prentice Hall, NJ, 550

(44) Turner, J. A. and Rice, E. L., **1975**. Microbial decomposition of ferulic acid in soil. *J. Chem. Ecol.* **1**, 41-58

(45) Vaughan, D., Sparling, G. P., and Ord, B. G., **1983**. Amelioration of the phytotoxicity of phenolic acids by some soil microbes. *Soil Biol. Biochem.* **15**, 613-614

(46) Wang, T. S. C., Huang, P. M., Chou, C.-H. and Chen, J.-H., **1986**. The role of soil minerals in the abiotic polymerization of phenolic compounds and formation of humic substances. In: Huang, P. M., Schnitzer, M., (Eds.), *Interactions of Soil Minerals with Natural Organics and Microbes*. Soil Science Society of America Publication No. 17. Soil Science Society of America, Madison, WI, 251-281

(47) Wang, T. S. C., Song, W. L. and Ferng, Y. L., **1978**. Catalytic polymerization of phenolic compounds by clay minerals. *Soil Sci.* **126**, 15-21

(48) Williamson. G. B., **1990**. Bacterial degradation of juglone. Evidence against allelopathy? *J. Chem. Ecol.* **16**, 1739-1742

(49) Willis, R. J., **1985**. The historical bases of the concept of allelopathy. *J. History Biology* **18**, 71-102

4 Benzoxazolin-2(3*H*)-ones – Generation, Effects and Detoxification in the Competition among Plants

D. Sicker, H. Hao and M. Schulz

Dedicated to Professor Horst Hennig on the occasion of his 65th birthday

CONTENT

Abstract 77
Introduction 78
Formation, Distribution and Biological Interactions in the Resistance
of Crop Plants 83
    Biosynthesis 83
    Chemical Synthesis of Aglucones and Acetal Glucosides 84
    Molecular Mode of Action 84
    Analytical Investigations 85
    Allelopathy 85
Results and Discussion 86
Methodology 92
References 94

ABSTRACT

Benzoxazolin-2(3*H*)-one is a phytotoxic allelochemical resulting from a two step degradation and root exudation process based upon an acetal glucoside of the 2,4-dihydroxy-2*H*-1,4-benz-oxazin-3(4H)-one skeleton. Such **benzoxazinoids** or **cyclic hydroxamic acids** occur in Acanthaceae, Poaceae, Ranunculaceae, and Scrophulariaceae. Their aglucones and resulting benzoxazolinones act as a plant's self resistance factors in case of pest attacks and as chemical weapons, in the soil, in competition with sensitive plant species. The driving force for investigations is to make agricultural use of these properties because benzoxazinoids are biosynthesized in cereals, like rye, wheat, and maize (Poaceae). However, the use of a derivative benzoxazolinone (BOA) as a natural herbicide may be rather limited, since numerous weeds, especially monocots, are able to detoxify the compound. At present we have identified four detoxification products: BOA-6-O-glucoside and the intermediate BOA-6-OH, as well as the nontoxic glucose carbamate and a gentiobioside carbamate. Carbamate is the major product in Poaceae over time. Efficient and fast induction of the enzymes involved in carbamate synthesis, together with exudation of the products, is regarded as a reason for a loss of pronounced BOA sensitivity. The detoxification effect found in weeds and crops reflects a membership in defined plant communities which seems to have been influenced by long term herbicides applications, a succession of cultivated crops, and the density of benzoxazinone containing species.

INTRODUCTION

In 1955, a first indication of the chemical basis for the increased resistance of rye plants towards pathogenic fungi was discovered: **benzoxazolin-2(3H)-one** (acronym: BOA).[78] Soon, it was found that BOA had two plant precursors, a benzoxazinoid acetal glucoside and its aglucone.[39,79] Thus, BOA is the final product of a degradation process of a (2R)-2-β-D-glucoside of 2,4-di-hydroxy-2H-1,4-benzoxazin-3(4H)-one (acronym: DIBOA-Glc) which undergoes enzymatic cleavage to its aglucone 2,4-dihydroxy-2H-1,4-benzoxazin-3(4H)-one (acronym: DIBOA) followed by a chemical fission into BOA and formic acid as shown in Figure 4.1.

These early findings, together with the ability to act as plant resistance factors towards pests, led to a cascade of some hundred papers dealing in an interdisciplinary manner with all aspects of **benzoxazinoids**. This name seems to be more correct than the name **cyclic hydroxamic acids**, which has often been used to name this class of compounds. However, it is not exact from the structural point of view, because it omits the fact that also a variety of acetal glucosides with lactam units was found in plants. They also may be split enzymatically to form lactam aglucones, but cannot undergo chemical decay into benzoxazolin-2(3H)-ones. Benzoxazinoids occur in Acanthaceae, Ranunculaceae, Scrophulariaceae, and last, but not least, Poaceae, mainly in cereals, like maize, rye, and wheat. In most cases, reported benzoxazinoids were discovered by chance. Usually, no systematic search has been undertaken for them.

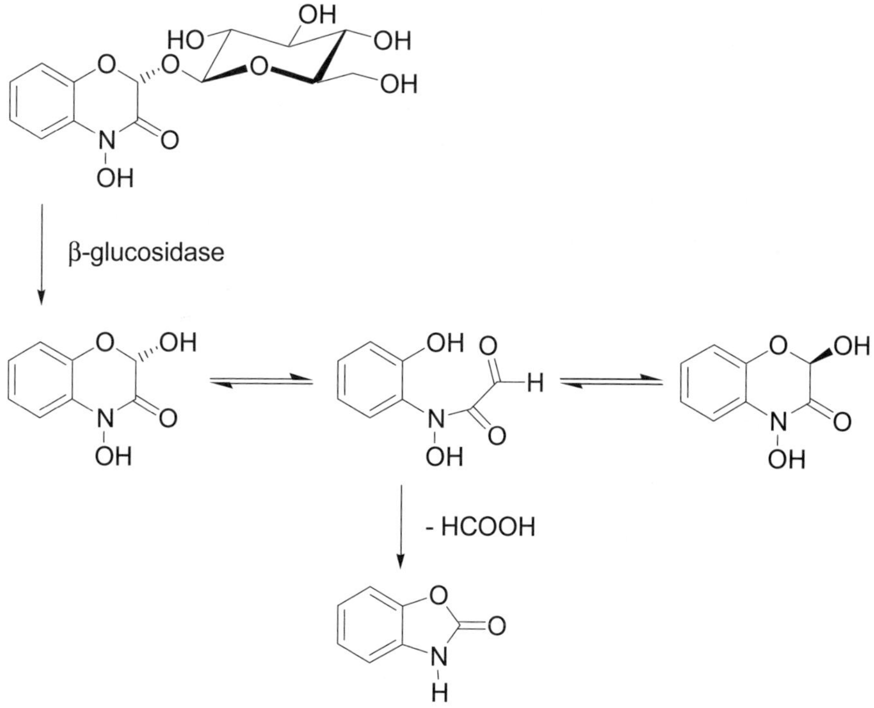

Figure 4.1

Enzymatic and chemical degradation of the benzoxazinoid acetal glucoside DIBOA-Glc.

The occurrence in some cereal species is a strong driving force for all investigations. Hence, the aim to make agricultural use of the results, either by breeding cereals with optimum benzoxazinoid content in the juvenile state, or by achieving a gene transfer for benzoxazinone biosynthesis into other plants of agricultural interest, is a possible future goal.

Because this paper deals in particular with benzoxazolinone chemistry, we will only present a survey of naturally occurring acetal glucosides with cyclic hydroxamic acid moieties (Fig. 4.2) as precursors of aglucones (Fig. 4.3) which may decompose to (substituted) benzoxazolin-2(3H)-ones (Fig. 4.4).

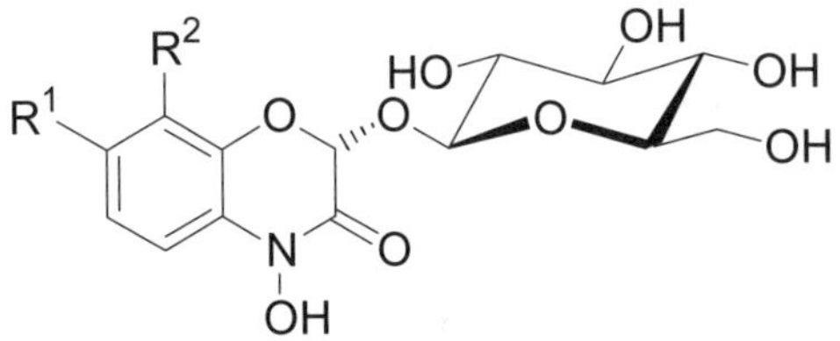

Figure 4.2
Natural benzoxazinone acetal glucosides with cyclic hydroxamic acid unit.

Acronym	R^1	R^2	Species (family), one reference cited for each plant
DIBOA-Glc	H	H	Secale cereale (Poaceae)[34] Triticum aestivum (Poaceae)[41] Consolida orientalis (Ranunculaceae)[60] Acanthus mollis (Acanthaceae)[84] Acanthus ebracteatus (Acanthaceae)[43]
DIMBOA-Glc	OMe	H	Zea mays (Poaeceae)[35] Triticum aestivum (Poaceae)[81] Secale cereale (Poaceae)[41] Coix lachryma jobi (Poaceae)[54]
DIM$_2$BOA-Glc	OMe	Ome	Zea mays[42]
7-Cl-DIBOA-Glc			Acanthus ebracteatus (Acanthaceae)[43]

The glucosides of interest belong to a special class of acetal glucosides which arises from the formal combination of two cyclic hemiacetals by dehydration. Principally, this offers the possibility to form four diastereomers by different combinations of the bond situated at the two anomeric centers. However, in all cases of the completely resolved natural structures, only the (2R)-2-β-D-glucosides of the 2-hydroxy-benzoxazinone skeleton have been found as a result of the stereoselective enzymatic glucosylation in plants. In exact drawings, this requires the backward orientation of the C-2-O-bond (Fig. 4.2). All glucosides are regarded as non-toxic precursors which are formed by the juvenile plant and stored in the vacuole until they are used by the plant for defense or attack.

Aglucones can reach the environment by several pathways. In case of a pest attack (insect, microbe, fungus) they are released by special stereospecific β-glucosidases in maize,[19,58] rye[73] and wheat[72] from their preinfectional glucosidic precursors. Furthermore, they can be set free passively by rotting plant material[89] or actively by root exudation.[63]

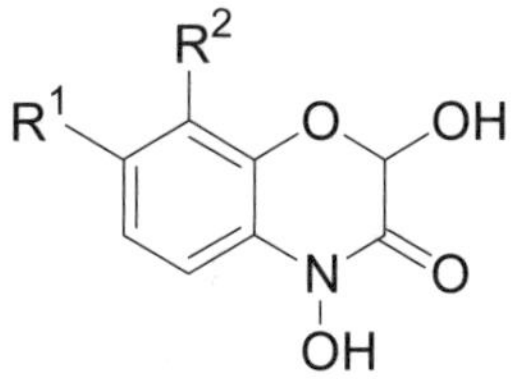

Acronym	R^1	R^2	Species (family), one reference cited for each plant
DIBOA	H	H	*Secale cereale*[79]
			Saccharum officinale (Poaceae)[48]
			Triticum aestivum[16]
			Zea mays[76]
			Hordeum vulgare (Poaceae)[4]
			Agropyron repens (Poaceae)[63]
			Acanthus mollis[84]
			Aphelandra tetragona (Acanthaceae)[77]
			Consolida orientalis[60]
DIMBOA	OMe	H	*Triticum aestivum*[31]
			Saccharum officinale[48]
			Zea mays[36]
			Agropyron repens[24]
DIM$_2$BOA-Glc	OMe	OMe	*Zea mays*[86]
TRIBOA	OH	H	*Zea mays*[86]
			Crossandra pungens (Acanthaceae)[61]

Figure 4.3
Natural benzoxazinone aglucones.

Aglucones are racemic compounds which cannot be split into pure enantiomers in preparative amounts of substance because of a fast equilibrium between both cyclic hemiacetal enantiomers (see Fig. 4.1). In drawings, this fact is expressed by a normal line between C-2 and OH. Figure 4.3 shows the aglucones DIBOA, DIMBOA and DIM$_2$BOA, which can be enzymatically released from their precursor glucosides (see Fig. 4.2). By far the best investigated benzoxazinoid compound is the cyclic hydroxamic acid DIMBOA. On the contrary, TRIBOA has only been found as an aglucone until now. It is expected to be the direct biosynthetic precursor of DIMBOA.

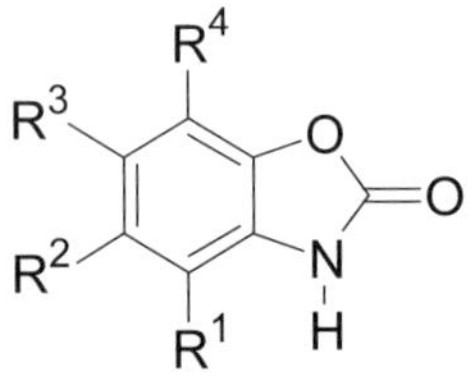

Acronym	R^1	R^2	R^3	R^4	Species (family), one reference cited for each plant
BOA	H	H	H	H	Secale cereale[78] Blepharis edulis (Acanthaceae)[14] Aphelandra tetragona[82] Zea mays[47]
MBOA	H	H	Ome	H	Zea mays[80] Coix lachryma jobi[54] Scoparia dulcis (Scrophulariaceae)[15] Aphelandra tetragona[82] Triticum aestivum[80]
DMBOA	H	H	Ome	OMe	Zea mays[47]
4-ABOA	Ac	H	H	H	Zea mays[22]
5-Cl-MBOA	H	Cl	Ome	H	Zea mays[44]
4-Cl-DMBOA	Cl	H	Ome	OMe	Zea mays[47]

Figure 4.4
Natural benzoxazolin-2(3H)-ones.

In juvenile maize plants the benzoxazinone glucoside content can reach values of ca. 1% of the dry mass, i.e., appreciable amounts. Hence, the plant's own defense and attack systems can liberate the toxic aglucone very rapidly and in large amounts. However, aglucones under discussion, which contain both a cyclic hydroxamic acid and a cyclic hemiacetal, are chemically unstable when in solution and decompose, with ring contraction, to form a benzoxazolin-2(3H)-one (Fig. 4.4) with elimination of formic acid. Such a transformation is shown from DIBOA-Glc to BOA in Figure 4.1. Analogously, MBOA and DMBOA arise from the two step degradation of DIMBOA-Glc and DIM$_2$BOA-Glc, respectively. The half life of DIMBOA in solution is about one day.[85] The rate of degradation is distinctively enhanced in alkaline medium. Several proposals for the ring contraction mechanism have been made.[3,8,10,70] Clearly, donor substitution, as in DIMBOA, accelerates the decomposition.[2] However, a glucosidic origin has only been proven for BOA, MBOA and DMBOA.

The biosynthetic origin of the other three benzoxazolinones mentioned in Figure 4.4 is not yet clear. At least other ways of formation than from the benzoxazinoid glucoside two step decay, like metabolic action of endogenous fungi on BOA, MBOA, or DMBOA, have to been considered.

Considerable effort has been made by us to detect the supposed precursor of 4-acetyl-BOA (4-ABOA) in plants of the Canadian maize variety Funks G 4106 from kernels of which it was isolated[22] and shown to be bioactive.[51,52] Though we eventually succeeded in the synthesis of the supposed corresponding 5-acetyl-2,4-dihydroxy-2H-1,4-benzoxazin-3(4H)-one, which indeed rapidly decomposed to form 4-ABOA in solution,[21] we were unable to detect this aglucone or its glucoside in juvenile plant material.

Finally, some leading references and reviews on other research topics of the benzoxazinoids shall be cited to allow an entrance in these related fields.

FORMATION, DISTRIBUTION AND BIOLOGICAL INTERACTIONS IN THE RESISTANCE OF CROP PLANTS

The entire subject has been reviewed in detail several times.[32,56,57,65] Recently, a new adsorptive method for the isolation of DIMBOA has been reported.[49] The impact of benzoxazinoids on the western corn rootworm (*Diabrotica virgifera* LeConte) development has been studied.[17] The allocation of a hydroxamic acid and biomass during vegetative development of rye has been investigated.[27] Effects of benzoxazinoids from maize on survival and fecundity of aphids have been explored.[13] DIMBOA concentrations have been measured in various isolines of wheat and corresponding plant introduction lines.[55] The variation of the content of several benzoxazinoids in relation to the age and plant organ has been determined in maize plants.[12]

BIOSYNTHESIS

In the late 1990's, a breakthrough in biosynthesis was reached based on the finding that benzoxazinone and tryptophane biosynthesis have a formal branch point, and indole is a precursor of the benzoxazinones.[18] The biosynthetic pathway from indole as the precursor of the benzoxazinone moiety was elucidated both on the levels of the responsible gene cluster discovered and the corresponding enzymes.[23] The stepwise oxidation of indole with molecular oxygen by means of cytochrome P450 monooxygenases in maize was found to be the biosynthetic route.[29] The specificity and conservation among grasses was studied for the cytochrome P450 monooxygenases of the DIBOA biosynthesis.[30] The most exciting step, the ring expansion of the indolinone to the benzoxazinone moiety was recently investigated in detail.[71] In the light of these new findings, it was necessary to retract former assumptions on the origin of benzoxazinones. The role of natural benzoxazinones in the survival strategies of plants was summarized with special emphasis on their biosynthesis, mode of action and allelopathy.[66] A current topic of biosynthesis studies is the induced accumulation of the 4-methylated DIMBOA-Glc in maize.[59]

CHEMICAL SYNTHESIS OF AGLUCONES AND ACETAL GLUCOSIDES

Two reviews cover the whole synthetic literature in the field until 1997.[1,67] Suitable syntheses for the main aglucones DIBOA[68] and DIMBOA[64,74] have been reported, as well as a general diastereoselective glucosidation method to form DIBOA-Glc, DIMBOA-Glc and other benzoxazinone glucosides.[46]

MOLECULAR MODE OF ACTION

The chemical mechanisms supporting the biological effects of the actions elicited by benzoxazinones and benzoxazolinones are not well understood. However, a lot of work has been invested in this field, and some reasonable hypotheses exist.[37] It has been shown that the combination of both cyclic hemiacetal and cyclic hydroxamic acid is a requirement for high bioactivity, which can be enhanced by a 7-methoxy donor substituent.

Most likely, the two main pathways consist of reactions of the electrophilic ring-opened aldehyde form of the hemiacetal with bionucleophiles and of reactions of a unique multi-centered cationic nucleophile that can be generated from DIMBOA by N-O-fission.[37] Recently, we reported a novel hypothesis for the mode of bioactivity based on the formation of 3-formyl-6-methoxybenzoxazolin-2(3H)-one (FMBOA) by formal dehydration of DIMBOA.[40] FMBOA was proven to be a potent formyl donor towards typical nucleophiles occurring in biomolecules and could, if formed under natural conditions, also lead to biological effects by formylation of biomolecules. The mechanism of this new dehydration was elucidated with [2-^{13}C]-DIMBOA.[33]

ANALYTICAL INVESTIGATIONS

Benzoxazinoids are of interest for analytical investigations for two reasons. First, there is need for clear identification and analytical separation of benzoxazinone glucosides, as well as aglucones and benzoxazolinones. To fulfill this task, mainly mass spectrometric, liquid and gas chromatographic techniques have been developed.[6,11,87] A sophisticated procedure for the first simultaneous HPLC separation of all three classes of compounds has been reported.[5] Second, racemic benzoxazinone aglucones have two chiral cyclic hemiacetals and make interesting targets to study, if their enantioseparation is possible. It was proven that separation of enantiomers is impossible by HPLC[50] or HPCE.[75] Only enantiodifferentiation was possible by a chiral NMR method.[45] The whole analytical topic has been reviewed.[20]

ALLELOPATHY

As already mentioned DIBOA, DIMBOA and their derivatives, the benzoxazolinones BOA and MBOA are highly bioactive compounds. In addition to their role in pest resistance, they are phytotoxic against a number of crops and weeds.[3,7,9,28,66] Several prerequisites are important, however, for manifestation of the phytotoxic effects. Besides abiotic, edaphic and climatic parameters, as well as genetic disposition of acceptor plants, the compounds must be present in sufficient amounts. In most cases traces have no inhibitory effects. Low concentrations are degraded rapidly by microorganisms to 2-aminophenol and other compounds, e.g., phenoxazinones and N-(2-hydroxylphenyl)-malonamic acid, or they may be adsorbed by soil particles. Active release by donor plants may compensate the disappearance resulting from the activity of microorganisms.

Root exudation of benzoxazinones has been described for wheat, rye, corn, and quackgrass,[24, 63] which may increase the concentration within the rhizosphere. The major part of the compounds present in the soil of rye or corn fields, however, originate from rotting plant material.[3] Another important prerequisite is the genotype. Visible effects of phytotoxicity imply sensitive acceptor species. Weed control by benzoxazolinones is only possible when the seedlings of a given species respond to the compounds with reduced viability that finally results in severe damage and death.

Adapted species may have developed, however, strategies which enable them to survive allelopathic attacks. One of those strategies certainly includes detoxification of absorbed allelochemicals by constitutive or inducible pathways. Metabolization and detoxification are known reactions in a number of crops upon application of diverse synthetic herbicides.[38] Enhanced herbicide detoxification is an important factor in the development of nontarget-site cross-resistance and multiple resistance. It is reasonable to expect comparable strategies in plants that are relatively resistant to allelochemicals such as DIBOA, DIMBOA, and their derivatives. Especially in ecosystems where co-existing species have to be adapted to each other, detoxification of absorbed allelochemicals may play a crucial role under defined circumstances.

In our studies we used model systems to elucidate detoxification capacities of weeds and crops for benzoxazolin-2(3H)-one. Model systems have the advantage of demonstrating biochemical events that may not show under complex natural field conditions. The purpose of these studies is to gain insights in the possible participation of detoxification in the often observed difference of monocots and dicots against allelochemicals. Second, we questioned whether detoxification capacities reflect the membership of species to defined plant communities.

RESULTS AND DISCUSSION

Earlier, we analyzed oat roots after BOA incubation of seedlings for at least 24 h and up to 72 h. Two new products were present in the extracts, which were identified as BOA-6-OH and BOA-6-O-glucoside.[83] In addition, a third product was found that increased with incubation time. It was slightly more hydrophobic than BOA-6-OH with an UV scan very similar to BOA. The compound was purified, subjected to chemical identification and characterized as BOA-N-glucoside. This product was not found in roots of *Vicia faba* var. Alfred, but this seems to depend on variety; e.g., *Vicia faba* var. Dreifach Weisse, was able to produce the compound. Both varieties contained BOA-6-OH and the corresponding glucoside.

Several members of the Poaceae were checked for BOA detoxification capacity; the results are presented in Table 4.1. All species, with one exception, contained the N-glucoside, although there were remarkable differences in the accumulated amounts following incubation at 24 h and 48 h. *Zea mays* exhibited the most effective detoxification capacity, as BOA-6-O-glucoside was synthesized in traces only or was a minor product. The N-glucosylated compound already appeared after 8 h incubation in the presence of 500 µM BOA, and it was the major metabolite after 24 h. We found a fourth compound which started to accumulate after 18 h. The substance was isolated for structural elucidation.

Table 4.1

BOA detoxification of several Poaceae; species containing benzoxazinones are bold printed.

Species	BOA-6-O-glucoside	Glucose carbamate	Gentiobioside carbamate
Zea mays	14	378	27
Triticum aestivum	111	280	n.a.
Lolium perenne	109	246	n.a.
Avena sativa	540	530	n.a
Avena fatua	1510	300	n.a.
Digitaria sanguinalis	35	310	n.a.

Initially, the compound was treated with ß-glucosidase from almonds. The hydrolysis resulted in a product which was present in a sample of synthetic BOA-N-glucoside dissolved in methanol and which was first regarded as a byproduct. In the aged solution, the assumed byproduct was the major constituent. At that time, it became obvious to reinvestigate the structure of the BOA-N-glucoside and the supposed derivative. The chemical analysis unequivocally showed that synthetic BOA-N-glucoside was not stable over time but isomerized yielding a glucoside carbamate. Thus, the natural product found in plant extracts, first thought to be BOA-N-glucoside, is identical with this glucoside carbamate. The fourth product from corn roots was identified as gentiobioside carbamate[69] (Fig. 4.5).

The synthesis of BOA-6-O-glucoside is catalyzed by constitutive enzymes that may be upregulated. A glucosyltransferase that accepted BOA-6-OH as a substrate was measurable in protein extracts of corn roots harvested from control plants. Detoxification via glucoside carbamate synthesis is inducible and seems to be more complicated than simple N-glucosylation (data unpublished). The biosynthesis of the compounds is part of our ongoing research.

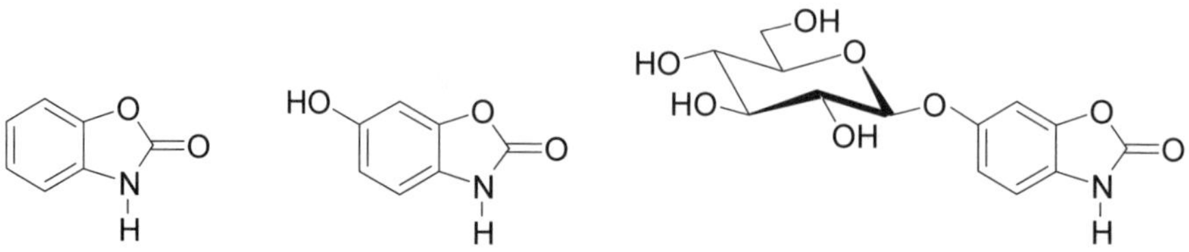

BOA BOA-6-OH BOA-6-*O*-glucoside

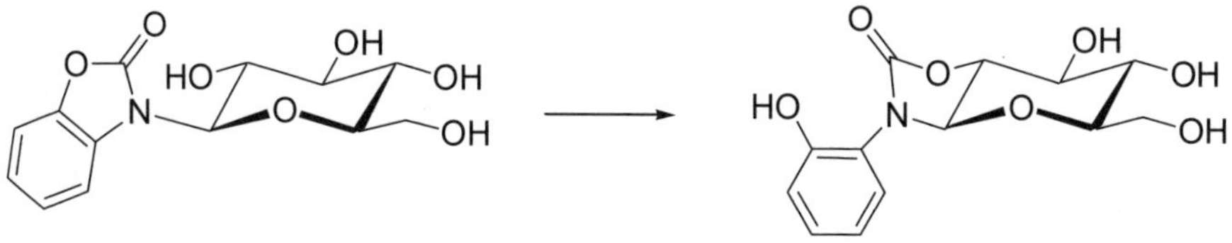

Isomerisation of synthetic BOA-*N*-glucoside to glucoside carbamate

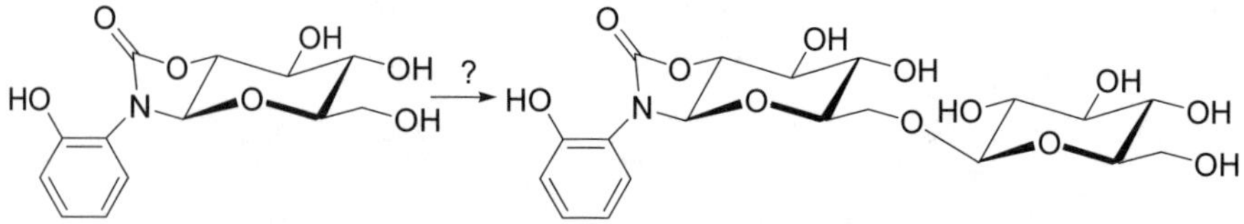

Supposed biosynthesis of gentiobioside carbamate by glucosylation of glucoside carbamate.

Figure 4.5

Detoxification of BOA and resulting products in higher plants.

BOA-6-OH, BOA-6-*O*-glucoside and glucoside carbamate were tested for phytotoxicity using the cress test. Only the carbamate had no inhibitory influence on radicle growth up to 1 mM, BOA-6-*O*-glucoside was still slightly toxic, but BOA-6-OH was more toxic than the original compound. Thus, accumulation of free BOA and, moreover, that of BOA-6-OH should be fatal to plants, as indicated by blackening of root tips in BOA-sensitive *Vicia faba* var. Alfred.[62]

BOA-metabolites are not stable constituents of incubated plants but are exuded, at least partially, as found with corn roots.[69] Corn seedlings transferred to soil filled pots after BOA incubation lost the extractable metabolites within 5-6 days. In other experiments, the seedlings were transferred to tap water basins after incubation. The water was removed every day and evaporated to dryness. The residue contained relatively high amounts of glucoside carbamate and low concentrations of gentiobioside carbamate. Since the experimental conditions were nonsterile, it is rather likely that some gentiobioside carbamate was hydrolysed by microorganisms to glucose carbamate and glucose.

Whereas, *Acinetobacter calcoaceticus* strains (*Waksmania aerata* and *Pseudomonas iodina* as well) are able to cleave benzoxazolinone to 2-aminophenol resulting in phenoxazinone (Fig. 4.6), bacteria belonging to this group are unable to metabolize glucose carbamate and BOA-6-OH[25,26] (Burziak et al. unpublished).

Also, the dicots *Coriandrum sativum* and *Galinsoga ciliata* do not accumulate BOA metabolites. Incubated seedlings lost BOA-6-O-glucoside over a period of 5 to 7 days. The further fate of BOA metabolites is still unknown but currently under investigation.

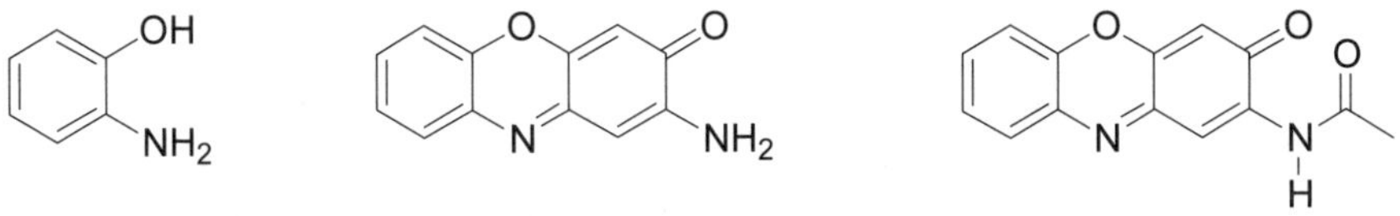

2-aminophenol   2-amino-3*H*-phenoxazin-3-one   2-acetylamino-3*H*-phenoxazin-3-one

Figure 4.6

The microbial degradation product 2-aminophenol as precursor for phenoxazinones.

We tested a number of dicot species for their ability to detoxify BOA. Most of them were incubated as seedlings but some as adult plants when germination failed under laboratory conditions. Applied BOA concentration was 100 and 500 µM, incubation time was 24 h. The plants were separated into roots and shoots after incubation and the material extracted with 30% methanol. HPLC analyses of the extracts revealed remarkable differences in the detoxification capacities. Major detoxification appeared within the roots, whereas shoots were only involved when roots seemed to be overtaxed. All species tested were able to perform BOA-6-O-glucoside synthesis, but with regard to glucoside carbamate synthesis, differences became obvious (Table 4.2).

Two facts seem to be of special importance – first, of course, the existence of the genes necessary for carbamate synthesis and second, the mode of gene induction, the velocity by which the corresponding proteins are present. We found species that were unable to synthesize glucoside carbamate within 24 h, among them two species of American origin, *Helianthus annuus* and *Galinsoga ciliata*. The same result was obtained with *Polygonum aviculare* and *Urtica urens*, both belonging to Chenopodietea. Detoxification of *Chenopodium album* broke down when 500 µM BOA was applied. *Plantago major*, existing in disturbed habitats, was the only dicot species tested with an excellent detoxification capacity, comparable to corn.

Table 4.2

BOA detoxification capacity of several weeds with regard to glucoside carbamate synthesis within 24 h.

Metabolization capacity	Species	Family	Vegetation Class
High	*Plantago major*	Plantaginaceae	*Agrostietea stoloniferae* communities
	Coriandrum sativum (crop)	Apiaceae	Secalietea
	Centaurea cyanus	Asteraceae	Secalietea
	Carduus nutans	Asteraceae	*Artemisietea vulgaris* communities
	Papaver rhoeas	Papveraceae	Secalietea
	Matricaria chamomilla	Asteraceae	Secalietea
	Daucus carota	Apiaceae	*Artemisietea vulgaris* communities
moderate			
	Consolida regalis	Ranunculaceae	Secalietea
	Agrostemma githago	Caryophyllaceae	Secalietea
	Capsella bursa pastoris	Brassicaceae	Secalietea Chenopodietea
	Legousia speculum veneris	Campanulaceae	Secalietea
	Chenopodium album	Chenopodiaceae	Chenopodietea
	Polygonum aviculare	Polygonaceae	Chenopodietea
	Urtica urens	Urticaceae	Chenopodietea
Low	*Galinsoga ciliata*	Asteraceae	Neophyte

Another group of common weeds that produces lower amounts of carbamate consists of plants belonging to Secalietea and *Artemisietea vulgaris* communities. They are naturally occurring with character species (*Triticum aestivum*, *Secale cereale*, *Agropyron repens*) containing benzoxazinones. There were also species belonging to the Secalietea communities with only a low capacity to synthesize glucoside carbamate (Table 4.2). Interestingly, they are endangered species, or in the case of *Agrostemma githago* already extinct. *Capsella bursa pastoris*, which also fit in the latter group, is a species occurring in both vegetation classes, Secalietea and Chenopodietea.

It is clear that BOA detoxification, mainly via glucoside carbamate production of the species tested, is not a feature of certain plant families but is rather combined with their occurrence in defined plant communities where benzoxazinone containing character species exist with high density covers. *Plantago major* is regarded as an exception.

In a further approach, we checked some species belonging to Amarantho-Chenopodion communities (syn. Consolido-Eragrostion poaeoidis) for their detoxification capacities (Table 4.3).

In those communities *Consolida orientalis* is presented as a character species, together with *Heliotropium europaeum*, *Hibiscus trionum* and accompanied by *Amaranthus albus*, *Portulacca oleracea*, *Digitaria sanguinalis*, *Diplotaxis tenuifolia* and other additional species.[53] *Consolida orientalis* is the only Ranunculaceae known to contain benzoxazinones, especially in the flowers.[60] But in contrast to Secalietea communities, with the character species wheat and rye, *C. orientalis* is only presented with a few, scattered distributed individuals (W. Nezedal, University of Erlangen, personal communication). Table 4.3 demonstrates striking differences for the detoxification capacity observed in some species belonging to Amarantho-Chenopodion communities after 24 h of incubation.

Digitaria sanguinalis contained glucoside carbamate as the major detoxification product and low amounts of BOA-6-O-glucoside, which agrees with results obtained from other Poaceae. *Consolida orientalis* is able to produce glucose carbamate, but surprisingly, the induction of the pathway obviously takes longer than in *Digitaria sanguinalis*, although the species contain benzoxazinone. *Amaranthus albus* contained only BOA-6-O-glucoside after 24 h of exposure to BOA, whereas *Diplotaxis tenuifolia* accumulated harmful BOA-6-OH as a major product. A still unidentified product was found in *Portulacca oleracea*. The compound was isolated and is currently under structural analysis.

One reason why species belonging to plant communities of the vegetation class Amarantho-Chenopodion behave so differently in comparison to those belonging to Secalietea communities may be due to the fact that *Consolida orientalis* is only sporadically distributed with a few individuals. The amount of benzoxazinone and the resulting benzoxazolinone released into the environment should be very low, too low for the development of a certain selection pressure on other species. Thus, with regard to Secalietea communities, a completely different situation is given. On the other hand, it cannot be excluded that in some managed ecosystems several weeds developed cross resistances against defined synthetic herbicides and related molecules. As a consequence, they are able to detoxify a large variety of compounds including benzoxazolinones. Moreover, different varieties of the same crop exhibit variations in their detoxification capacity, perhaps as a result of differences in cultivation progress and success. Thus, the alternative use of benzoxazinones as natural herbicides may be rather limited.

Table 4.3

Detoxification capacities in species belonging to the Amarantho-Chenopodion vegetation class; species containing benzoxazinones are bold printed.

Detoxification Capacity	Species	Detoxification Products after 24 h of Incubation	Family	Vegetation Class
High	*Digitaria sanguinalis*	BOA-6-O-glucoside glucoside carbamate (major product)	Poaceae	Amarantho-Chenopodion
High-moderate	*Consolida orientalis*	BOA-6-O-glucoside (major product) glucoside carbamate	Ranunculaceae	Amarantho-Chenopodion
Low	*Diplotaxis tenuifolia*	BOA-6-O-glucoside BOA-6-OH (major product)	Brassicaceae	Amarantho-Chenopodion
Moderate	*Amaranthus albus*	BOA-6-O-glucoside	Amaranthaceae	Amarantho-Chenopodion
?	*Portulacca oleracea*	unknown product	Portulaccaceae	Amarantho-Chenopodion

METHODOLOGY

Plant material

Seedlings and adult plants of the following species were used for the studies:[62,69,83] Amaranthaceae: *Amaranthus albus*; Apiaceae: *Daucus carota, Coriandrum sativum*; Asteraceae: *Matricaria chamomilla, Centaurea cyanus, Galinsoga ciliata, Helianthus annuus, Carduus nutans*; Brassicaceae: *Capsella bursa pastoris* (adult), *Diplotaxis tenuifolia*; Campanulaceae: *Legousia speculum veneris* (adult); Caryophyllaceae: *Agrostemma githago*; Chenopodiaceae: *Chenopodium album*; Fabaceae: *Vicia faba*; Malvaceae: *Hibiscus trionum*;

Papaveraceae: *Papaver rhoeas* (adult); Plantaginaceae: *Plantago major* (adult); Poaceae: *Avena sativa, Avena fatua, Zea mays, Triticum aestivum, Lolium perenne, Digitaria sanguinalis*; Portulaccaceae: *Portulacca oleracea* ssp. *sativa*; Ranunculaceae: *Consolida regalis* (adult), *Consolida orientalis*; Polygonaceae: *Polygonum aviculare* (adult); Urticaceae: *Urtica urens* (adult).

Plants were incubated with (30 mL/g fresh weight) 500 µM BOA in MES/KOH buffer supplemented with 0.5 mM $CaSO_4$, 1 mM KCl and 1% Na-ascorbate (w/v) for 24 and 48 h at room temperature.[88] During the incubation, the medium was aerated. After incubation, plants were washed with water, separated into roots and shoots, and homogenized with 30% methanol (1mL / g fresh weight) in the presence of quartz. The homogenate was filtered through miracloth, the filtrate centrifuged at 10.000 g, and the supernatant used for analysis. When concentration was necessary, the supernatants were evaporated to dryness and the residue dissolved in a sufficient quantity of 30% methanol. The solution was centrifuged, then used for analysis.

Analyses

HPLC analysis was performed with a Beckman system equipped with DA detector module using a RP 18 column. Detection was performed at 280 nm. Calculation of concentrations was based on external standard curves with the corresponding compounds. Each experiment was triplicated.

New compounds were collected and purified via HPLC. Fractions containing the compounds were combined and evaporated to dryness using a speed vacuum centrifuge. Purified samples were identified by MS (Bruker Daltonics 7T APEX II FT-ICR mass spectrometer with electron spray ionization, positive mode) and by NMR (1H-NMR, 1H-1H COSY, HMBC, and HMQC, Bruker DRX 600 spectrometer). NMR analyses were performed by Bernd Schneider (Max Planck-Institut für Chemische Ökologie, Jena) and Lothar Hennig (Institut für Organische Chemie, Universität Leipzig), and MS analysis by Sabine Giesa (Institut für Analytische Chemie, Universität Leipzig). Aliquots of the purified compounds were incubated in presence of ß-glucosidase from almonds (Sigma) according to the recommendation of the company. Enzyme assays were stopped by boiling, followed by centrifugation of the mixture at 10.000 g.

ACKNOWLEDGMENTS

D.S. thanks the Fonds der Chemischen Industrie for financial support of this work. M.S. thanks the DFG for financial support. Prof. Dr. Huang Hao thanks the Alexander von Humboldt Foundation for research fellowship granting (January 1. - June 30. 2001: University of Leipzig, D.S; July 1. - December 31. 2001 University of Bonn, M.S.). We thank Drs. S. Giesa, L. Hennig, and P. D. B. Schneider for their analytical measurements.

REFERENCES

(1) Atkinson, J., Arnason, J., Campos, F., Niemeyer, H. M., and Bravo, H. **1992**. Synthesis and reactivity of cyclic hydroxamic acids – resistance factors in the Gramineae. In: Baker, D. R., Fenyes, J. G., and Steffens, J. J. (Eds.), *Synthesis and Chemistry of Agrochemicals III*; ACS Symposium Series Vol. **504**. American Chemical Society, 349-359

(2) Atkinson, J., Morand, P., Arnason, J. T., Niemeyer, H. M. and Bravo, H. R. **1991**. Analogues of the cyclic hydroxamic acid 2,4-dihydroxy-2H-1,4-benzoxazin-3-one: decomposition to benzoxazolinones and reaction with □-mercaptoethanol. *J. Org. Chem.* **56**, 1788-1800

(3) Barnes, J. P., Putnam, A. R., and Burke, B. A. **1986**. Allelopathy activity of rye (*Secale cereale* L.). In: Putnam, A. R. and Tang, C. S. (Eds.). *The Science of Allelopathy*. Wiley-Interscience, New York, 271-286

(4) Barria, B. N., Copaja, S. V. and Niemeyer, H. M. **1992**. Occurrence of DIBOA in wild Hordeum species and its relation to aphid resistance. *Phytochemistry* **31**, 89-91

(5) Baumeler, A., Hesse, M., and Werner, C. **2000**. Benzoxazinoids – cyclic hydroxamic acids, lactams and their corresponding glucosides in the genus *Aphelandra* (Acanthaceae). *Phytochemistry* **53**, 213-222

(6) Bigler, L., Baumeler, A., Werner, C., and Hesse, M. **1996**. Detection of noncovalent complexes of hydroxamic-acid derivatives by means of electrospray mass spectrometry. *Helv. Chim. Acta* **79**, 1701-1709

(7) Bravo, H. R. and Lazo, W. **1996**. Antialgal and antifungal activity of natural hydroxamic acids and related compounds. *J. Agric. Food Chem.* **44**, 1569-1571

(8) Bravo, H. R. and Niemeyer, H. M. **1985**. Decomposition in aprotic solvents of 2,4-dihydroxy-7-methoxy-1,4-benzoxazin-3-one, a hydroxamic acid from cereals. *Tetrahedron* **41**, 4983-4986

(9) Bravo, H. R., Copaja, S. V. and Lazo, W. **1997**. Antimicrobial activity of natural 2-benzoxazolinones and related derivatives. *J. Agric. Food Chem.* **45**, 3255-3257

(10) Brendenberg, J.-B., Honkanen, E. and Virtanen, A. I. **1962**. The kinetics and mechanism of the decomposition of 2,4-dihydroxy-1,4-benzoxazin-3-one. *Acta Chem. Scand.* **16**, 135-141

(11) Cambier, V., Hance, T. and de Hoffmann, E. **1999**. Non-injured maize contains several 1,4-benzoxazin-3-one related compounds but only as glucoconjugates. *Phytochem. Anal.* **10**, 119-126

(12) Cambier, V., Hance, T. and de Hoffmann, E. **2000**. Variation of DIMBOA and related compounds content in relation to the age and plant organ in maize. *Phytochemistry* **53**, 223-229

(13) Cambier, V., Hance, T. and de Hoffmann, E. **2001**. Effects of 1,4-benzoxazin-3-one derivatives from maize on survival and fecundity of *Metopolophium dirhodium* (Walker) on artificial diet. *J. Chem. Ecol.* **27**, 359-370

(14) Chatterjee, A., Sharma, N. J., Basserji, J., and Basa, S. C. **1990**. Studies on Acanthaceae - Benzoxazine glucoside and benzoxazolone from *Blepharis edulis* Pers. *Ind. J. Chem.* **29B**, 132-134

(15) Chen, C. M. and Chen, M. T. **1976**. 6-Methoxybenzoxazolinone and triterpenoids from roots of *Scoparia dulcis*. *Phytochemistry* **15**, 1997-1999

(16) Copaja, S. V., Barria, B. N. and Niemeyer, H. M. **1991**. Hydroxamic acid content in perennial Triticeae. *Phytochemistry* **30**, 1531-1534

(17) Davis, C. S., Ni, X. Z., Quisenberry, S. S. and Foster, J. E. **2000**. Identification and quantification of hydroxamic acids in maize seedling root tissue and impact on western corn rootworm (Coleoptera: Chrysomelidae) larval development. *J. Econ. Entomol.* **93**, 989-992

(18) Desai, R. S., Kumar, P. and Chilton, W. S. **1996**. Indole is an intermediate in the biosynthesis of cyclic hydroxamic acids in maize. *Chem. Commun.*, 1321

(19) Ebishi, K., Ishihara, A., Hirai, N. and Iwamura, H. Z. **1998**. Occurrence of 2,4-dihydroxy-7-methoxy-1,4-benzoxazin-3-one (DIMBOA) and a β-glucosidase specific for its glucoside in maize seedlings. *Z. Naturforsch. C: Biosci.* **53**, 793-798

(20) Eljarrat, E. and Barcelo, D. **2001**. Sample handling and analysis of allelochemical compounds in plants. *Trac-Trends Anal. Chem.* **20**, 584-590

(21) Escobar, C. A., Kluge, M. and Sicker, D. **1997**. Biomimetic synthesis of 4-acetylbenzoxazolin-2(3*H*)-one isolated from *Zea mays*. *J. Heterocycl. Chem.* **34**, 1407-1414

(22) Fielder, D. A., Collins, F. W., Blackwell, B. A., Bensimon, C., and ApSimon, J. W. **1994**. Isolation and characterization of 4-acetyl-benzoxazolin-2-one (4-ABOA), a new benzoxazolinone from *Zea mays*. *Tetrahedron Lett.* **35**, 521-524

(23) Frey, M., Chomet, P., Glawischnig, E., Stettner, C., Grün, S., Winklmair, A., Eisenreich, W., Bacher, A., Meeley, R. B., Briggs, S. P., Simcox, K., and Gierl, A. **1997**. Analysis of a chemical plant defence mechanism in grasses. *Science* **277**, 696-699

(24) Friebe, A., Schulz, M., Kück, P., and Schnabl, H. **1995**. Phytotoxins from shoot extracts of *Agropyron repens* seedlings. *Phytochemistry* **38**, 1157-1159

(25) Friebe, A., Wieland, I., and Schulz, M. **1996**. Tolerance of *Avena sativa* to the allelochemical benzoxazolinone. Degradation of BOA by root colonizing bacteria. *Angew. Bot.* **70**, 150-154

(26) Gerber, N. N. and LeChevalier, M. P. **1964**. Phenazines and phenoxazinones from *Waksmania aerata* sp. nov. and *Pseudomonas iodina*. *Biochemistry* **3**, 598-602

(27) Gianoli, E., Rios, J. M. and Niemeyer, H. M. **2000**. Allocation of a hydroxamic acid and biomass during vegetative development in rye. *Acta Agric. Scand. Sect. B – Soil Plant Sci.* **50**, 35-39

(28) Gierl, A. and Frey, M. **2001**. Evolution of benzoxazinone biosynthesis and indole production in maize. *Planta* **213**, 493-498

(29) Glawischnig, E., Eisenreich, W., Bacher, A., Frey, M., and Gierl, A. **1997**. Biosynthetic origin of oxygen atoms in DIMBOA from maize: NMR studies with $^{18}O_2$. *Phytochemistry* **45**, 715-718

(30) Glawischnig, E., Grün, S., Frey, M. and Gierl, A. **1999**. Cytochrome P450 monooxygenases of DIBOA biosynthesis: specificity and conservation among grasses. *Phytochemistry* **50**, 925-930

(31) Grambow, H. J., Lückge, A., Klausener, E., and Müller, Z. **1986**. Occurrence of 2-(2-Hydroxy-4,-7-dimethoxy-2H-1,4-benzoxazin-3-one)-β-D-glucopyra-

noside in *Triticum aestivum* leaves and its conversion into 6-methoxybenzoxazolinone. *Z. Naturforsch.* **41c**, 684-690

(32) Gross, D. **1989**. Antimicrobial defense compounds in the Gramineae. *J. Plant Dis. Prot.* **96**, 535-553

(33) Hao, H., Sieler, J. and Sicker, D. **2002**. A mechanistic dehydration study with [2-^{13}C]-DIMBOA. *J. Nat. Prod.* **65**. In press

(34) Hartenstein, H. and Sicker, D. **1994**. (2R)-2-☐-D-Glucopyranosyloxy-4-hydroxy-2H-1,4-benzox-azin-3(4H)-one from *Secale cereale. Phytochemistry* **35**, 827-828

(35) Hartenstein, H., Klein, J. and Sicker, D. **1993**. Efficient isolation procedure for (2R)-2-β-D-Glucopyranosyloxy-4-hydroxy-7-methoxy-2H-1,4-benzoxazin-3(4H)-one from maize. *Ind. J. Heterocycl. Chem.* **2**, 151-153

(36) Hartenstein, H., Lippmann, T. and Sicker, D. **1992**. An efficient procedure for the isolation of pure 2,4-dihydroxy-7-methoxy-2H-1,4-benzoxazin-3(4H)-one (DIMBOA) from maize. *Ind. J. Heterocycl. Chem.* **2**, 75-76

(37) Hashimoto, Y. and Shudo, K. **1996**. Chemistry of biologically active benzoxazinoids. *Phytochemistry* **43**, 551-559

(38) Hatzios, K. K. (Ed.) **1997**. *Regulation of Enzymatic Systems Detoxifying Xenobiotics in Plants.* NATO ASI Series 37. Kluwer Academic Publisher Group. Dordrecht, The Netherlands

(39) Hietala, P. K. and Virtanen, A. I. **1960**. Precursors of benzoxazolinone in rye plants: II. Precursor I, the glucoside. *Acta Chem. Scand.* **14**, 502-504

(40) Hoffmann, A. and Sicker, D. **1999**. A formylating agent by dehydration of the natural product DIMBOA. *J. Nat. Prod.* **62**, 1151-1153

(41) Hofman, J. and Hofmanova, O. **1969**. 1,4-Benzoxazine derivatives in plants. Sephadex fractionation and identification of a new glucoside. *Eur. J. Biochem.* **8**, 109-112

(42) Hofman, J. and Masojidkova, M. **1973**. 1,4-Benzoxazine glucosides from *Zea mays. Phytochemistry* **12**, 207-208

(43) Kanchanapoom, T., Kasai, R., Picheansoonthon, C., and Yamasaki K. **2001**. Megastigmane, aliphatic alcohol and benzoxazinoid glycosides from *Acanthus ebracteatus. Phytochemistry* **58**, 811-817

(44) Kato-Noguchi, H., Kosemura, S. and Yamamura, S. **1998**. Allelopathic potential of 5-chloro-6-methoxy-2-benzoxazolinone. *Phytochemistry* **48**, 433-435

(45) Klein, J., Hartenstein, H. and Sicker, D. **1994**. First discrimination of enantiomeric cyclic hemiacetals and methyl acetals derived from hydroxamic acids and lactams of Gramineae by means of ^{1}H NMR using various chiral solvating agents. *Magn. Reson. Chem.* **32**, 727-731

(46) Kluge, M. and Sicker, D. **1996**. Double diastereoselective glucosidation of cyclic hemiacetals: synthesis of the 1,4-benzoxazinone acetal glucosides GDIBOA and GDIMBOA from Gramineae. *Tetrahedron* **52**, 10389-10398

(47) Kosemura, S., Emori, H., Yamamura, S., Anai, T., and Aizawa, H. **1995**. Isolation and characterization of 4-chloro-6,7-dimethoxybenzoxazolin-2-one, a new auxin-inhibiting benzoxazolinone from *Zea mays*. *Chem. Lett.*, 1053-1054

(48) Kumarasinghe, N. C. and Wratten, S. D. **1998**. Hydroxamic acids in sugarcane and their effect on the sugarcane plant hopper *Pyrilla perpusilla*. *Sugar Cane* **5**, 18-22

(49) Larsen, E. and Christensen, L. P. **2000**. Simple method for large scale isolation of the cyclic arylhydroxamic acid DIMBOA from maize (*Zea mays* L.). *J. Agric. Food Chem.* **48**, 2556-2558

(50) Lippmann, T., Hartenstein, H. and Sicker, D. **1993**. Enantiomeric separation of close analogs to naturally occurring 1,4-benzoxazin-3-ones by liquid chromatography using β-cyclodextrin-modified stationary phase. *Chromatographia* **35**, 302-304

(51) Miller, D. J., Fielder, D. A., Dowd, P. A., Norton. R. A., and Collins, F. W. **1996**. Isolation of 4-acetyl-benzoxazolin-2-one (4-ABOA) and diferuloylputrescine from an extract of gibberella ear rot-resistant corn that blocks mycotoxin biosynthesis, and the insect toxicity of 4-ABOA and related compounds. *Biochem. Syst. Ecol.* **24**, 647-658

(52) Miller, D. J., Miles, M. and Fielder, D. A. **1997**. Kernel concentrations of 4-acetylbenzoxazolin-2-one and diferuloylputrescine in maize genotypes and gibberella ear rot. *J. Agric. Food Chem.* **45**, 4456-4459

(53) Mucina, L., Grabherr, G. and Ellmauer, T. (Eds.) **1993**. *Die Pflanzengesellschaften Österreichs. I. Anthropogene Vegetation.* Gustav Fischer Verlag, Jena

(54) Nagao, T., Otsuka, H., Kohda, H., Sato, T., and Yamasaki, K. **1985**. Benzoxazinones from *Coix lachryma jobi* Var. Ma-Yuen. *Phytochemistry* **24**, 2959-2962

(55) Ni, X. Z. and Quisenberry, S. S. 2000. Comparison of DIMBOA concentrations among wheat isolines and corresponding plant introduction lines. *Entomol. Exp. Appl.* **96**, 275-279

(56) Niemeyer, H. M. 1988. Hydroxamic acids (4-hydroxy-1,4-benzoxazin-3-ones), defense chemicals in the Gramineae. *Phytochemistry* **27**, 3349-3358

(57) Niemeyer, H. M., and Perez, F. J. 1995. Potential of hydroxamic acids in the control of cereal pests, diseases, and weeds. In: Inderjit, Dakshini, K. M. M., and Einhellig, F. A. (Eds.). *Allelopathy: Organisms, Processes, and Applications.* ACS Symposium Series, Vol. 582. American Chemical Society, 260-270

(58) Oikawa, A., Ebisui, K., Sue, M., Ishihara, A., and Iwamura, H. 1999. Purification and characterization of a β-glucosidase specific for 2,4-dihydroxy-7-methoxy-1,4-benzoxazin-3-one (DIMBOA) glucoside in maize. *Z. Naturforsch. C: Biosci.* **54**, 181-185

(59) Oikawa, A., Ishihara, A., Hasegawa, M., Kodama, O., and Iwamura, H. 2001. Induced accumulation of 2-hydroxy-4,7-dimethoxy-1,4-benzoxazin-3-one glucoside (HDMBOA-Glc) in maize leaves. *Phytochemistry* **56**, 669-675

(60) Özden, S., Özden T., Attila, J., Kücükislamoglu, M., and Okatan, A. 1992. Isolation and identification via high performance liquid chromatography and thin layer chromatography of benzoxazolinone precursors from *Consolida orientalis* flowers. *J. Chromatogr.* **609**, 402-406

(61) Pratt, K., Kumar, P., and Chilton, W. S. 1995. Cyclic hydroxamic acids in dicotyledonous plants. *Biochem. Syst. Ecol.* **23**, 781-785

(62) Schulz, M. and Wieland, I. 1999. Variations in metabolism of BOA among species in various field communities – biochemical evidence for co-evolutionary processes in plant communities? *Chemoecology* **9**, 133-141

(63) Schulz, M., Friebe, A., Kück, P., Seipel, M., and Schnabl, H. 1994. Allelopathic effects of living quackgrass (*Agropyron repens* L.). Identification of inhibitory allelochemicals exuded from rhizome borne roots. *Angew. Bot.* **68**, 195-200

(64) Sicker, D. and Hartenstein, H. 1993. A new general approach to the 2-hydroxy-2*H*-1,4-benzoxazin-3(4*H*)-one skeleton *via* diisobutylaluminum hydride reduction of 2,3-dioxo-1,4-benzoxazines. *Synthesis*, 771-772

(65) Sicker, D. and Schulz, M. 2002, Benzoxazinones in plants: occurrence, synthetic access, and biological activity. *Studies in Natural Product Chemistry* **27**. In press

(66) Sicker, D., Frey, M., Schulz, M., and Gierl, A. **2000**. Role of natural benzoxazinones in the survival strategies of plants. In: Jeong, K. W. (Ed.), *International Review of Cytology – A Survey of Cell Biology*, Vol. **198**. Academic Press, San Diego, CA, 319-346

(67) Sicker, D., Hartenstein, H. and Kluge, M. **1997**. Natural benzoxazinoids - Synthesis of acetal glucosides, aglucones, and analogues. In: Pandalai, S. G. (Ed.), *Recent Research Developments in Phytochemistry*, Vol. **1**. Research Signpost, Trivandrum, 203-223

(68) Sicker, D., Prätorius, B., Mann, G., and Meyer, L. **1989**. A convenient synthesis of 2,4-dihydroxy-2*H*-1,4-benzoxazin-3(4*H*)-one. *Synthesis*, 211-212

(69) Sicker, D., Schneider, B., Hennig, L., Knop, M., and Schulz, M. **2001**. Glucoside carbamate from benzoxazolin-2(3*H*)-one detoxification in extracts and exudates of corn roots. *Phytochemistry* **58**, 819-825

(70) Smissman, E. E., Corbett, M. D., Jenny, N. A., and Kristiansen, O. **1972**. Mechanism of transformation of 2,4-dihydroxy-1,4-benzoxazin-3-ones and 2-hydroxy-2-methyl-4-methoxy-1,4-benzoxazin-3-one to 2-benzoxazolinone. *J. Org. Chem.* **37**, 1700-1704

(71) Spiteller, P., Glawischnig, E., Gierl, A., and Steglich, W. **2001**. Studies on the biosynthesis of 2-hydroxy-1,4-benzoxazin-3-one (HBOA) from 3-hydroxyindolin-2-one in *Zea mays*. *Phytochemistry* **57**, 373-376

(72) Sue, M., Ishihara, A. and Iwamura, H. **2000**. Purification and characterization of a hydroxamic acid glucoside beta-glucosidase from wheat (*Triticum aestivum* L.) seedlings. *Planta* **210**, 432-438

(73) Sue, M., Ishihara, A. and Iwamura, H. **2000**. Purification and characterization of a β-glucosidase from rye (*Secale cereale* L.) seedlings. *Plant Sci.* **155**, 67-74

(74) Tays, K. and Atkinson, J. **1998**. An improved synthesis of cyclic hydroxamic acids from Gramineae. *Synth. Commun.* **28**, 903-912

(75) Thunecke, F., Hartenstein, H., Sicker, D., and Vogt, C. **1994**. Separation of enantiomers and diastereomers of 4-hydroxy-2*H*-1,4-benzoxazin-3(4*H*)-one derivatives by capillary electrophoresis. *Chromatographia* **38**, 470-474

(76) Tipton, C. L., Klun, J. A., Husted, R. R., and Pierson, M. D. **1967**. Cyclic hydroxamic acids and related compounds from maize. Isolation and characterization. *Biochemistry* **6**, 2866-2870

(77) Todorova, M., Werner, C. and Hesse, M. **1994**. Enzymatic phenol oxidation and polymerization of the spermine alkaloid aphelandrine. *Phytochemistry* **37**, 1251-1256

(78) Virtanen, A. I. and Hietala, P. K. **1955**. 2(3*H*)-Benzoxazolinone, an antifusarium factor in rye seedlings. *Acta Chem. Scand.* **9**, 1543-1545

(79) Virtanen, A. I. and Hietala, P. K. **1960**. Precursors of benzoxazolinone in rye plants: I. Precursor II, the aglucone. *Acta Chem. Scand.* **14**, 499-502

(80) Wahlroos, Ö. and Virtanen, A. I. **1959**. On the formation of 6-methoxybenzoxazolinone in maize and wheat plants. *Suomen Kemistilehti* **32B**, 139-140

(81) Wahlroos, Ö. and Virtanen, A. I., **1959**. The precursors of 6-methoxybenzoxazolinone in maize and wheat plants, their isolation and some of their properties. *Acta Chem. Scand.* **13**, 1906-1908

(82) Werner, C., Hedberg, C., Lorenzi-Riatsch, A., and Hesse, M. **1993**. Accumulation and metabolism of the spermine alkaloid aphelandrine in roots of *Aphelandra tetragona*. *Phytochemistry* **33**, 1033-1036

(83) Wieland, I., Friebe, A., Kluge, M., Sicker, D., and Schulz, M. **1999**. Detoxification of benzoxazolinone in higher plants. In: Macias, F. A., Galindo, J. C. G., Molinillo, J. M. G., and Cutler, H. G. (Eds.). *Recent Advances in Allelopathy - A Science for the Future*. Vol. 1, Servicio de Publicaciones – Universidad de Cadiz, Cadiz, Spain, 47-56

(84) Wolf, R. B., Spencer, G. F. and Plattner R. D. **1985**. DIBOA-Glc in *Acanthus mollis*. *J. Nat. Prod.* **48**, 59-62

(85) Woodward, M. D., Corcuera, L. J., Helgeson, J. P., and Upper, C. D. **1978**. Decomposition of 2,4-dihydroxy-7-methoxy-2*H*-1,4-benzoxazin-3(4*H*)-one in aqueous solutions. *Plant. Physiol.* **61**, 796-802

(86) Woodward, M. D., Corcuera, L. J., Schnoes, H. K., Helgeson, J. P., and Upper, C. D. **1979**. Identification of 1,4-Benzoxazin-3-ones in maize extracts by gas-liquid chromatography and mass spectrometry. *Plant Physiol.* **63**, 9-13

(87) Wu, H., Haig, T., Pratley, J., Lemerle, D., and An, M. **1999**. Simultaneous determination of phenolic acids and 2,4-dihydroxy-7-methoxy-1,4-benzoxazin-3-one in wheat (*Triticum aestivum* L.) by gas chromatography-tandem mass spectrometry. *J. Chromatogr. A* **864**, 315-321

(88) Yalpani, N., Schulz, M., Davis, M. P., and Balke, N. E. **1992**. Partial purification and properties of an inducible Uridine-5´-diphosphate-

glucose:salicylic acid glucosyltransferase from oat roots. *Plant Physiol.* **100**, 457-463

(89) Yenish, J. P., Worsham, A. D. and Chilton, W. S., **1995**. Disappearance of DIBOA-glucoside, DIBOA, and BOA from rye (*Secale cereale* L.) cover crop residue. *Weed Sci.* **43**, 18-20

5 Heliannanes – a Structure-Activity Relationship (SAR) Study

F. A. Macías, J. M. G. Molinillo, D. Chinchilla and J. C. G. Galindo

CONTENT

Abstract 103
Introduction 103
Heliannanes – Structure and Biogenesis 106
Heliannanes – Synthetic Studies 109
Heliannanes – Bioactivity 115
Methodology 117
References 121

ABSTRACT

Heliannanes constitute a new type of sesquiterpenes isolated from terrestrial (*Helianthus annuus*) and marine (*Haliclona ?fascigera*) organisms. They share as a common structural feature a substituted aromatic ring fused with an oxygen-containing heterocycle of variable size. The novelty of this structure and the phytotoxic properties of some members of this family led us to accomplish a Structure-Activity Relationship (SAR) study. At the same time, the interest of several research groups focused in this family of compounds, and several synthetic approaches to racemic or enantiopure heliannuols have been accomplished and published. The scope of this review is to present together all the literature published on the isolation, structural characterization, bioactivity and synthesis of heliannuols.

INTRODUCTION

We have been interested for a long time in the study of allelopathic crops such as sunflower, wheat, rice, and red clover. The reason for such an interest is that the knowledge of the chemicals involved in the allelopathic interactions will lead to a better understanding of the complex relationships that occur in nature. This is not just an academic question: this knowledge can be used to develop new low-input, environmentally friendly agricultural practices. Several scenarios can be envisioned in these studies:

(I) The most obvious is the direct use of the allelochemicals as new herbicides, targeting new sites of action. Few successful examples can be mentioned but, among them, glyphosate (**1**) [the synthetic version of phosphinothricin (**2**), a metabolic breakdown product of the *Streptomyces* spp. toxin bialaphos (**3**)] is a worldwide-marketed herbicide that might be claimed as a paradigm of this approach (Fig. 5.1). Phosphinothricin also is a potent inhibitor of glutamine synthetase.[14] However, this is not the common situation. Main problems concerning the direct use of allelopathic agents as herbicides are the low amounts of compound usually obtained (excepting in those coming from bacterial cultures, that can be continuously produced) and their complex structures, which make difficult and expensive their total synthesis at multigram scale.

(II) A second option is using the allelochemicals as lead compounds. In many cases, the activities obtained for the allelochemicals (especially those coming from plants) are low in comparison with the commercial herbicides. Should we throw away them? Our opinion is no, as they give us important information leading to the discovery of new sites of action and widening our scope with new structural types. In this case, the development of new herbicides is accomplished through Structure-Activity Relationships (SAR) studies. The introduction of chemical changes in the original structure is correlated with the results obtained in the bioassays. A picture of those factors affecting the activity should arise from this approach, which, thus, allows drawing the better molecule to be used as herbicide. This approach is commonly used in "classical" synthetic herbicide production, drug design, etc., but it has been little exploited using natural products as leads for herbicides. Cinmethylin (**4**) could represent an example, even though not a good one. The structure of this compound is closely related to the monoterpene 1,4-cineole (**5**), but apparently it is not the result of this strategy (Fig. 5.1). However, recent findings show that cinmethylin is a proherbicide that is cleaved by the benzyl-ether side chain, giving rise to the real herbicide. 1,4-Cineole targets asparagine synthetase, a key enzyme in the biosynthesis of the aminoacid asparagine.[38]

(III) Transgenic crops are another choice. It is obvious that transfer of allelopathic traits from allelopathic plants to nonallelopathic crops is an attractive approach to the problem of weed control. This transfer will enhance the plant resistance to weed attack and, probably, to other pests. However, the genes encoding the biogenetic pathway of plant toxins are not well understood, and few of them have been cloned. Another problem arise from ethic concerns. In our opinion, it has not been demonstrated yet that the use of transgenic crops does not have environmentally adverse effects. In Europe the use of transgenic crops is severely limited, and public opinion it is not favorable regarding the use of foods based on transgenic crops. Another possibility is the use of transgenic microorganisms producing the toxin by fermentation. However, problems coming from autotoxicity can be envisioned, unless the allelochemical is produced as a harmless pro-toxin (e.g., in glucosilated form).

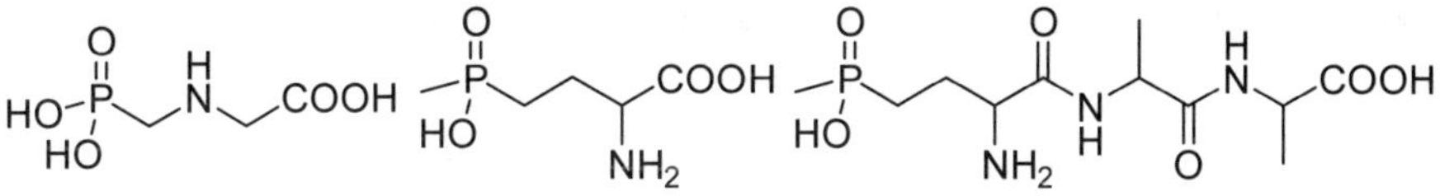

glyphosate (**1**)   phospinothricin (**2**)   bialaphos (**3**)

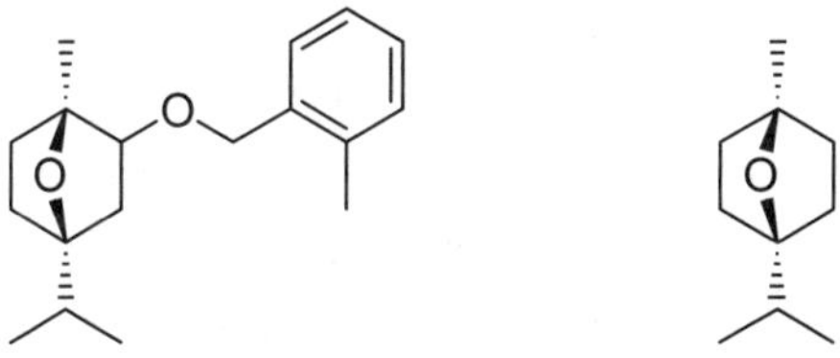

cinmmethylin (**4**)   1,4-cineole (**5**)

Figure 5.1
Some natural toxins and their structurally related commercial herbicides.

The present study falls into the second class mentioned above. Heliannuols are a new type of plant toxins with a good activity, but they have been isolated in small amounts which are not good enough to be used directly as herbicide. Consequently, chemical methods to get access to multi-gram amounts of heliannuols and to improve their solubility, stability, and phytotoxicity are needed.

Sunflower is a rich source of sesquiterpenes,[21] especially sesquiterpene lactones.[23,28,31,44] Due to its economic importance, sunflower has been extensively studied, thus leading to the isolation and chemical characterization of phenolic compounds (benzoic acid derivatives,[9,39,51] coumarins,[51] and flavonoids[6,10,19]), diterpenes,[4,33,34,36] and triterpenes.[12,37] Most of these studies were performed using a classical methodology where the extracts were obtained by soxhlet extraction using organic solvents.[6,4] When we started our studies on sunflower, a different approach was used. We performed water extracts of whole leaves trying to simulate rain natural conditions. Then, the aqueous extracts obtained were re-extracted with organic solvents (chloroform and ethyl acetate, sequentially) to obtain an easy to handle organic crude extract. In the results, the chemical composition of these extracts were extremely different from that reported in previous studies for sunflower. Two new families of sesquiterpenes, heliannuols and heliespirones, and a number of new sesquiterpenes and sesquiterpene lactones were described.

HELIANNANES – STRUCTURE AND BIOGENESIS

Heliannuol A (**7**) is the first heliannane reported in the literature.[27] It was isolated from sunflower leaves and all the following members of the family have been isolated from the same source, but in different sunflower varieties. Surprisingly, no heliannuols have been isolated from other terrestrial sources so far. However, the basic heliannane skeleton (**6**) has been isolated lately from a marine organism, the Indo-Pacific sponge *Haliclona ?fascigera*.[8] Heliannanes isolated from different sources are summarized in Table 5.1 and Figure 5.2; new heliannanes have been isolated from other sunflower varieties (unpublished results).

Heliannane (**6**) has a different origin from the rest of the members of the heliannane family. Thus, two important and closely connected points of discussion arose regarding the absolute stereochemistry and the biogenetic pathway.

The absolute stereochemistry is not just an academic question; it has strong implications in the biogenetic pathway and also in the bioactivity. This is of special importance when the ultimate goal is to find a practical application for such bioactivity. It is well known that only one member corresponding to one of the two possible *d/l* optical rotation values exists in nature. Sometimes, when the natural product is active, the other is inactive or, what is even worse, has unexpected detrimental effects.

Heliannanes – a Structure-Activity Relationship (SAR) Study

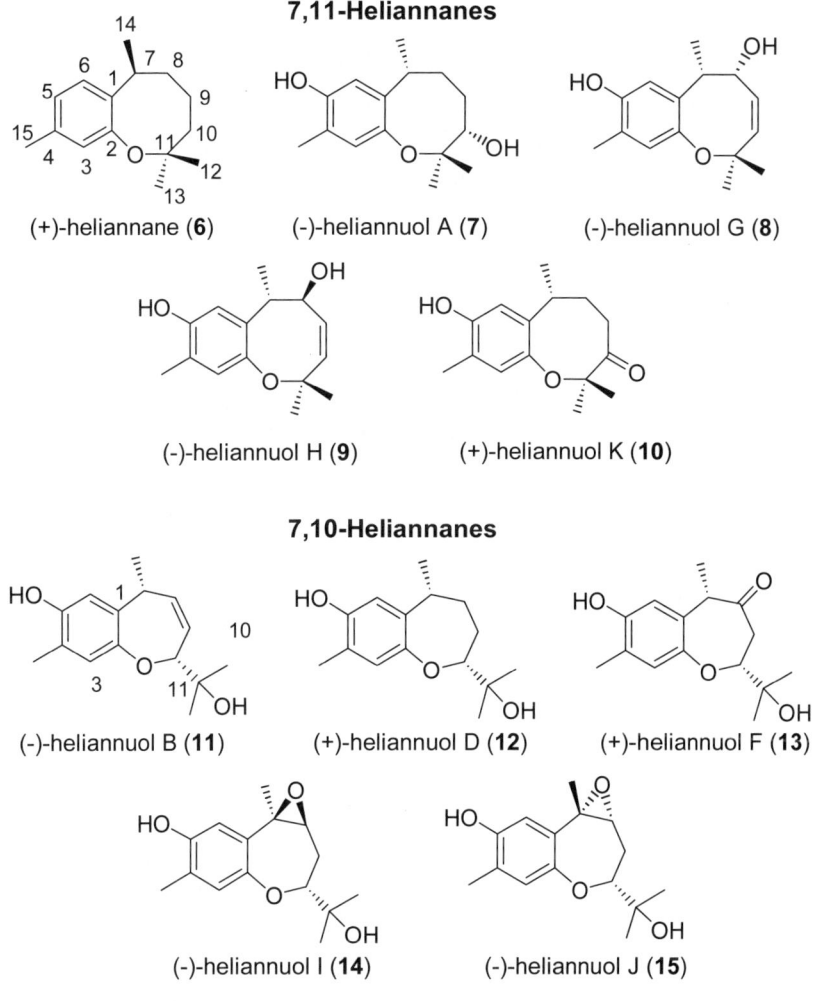

Figure 5.2

Natural heliannanes isolated from marine (**6**) and terrestrial (**7-17**) organisms.

Table 5.1

Heliannanes isolated from terrestrial and marine organisms.

Heliannanes	Origin	Reference
Heliannane (6)	*Haliclona ?fascigera*	8
Heliannuol A (7)	*H. annuus* cv. SH-222	20,27
Heliannuol B (11)	*H. annuus* cv. SH-222	27
Heliannuol C (16)	*H. annuus* cv. SH-222	27
Heliannuol D (12)	*H. annuus* cv. SH-222	27
Heliannuol E (17)	*H. annuus* cv. SH-222	26
Heliannuol F (13)	*H. annuus* cv. SH-222, and cv. VYP	29
Heliannuol G (8)	*H. annuus* cv. SH-222	29
Heliannuol H (9)	*H. annuus* cv. SH-222, and cv. VYP	29
Heliannuol I (14)	*H. annuus* cv. SH-222	29
Heliannuol J (15)	*H. annuus* cv. VYP, and cv. Peredovick	22,29
Heliannuol K (10)	*H. annuus* cv. SH-222	29

Absolute configuration of Heliannuol A (7) has been established as (7R,10S)[30] by using the modified Mosher methodology.[35,43] This assignation has been further confirmed by asymmetric synthesis of the enantiomer (+)-heliannuol A.[47] Biogenetic considerations allow us to establish the absolute stereochemistries to the rest of the members of the heliannane family (Fig. 5.1), excepting heliannuol C.[30] Recently, the total synthesis of (-)-heliannuol E (17) has been reported[40] and, again, confirms the previous assignment of absolute stereochemistry. The absolute stereochemistry of the marine heliannane (6) is proposed as just the opposite (7S) based on the same biogenetic arguments.[8,30]

Regarding the biogenetic pathway, a common γ-bisabolene precursor is proposed for marine and terrestrial heliannuols. The route then diverts through (+)-curcuphenol to (+)–heliannane in marine organisms,[8] and through (-)-curcuquinone to (-)-heliannuol A in plant heliannuols.[20,30] The differentiation arose from reduction of the double bond at C7-C8, thus giving rise to the stereogenic centre at C7: 7R in terrestrial heliannuols and 7S in marine heliannuols. The isolation of (+)-curcuphenol from marine sponges strengths this hypothesis.[52] Then, the biogenetic route proposes as a key intermediate an epoxide derivative of curcuhydroquinone. The basic or acid catalyzed nucleophilic opening of this epoxide should lead to the head members of 7,11- and 7-10-heliannanes (heliannuols A and D, respectively). More recently, we have isolated from *H. annuus* a new sesquiterpene;

helibisabonol A.[24] This compound should correspond to the hypothetic epoxide ring opening reaction and thus, modifies the cyclization step from a nucleophilic attack of the alcohol to an etherification reaction.

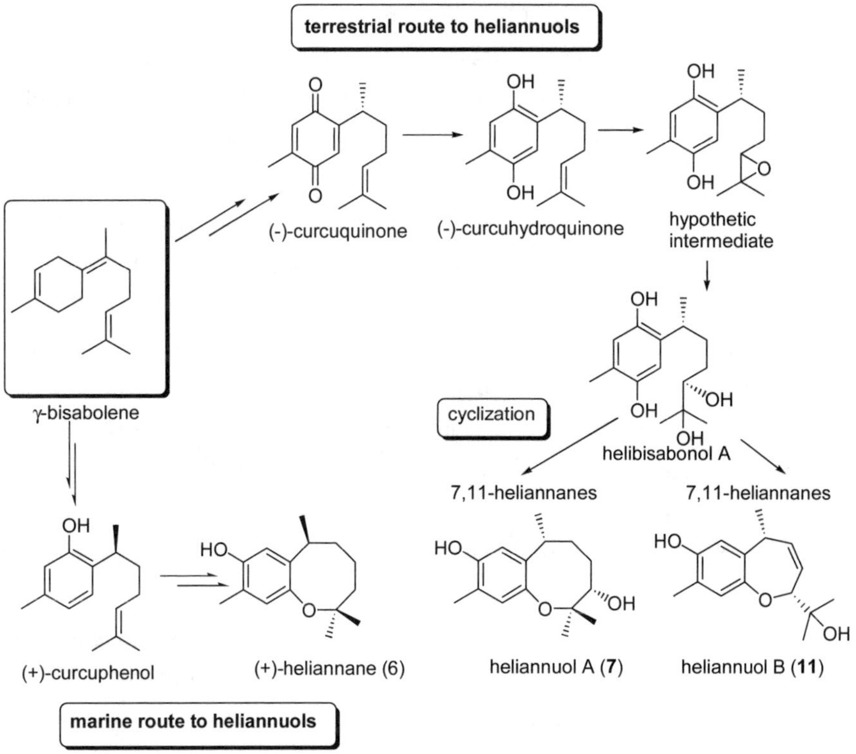

Figure 5.3

Biogenetic pathways proposed for marine and terrestrial heliannanes.

HELIANNANES – SYNTHETIC STUDIES

Once the absolute stereochemistry of the natural products is well established, it comes time for the synthetic approaches. In fact, most of the time the elucidation process and the development of an efficient synthetic methodology to get access to the compound occur almost at the same time. Competition among synthetic chemists is very strong and, if the natural product is novel enough, a race towards the "first synthesis" usually starts as soon as the structure of the compound is disclosed.

Allelopathy scientists usually disregard "hard" chemistry matters. This is probably due to how allelopathy was born and evolved. Organic chemists became involved late in the development of allelopathy, once plant physiologists, ecologists and agronomists established the "core" of the discipline. However, the definition of allelopathy itself as the study of the chemical interactions among plants and their environment[11] implies that chemical aspects cannot be neglected but also that they are crucial to the understanding of the complex inter-relationships observed in every ecological niche. Moreover, a true understanding of the allelopathic interactions is not possible without the knowledge of the chemistry involved between the allelopathic agents and their targets at the molecular level. Most of the time, the first pebble in the scientist's shoe is the small amount obtained for most of natural products. Such small quantities make it extremely difficult to accomplish mode of action studies. Consequently, the development of synthetic methodologies to get access to enough of the compound to perform mode of action studies is imperative. At the same time, during the course of these synthetic studies, many chemical analogues arise that can be also used for SAR studies.

Heliannuols are not an exception. Only one year later following the publication of heliannuol A, the first total synthesis of the racemic (±)-heliannuol A using the coumarin **18** as starting material was published.[7] A key step in this synthesis was the ring closure achieved by two different methods: intramolecular Julia coupling of the sulfone **19** and sulfone ester cyclization of the alternative sulfone **20** (Fig. 5.4.A). Desulfonation and demethylation yielded the desired (±)-heliannuol A.

An elegant synthesis of (±)-heliannane connecting ortho metalation and olefin metathesis as key steps has been reported.[46] Ring closing olefin metathesis (RCM) constitutes one of the more powerful synthetic tools recently developed to obtain medium and large size rings.[41] In this case, the starting material was the *m*-cresol O-carbamate **21**, which suffered selective regioselective lithiation, transmetalation to the cuprate, Michael addition, and Wittig olefination. Deprotection of the carbamoyl moiety led to the appropriate substrate (**22**) for etherification and introduction of the second alkenyl side chain (compound **23**). RCM using the ruthenium Grubb's catalyst followed by palladium catalyzed hydrogenation afforded the desired heliannane in racemic form (Fig. 5.4.B).

First total synthesis of (±)-heliannuol D, bearing a seven-member heterocyclic ring, has been recently reported.[50] The most remarkable aspect of the retro-synthetic analysis is the biomimetic approach, in which the key step is the base-catalyzed ring opening of the hypothetic epoxide intermediate **26** (Fig. 5.5).

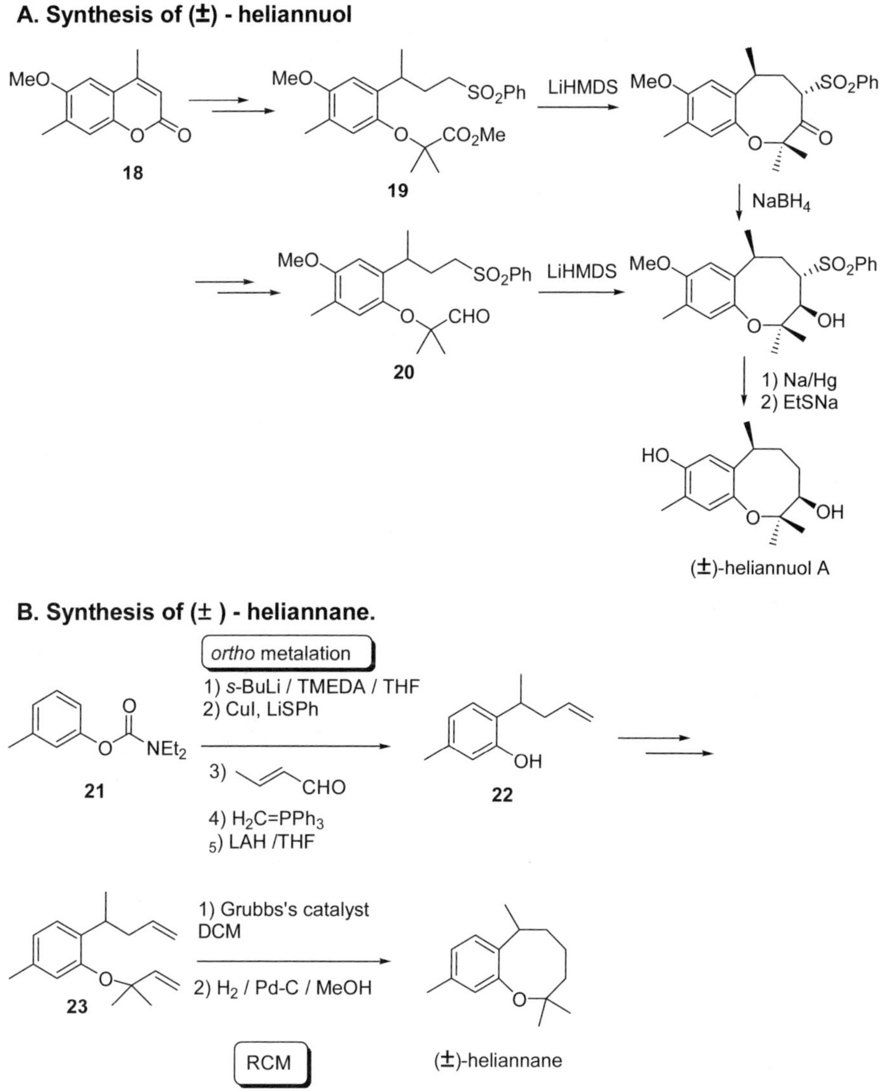

A. Synthesis of (±) - heliannuol

B. Synthesis of (±) - heliannane.

Figure 5.4

A. First total synthesis of (±)-heliannuol A. **B.** First total synthesis of (±)-heliannane.

All the synthesis mentioned end up with racemic mixtures. In spite of the high importance of getting enantiopure compounds, the first report of an enantioselective total synthesis of heliannuols A and D have not been published until recently.[47] The key point of the synthetic strategy was to get access to the enantiopure epoxide **35**. Base-catalyzed intramolecular cyclization led to the desired heliannuols A and D in enantiopure forms.

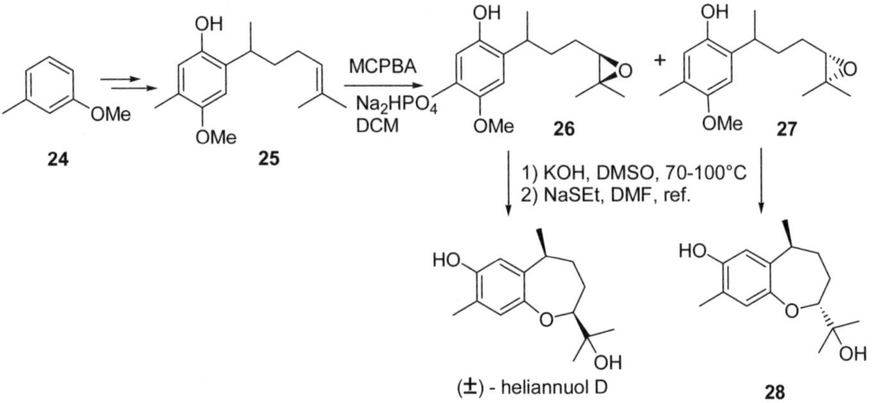

Figure 5.5

Biomimetic synthesis of (±)–heliannuol D and its C-10 epimer **28**.

The proper stereochemistry was achieved by enzyme catalyzed desymmetrization of the prochiral 1,3-diol **30**. *Candida antarctica* lipase (CAL)-catalyzed transesterification yielded the monoacetate **31**, which gave rise to the methyl with the proper stereochemistry **32**. The generation of the desired chiral epoxide **35** was achieved by asymmetric dihydroxylation employing AD-mix-α,[42] followed by epoxide formation. Base-catalyzed etherification yielded the mixture of the enantiopure (+)-heliannuol A and (-)-heliannuol D. Unfortunately these compounds correspond to the opposite *d/l* series and correspond to the enantiomers of the natural products (-)–heliannuol A and (+)-heliannuol D (Fig. 5.6.A).

(-)-Heliannuol E, the only representative of six-member ring heliannuols, is the first heliannuol synthesized belonging to the natural *d/l* series.[40] The synthesis has been developed by Shishido's group, which also published the enantioselective synthesis of (+)-heliannuol A and (-)-heliannuol D. Following a similar strategy, key steps are the desymmetrization of the diol derivative **37** and cyclization of the epoxide **39** (Fig. 5.6.B). This time, the use of lipase AK transformed the diacetate **35** in the desired monoacetate **36**, bearing the correct stereochemistry at the carbon supporting the future vinyl group. Unfortunately it was not possible to obtain the epoxide in an enantioselective form. Consequently, the mixture of epoxides **39** was subjected to cyclization basic conditions (K$_2$CO$_3$ in MeOH), the epoxides (**40**) separated by preparative HPLC, and the desired isomer deprotected and dehydrated to yield (−)-heliannuol E.

A. First enantioselective synthesis of (+)-heliannuol A and (−)-heliannuol D.

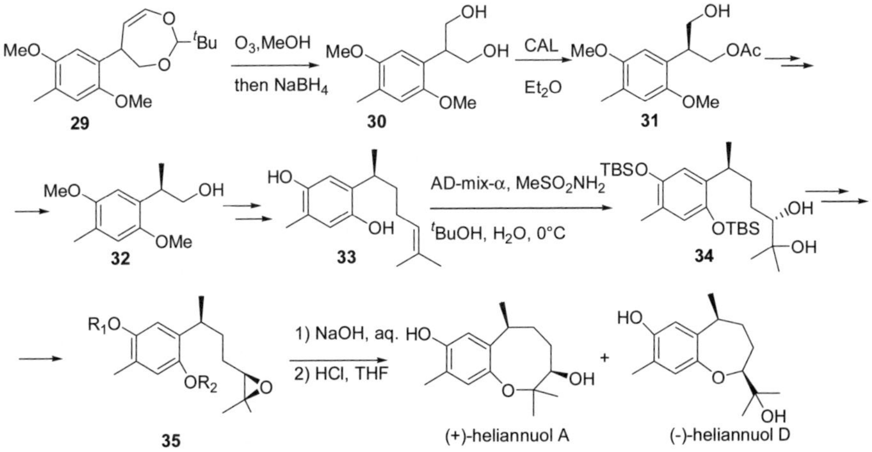

B. First enantioselective synthesis of (−)-heliannuol E.

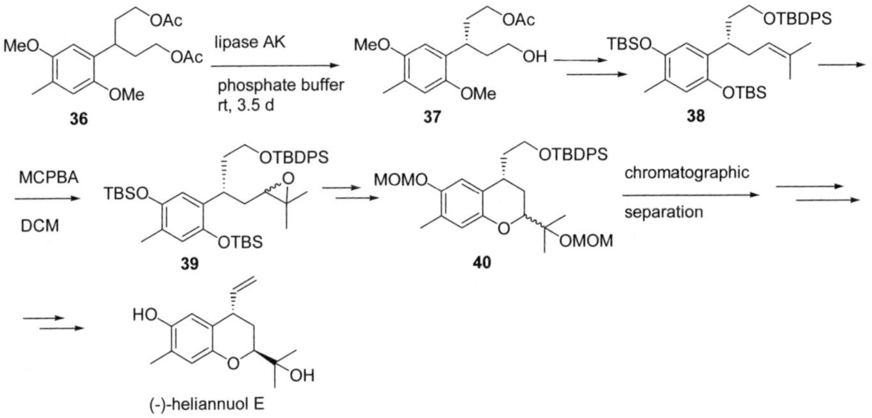

Figure 5.6

Enantioselective synthesis of heliannuols. **A**: (+)-heliannuol A and (−)-heliannuol D. **B**: (−)-heliannuol D.

Recently, we have published a high yield route to (±)-heliannuol D[16] based on biomimetic considerations. The synthesis starts with the readily available 2-

methyl-hydroquinone **41**; acetylation and Fries rearrangement (**42**), followed by Grignard chain elongation and epoxidation with *m*-CPBA lead to the two possible diastereoisomers **43** and **44**. The key step in the synthesis is the basic-catalyzed epoxide ring opening etherification that yields both possible diastereoisomers (±)-heliannuol D and **45** with an overall yield of 60.7 % for the whole process (Fig. 5.7).

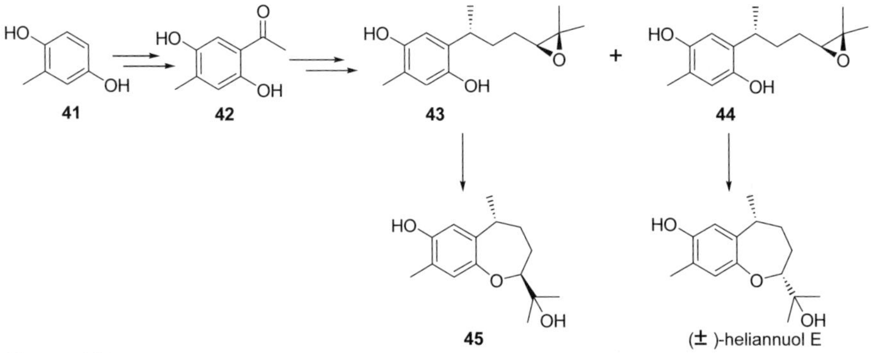

Figure 5.7

Biomimetic total synthesis of (±)-heliannuol D.

The interest in these compounds is becoming more obvious as more synthetic strategies are being disclosed. Heliannuol A has been synthesized by first getting access to the oxepane ring, followed by cyclopropanation and ring expansion using 4,7-dimethylcoumarin as starting material.[48] (±)-Heliannuol E has also been obtained through a ring expansion strategy using the spirodienone **46** as key intermediate (Fig. 5.8A).[3] Finally, a enantioselective total synthesis of (-)-heliannuol A has been recently disclosed[13] in which the key steps are an enzymatic desymmetrization using porcin pancreatic lipase (PPL) to obtain the proper enantiomer (78% ee) and ring closing metathesis (RCM) using the second generation Grubbs catalyst to get access to the eight-membered ring (Fig. 5.8B).

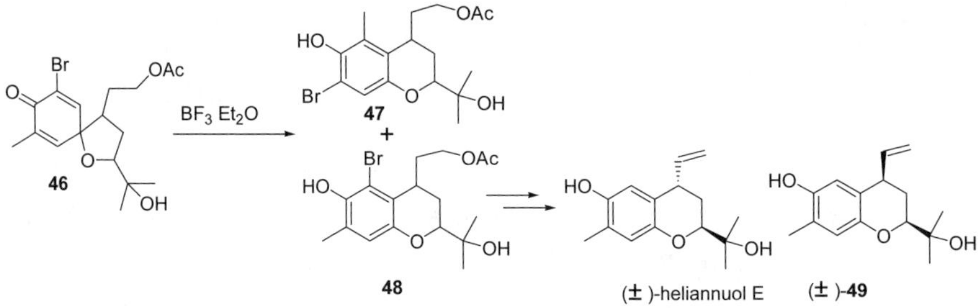

Figure 5.8.A

Synthesis of (±)-heliannuol E using a ring expansion strategy.

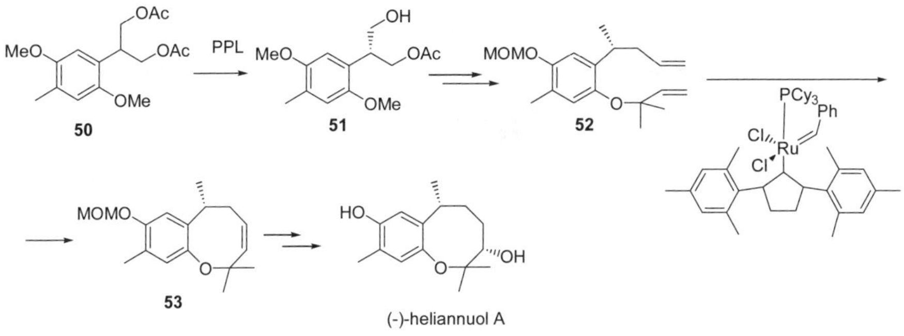

Figure 5.8.B

Synthesis of natural (−)–heliannuol A using a desymmetrisation-RCM strategy.

HELIANNANES – BIOACTIVITY

We have evaluated the activities of the heliannuols isolated from sunflower (**7-17**) using Petri dish bioassays with monocot and dicot standard target species (STS), as previously proposed by our group.[15] As we had no access to a sample of Heliannane (**6**) our study was limited to the heliannuols isolated by us. We also had access to several possible biogenetic precursors of heliannuols (helibisabonol A, Fig. 5.3), isolated[24] and lately synthesized[18] by us, which were also included in these studies. More recently, we have adapted to our needs another bioassay based on the use of etiolated wheat coleoptiles, as proposed by Cutler et al.[2] The discussion presented herein about the structural requirements for bioactivity is based on all these data. It can be claimed that bioactivity data correspond to bioassays performed at different moments and with different batches of seeds. However, to assure the reproducibility of the bioassays, we introduce in every assay an internal standard of known activity (in our case, the commercial herbicide Logran®). In all cases, the internal standard showed similar values of activity. Consequently, comparison among data from different bioassays can be done if we make the analysis in a broad sense.

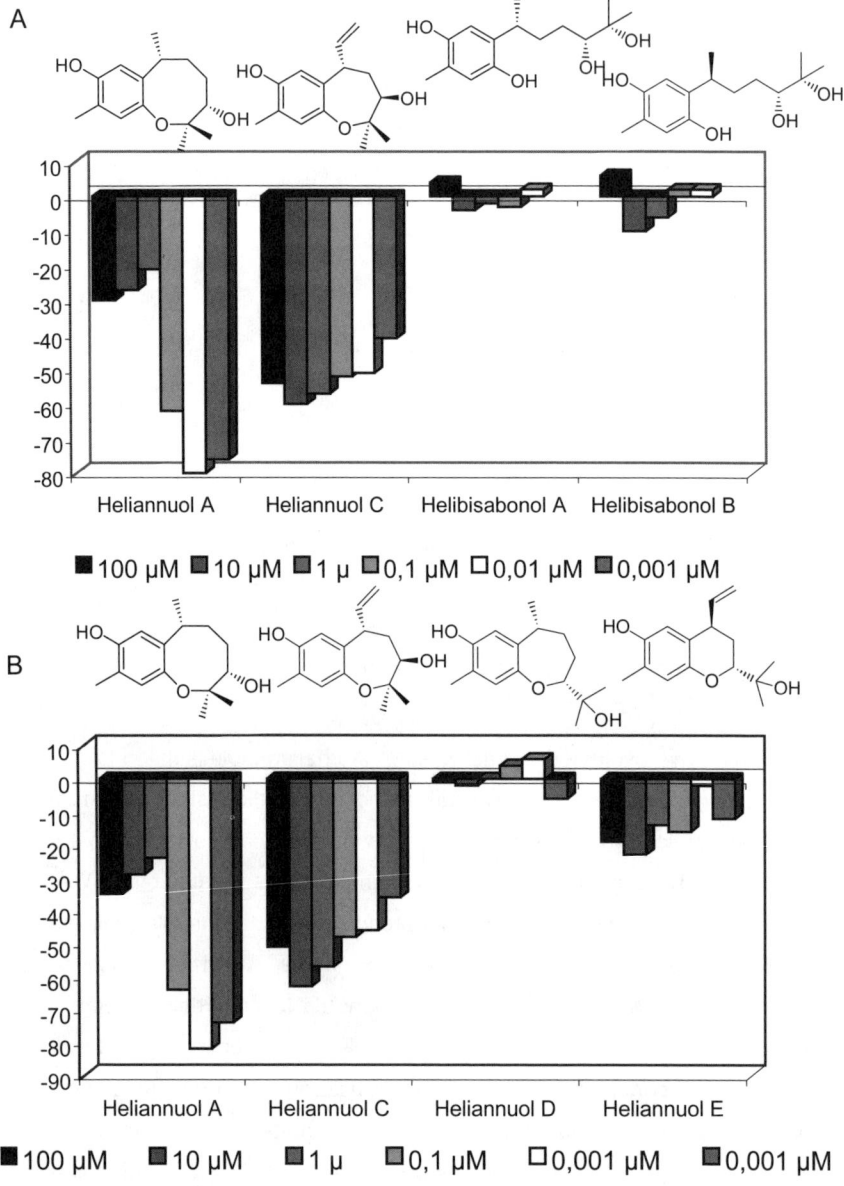

Figure 5.9

A: Influence of the heterocycle in the germination of le ttuce. Note how heterocyclic compounds Heliannuol A and C are active, whereas open chain compounds Helibisabonol A and B are inactive. B: Influence of the ring size in the germination of lettuce. Note how the bioactivity decreases with the size of the ring: 8 > 7 > 6

According to these data, several conclusions arose from the comparison of the bioactivity:

a) As a general behavior, inhibitory effects are observed usually in the dicots lettuce (*Lactuca sativa*) and cress (*Lepidium sativum*). On the other hand, monocot growth (barley, *Hordeum vulgare*, and onion, *Allium cepa*) is inhibited.[20,29,30]

b) The presence of a heterocyclic ring is crucial for the activity. Most heliannuols are active on dicot and/or monocot species.[20,29,49] They are also active in the etiolated wheat coleoptile bioassay.[24] However, when we assayed the corresponding open-chain precursors Helibisabonol A and B and other related compounds obtained by synthesis (Fig. 5.9A), they were found to be inactive in the Petri dish bioassay (unpublished results)[1,17,32] and exhibited also lower activities in the coleoptile biosassay (unpublished results).[5]

c) The ring size of the heterocycle is a key factor: the decreasing order of activity in the Petri dish bioassay is eight members > seven members > six members (Fig. 5.9B). This is true also in the wheat coleoptile bioassay.[50]

d) The position of the hydroxyl group also influences the activity. Those compounds with the hydroxyl group located in the heterocyclic ring (7,11-helinannuols and 8,11-heliannuols) are more active that those with the hydroxyl group in the *iso*-propyl side chain (7,10-heliannuols).[49] (Fig. 5.10). When comparing 8,11-heliannuols with 7,11-heliannuols, the presence of a vinyl side chain also enhances the activity (Fig. 5.11).

e) Finally, the stereochemistry of each chiral centre is crucial for the bioactivity. Comparison of the effects of heliannuol G and its epimer at C-8 heliannuol H shows a great difference in the activity between both compounds (Fig. 5.12).

METHODOLOGY

Wheat coleoptiles bioassays[2,25]

Wheat seeds (*Triticum aestivum* L. cv. Cortex) were sown on 15 cm ⌀ Petri dishes filled with Whatman #1 filter paper and grown in the dark at 24°C for 4 days. The etiolated seedlings were removed from the dishes and selected for size uniformity. The selected etiolated seedlings were placed in a Van der Wij guillotine, and the apical 2 mm were cut off and discarded. The next 4 mm of the coleoptiles were removed for bioassay and kept in aqueous nutritive buffer for 1 h to synchronize the growth. Mother solutions of pure compounds were dissolved in DMSO and diluted to the proper concentration with a phosphate-citrate buffer containing 2% sucrose at pH 5.6 to a 0.5 % DMSO final maximum concentration.

Following concentrations were obtained by dilution. The bioassays were performed in 10 mL test tubes: five coleoptiles were added to each test tube containing 2 mL of the test solution. Three replicates were made for each test solution, and the experiments were run in duplicate. Test tubes were placed in a roller tube apparatus and rotated at 0.25 rpm for 24 h at 22°C in the dark. All manipulations were done under a green safelight. The coleoptiles were measured by digitalization of their photographic images and the data were statistically analyzed.

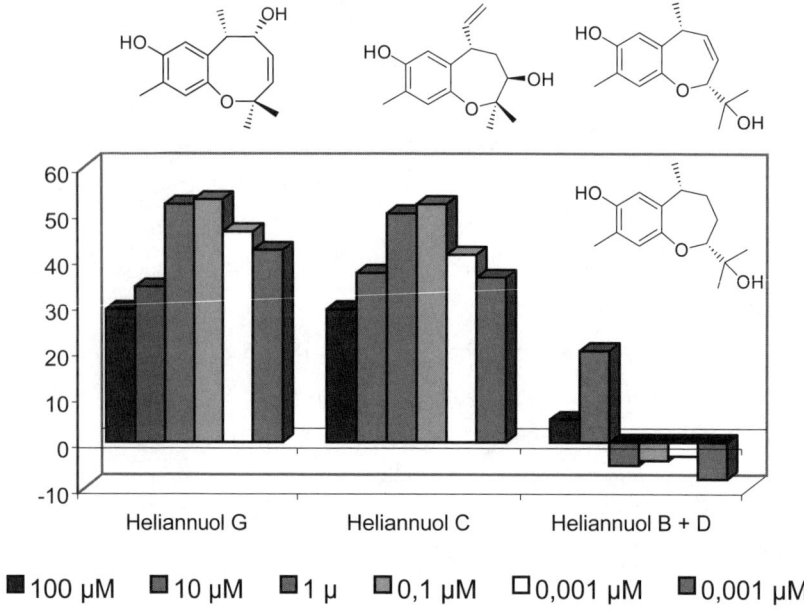

Figure 5.10

Influence of the position of the hydroxyl group in the root growth of barley. Note how those compounds with the hydroxyl group attached directly to the heterocycle (heliannuol G and C) are more active than those with the hydroxyl group located at the *iso*propyl side chain (mixture of heliannuols B and D).

Heliannanes – a Structure-Activity Relationship (SAR) Study

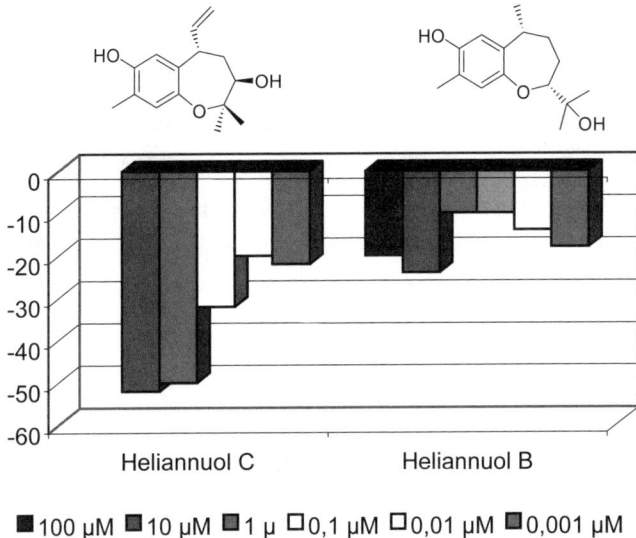

■ 100 μM ■ 10 μM ■ 1 μ ▫ 0,1 μM ▫ 0,01 μM ■ 0,001 μM

Figure 5.11

Influence of the presence of a vinyl side chain (heliannuol C) *versus* an isopropyl side chain (heliannuol B) in the germination of onion.

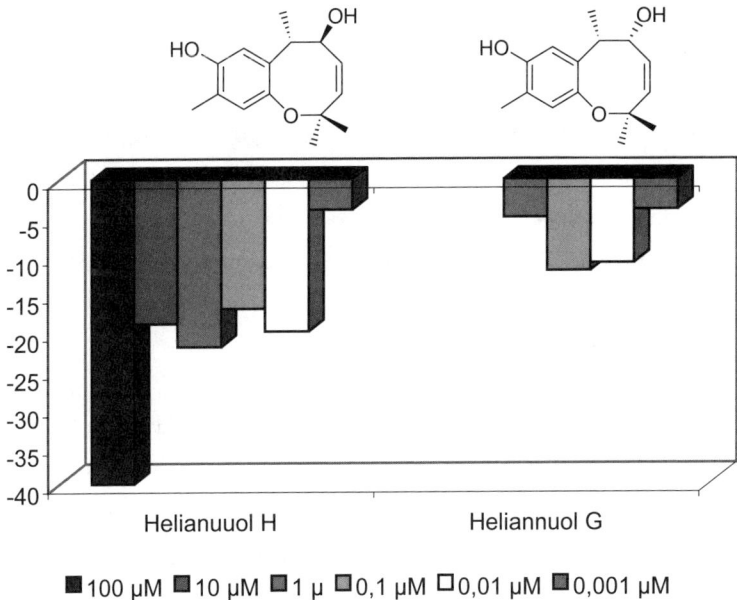

■ 100 μM ■ 10 μM ■ 1 μ ▫ 0,1 μM ▫ 0,01 μM ■ 0,001 μM

Figure 5.12

Influence of the stereochemistry of the chiral centre at C-8 in the root growth of lettuce.

Petri dish bioassays[43]

Seeds of lettuce (*Lactuca sativa* L. cv. Roman), cress (*Lepidium sativum* L. cv. Común), and onion (*Allium cepa* L. cv. Valenciana), were obtained from FITÓ, S.L. (Barcelona, Spain). Seeds of wheat and barley (*Hordeum vulgare* L.) were obtained from Rancho La Merced, Junta de Andalucía, Jerez, Spain. All undersized or damaged seeds were discarded, and the assay seeds were selected for uniformity. Bioassays were carried out in 9 cm Ø plastic Petri dishes, using Whatman #1 filter paper as support.

The general procedure for seedling bioassay was as follows: 25 seeds of each species were placed per dish, excepting *Hordeum vulgare* (10 seeds per dish), with 5 mL of the test solution, and incubated in the dark at 25°C. Four replicates for each concentration were set up. Germination and growth time varied for each plant species: *Lepidium sativum*, 3 days; *Lactuca sativa* and *Hordeum vulgare*, 5 days; and *Allium cepa*, 7 days.

Test mother solutions (10^{-2}M) were prepared using dimethyl sulfoxide (DMSO) and then diluted to 10^{-4}M using 10 mM MES (2-[*N*-morpholino]ethanesulphonic acid). Following solutions were obtained by dilution maintaining the 1% DMSO percentage. Parallel controls were performed. All pH values were adjusted to 6.0 before bioassay. All products were purified prior to the bioassay using HPLC equipped with a refractive index detector. Minimum degree of purity was 99% as extracted from the chromatograms.

Data are presented in figures where zero represents the control; positive values represent stimulation and negative values represent inhibition.

Statistical treatment

Germination and root and shoot length were tested by Welch´s test,[53] the differences between test solutions and controls being significant with $P<0.01$. Cluster analysis was performed using the Statistica package on a Pentium III PC.[45] The analysis was recorded to all compounds tested using as variables germination index and root and shoot growth.

ACKNOWLEDGMENTS

The authors are thankful for financial support from the Ministerio de

Educación y Ciencia, DGICYT, Spain. Project PB98-0575.

REFERENCES

(1) Chinchilla, D. **2001**. Biomimetic synthesis of heliannanes: 7,10-heliannane backbone. Ms. dissertation. University of Cadiz, Spain

(2) Cutler, H. G. **1984**. A fresh look at the wheat coleoptile bioassay. *Proceedings of the 11th Annual Meeting of the Plant Growth Regulator Society of America.* pp. 1-9

(3) Doi, F., Ogamino, T., Sugai, T., and Nishiyama, S. **2003**. Synthesis of bioactive sesquiterpene heliannuol E involving a ring-expansion reaction of spirodienones. *Synlett* **3**, 411-413

(4) Ferguson, G., McCrindle, R., Murphy, S. T., and Parvez, M. **1982**. Further diterpenoid constituents of *Helianthus annuus* I. Crystal and molecular-structure of methyl-*ent*-15-beta-hydroxy-trachyloban-19-oate. *J. Chem. Res.-S.*, **8**, 200-201

(5) Galindo, J. L. G. **2001**. Study of the polar fractions of *Helianthus annuus*. cv. Peredovick. Models for Natural Herbicides. Ms. dissertation. University of Cadiz, Spain

(6) Gao, F., Wang, H., and Mabry, T. J. **1987**. Sesquiterpene lactones and flavonoides from *Helianthus* species. *J. Nat. Prod.* **50**, 23-29

(7) Grimm, E. L., Levac, S., and Trimble, L. A. **1994**. Total synthesis of ($\pm$)-Heliannuol A. *Tetrahedron Lett.* **35**, 6847-6850

(8) Harrison, B. and Crews, P. **1996**. The structure and probable biogenesis of Helianane, a heterocyclic sesquiterpene, from the Indo-Pacific sponge *Haliclona ?fascigera*. *J. Org. Chem.* **62**, 2646-2648

(9) Herz, W. and Bruno, M. **1986**. Heliangolides, kauranes and other constituents of *Helianthus heterophyllus*. *Phytochemistry*, **25**, 1913-1916

(10) Herz, W., Kulanthaivel, P., and Watanabe, K., **1983**. *Ent*-kauranes and other constituents of three *Helianthus* species. *Phytochemistry* **22**, 2021-2025

(11) IAS Constitutions and Bylaws. **1999**. First International Symposium in Allelopathy. A Science for the Future. Cádiz, Spain

(12) Kasprzyk, Z. and Janiszowska, W. **1971**. Triterpenic alcohols from shoots of *Helianthus annuus*. *Phytochemistry* **10**, 1946-1947

(13) Kishuku, H., Shindo, M., and Shishido, K. **2003**. Enantioselective total synthesis of (-)-heliannuol A. *Chem. Commun.*, 350-351

(14) Lydon, J. and Duke, S. O. **1998**. Inhibitors of glutamine biosynthesis.In: Sing, B.K. (Ed.). *Plant Amino Acids: Biochemistry and Biotechnology*.

Marcel Dekker, New York, 445-464

(15) Macías, F. A., Castellano, D., and Molinillo, J. M. G. **2000**. Search for a standard phytotoxic bioassay for allelochemicals. Selection of standard target species. *J. Agric. Food Chem.* **48**, 2512-2521

(16) Macías, F. A., Chinchilla, D., Molinillo, J. M. G., Marín, D., Varela, R. M., and Torres, A. **2003**. Synthesis of heliannane skeletons. Facile preparation of (±)- heliannuol D. *Tetrahedron* **59**, 1679-1683

(17) Macías, F. A., Marín, D., Chinchilla, D., Galindo, J. L. G., and Molinillo, J. M. G. **2002**. First total synthesis of (±) helibisabonol A. Evaluation of its activity. Book of Abstracts of the III Encuentro Andaluz-Marroquí sobre la Química de Productos Naturales. Algeciras, Spain

(18) Macías, F. A., Marin, D., Chinchilla, D., and Molinillo, J. M. G. **2002**. First total synthesis of (±)-helibisabonol A. *Tetrahedron Lett.* **43**, 6417-6420

(19) Macías, F. A., Molinillo, J. M. G., Torres, A., Varela, R. M., and Castellano, D. **1997**. Bioactive flavonoides from *Helianthus annuus* cultivars. *Phytochemistry* **45**, 683-687

(20) Macías, F. A., Molinillo, J. M. G., Varela, R. M., and Torres, A. **1994**. Structural elucidation and chemistry of a novel family of bioactive sesquiterpenes: Heliannuols. *J. Org. Chem.* **59**, 8261-8266

(21) Macías, F. A., Molinillo, J. M. G., Varela, R. M., Torres, A., and Galindo, J. C. G. **1999**. Bioactive compounds from the genus *Helianthus*. In: Macías, F. A., Galindo, J. C. G., Molinillo, J. M. G., and Cutler, H. G. (Eds.). *Recent Advances in Allelopathy. A Science for the Future*. Vol. 1. Servicio de Publicaciones de la UCA. Cádiz, Spain, 124-148

(22) Macías, F. A., Oliva, R. M., Varela, R. M., Torres, A., and Molinillo, J. M. G. **1999**. Allelochemicals from sunflower leaves cv. Peredovick. *Phytochemistry* **52**, 613-621

(23) Macías, F. A., Torres, A., Molinillo, J. M. G., Varela, R. M., and Castellano, D. **1996**. Potential allelopathic sesquiterpene lactones from sunflower leaves. *Phytochemistry* **43**, 1205-1215

(24) Macías, F. A., Torres, A., Varela, R. M., Galindo, J. L. G., Álvarez, J. A., and Molinillo, J. M. G. **2002**. Bioactive terpenoids from sunflower leaves cv. Peredovick®. *Phytochemistry*, **61**, 687-692

(25) Macías, F. A., Varela, R. M., Simonet, A. M., Cutler, H. G., Cutler, S. J., Dugan, F. M., and Hill, R. A. **2000**. Novel bioactive Breviane spiroditerpenoids from *Penicillium brevicompactum* Dierckx. *J. Org. Chem.* **65**, 9039-9046

(26) Macías, F. A., Varela, R. M., Torres, A., and Molinillo, J. M. G. **1999**.

Heliannuol E. A novel bioactive sesquiterpene of the heliannane family. *Tetrahedron Lett.* **40**, 4725-4728

(27) Macías, F. A., Varela, R. M., Torres, A., and Molinillo, J. M. G. **1993**. Potential allelopathic guaianolides from cultivar sunflower leaves, cv. SH-222. *Phytochemistry* **34**, 669-674

(28) Macías, F. A., Varela, R. M., Torres, A., and Molinillo, J. M. G. **1993**. Novel sesquiterpene from bioactive fractions of cultivar sunflowers. *Tetrahedron Lett.* **34**, 1999-2002

(29) Macías, F. A., Varela, R. M., Torres, A., and Molinillo, J. M. G. **1999**. New bioactive plant heliannuols from cultivar sunflower leaves. *J. Nat. Prod.* **62**, 1636-1639

(30) Macías, F. A., Varela, R. M., Torres, A., and Molinillo, J. M. G. **2000**. Potential allelopathic activity of natural plant heliannanes: a proposal of absolute configuration and nomenclature. *J. Chem. Ecol.* **26**, 2173-2186

(31) Macías, F. A., Varela, R. M., Torres, A. Oliva, R. M., and Molinillo, J. M. G. **1998**. Bioactive norsesquiterpenoids from *Helianthus annuus* with potential allelopathic activity. *Phytochemistry*, **48**, 631-636

(32) Marín, D. **2001**. Biomimetic synthesis of sesquiterpenes from *Helianthus annuus*. Helibisabonol A. Ms. dissertation. University of Cadiz, Spain

(33) Martín Panizo, F. and Rodríguez, B. **1979**. Diterpene compounds of the sunflower (*Helianthus annuus*). *An. Quím.* **75**, 428-430

(34) Melek, F. R., Gage, D. A., Gershenzon, J., and Mabry, T. J. **1985**. Sesquiterpene lactone and diterpene constituents of *Helianthus annuus*. *Phytochemistry* **24**, 1537-1539

(35) Ohtani, I., Kusumi, T., Ishitsuka, M. O., and Kasikawa, H. **1989**. Absolute configuration of marine diterpenes possessing a xenicane skeleton. An application of an advanced Mosher's method. *Tetrahedron Lett.* **30**, 3147-3150

(36) Pyrek, J. S. **1970**. New pentacyclic diterpene acid trachyloban-19-oic acid from sunflower. *Tetrahedron* **26**, 5029

(37) Pyrek, J. S. and Baranowska, E. **1973**. Faradiol and arnidiol. Revision of structure. *Tetrahedron Lett.* **11**, 809-810

(38) Romagni, J. G., Duke, S. O. and Dayan, F. E. **2000**. Inhibition of plant asparagine synthetase by monoterpene cineoles. *Plant Physiol.* **123**, 303-313

(39) Saggese, E. J., Foglia, T. A., Leather, G., Thompson, M. P., Bills, D. D., and Hoagland, P. D. **1985**. Fractionation of allelochemicals from oilseed sunflowers and jerusalem artichokes. *ACS Symp. Ser.* **268**, 99-112

(40) Sato, K., Yoshimura, T., Shindo, Y., and Shishido, K. **2001**. Total synthesis

(41) Schuster, M. and Blechert, S. **1997**. Olefin metathesis in organic chemistry. *Angew. Chem., Int. Ed. Eng.* **36**, 2036-2056

(42) Sharpless, K. B., Amberg, W., Bennani, Y. L., Crispino, G. A., Hartung, J., Jeong, K. S., Kwong, H. L., Morizawa, K., Wand, Z. M., Xu, D., and Zhang, X. **1992**. The osmium-catalyzed asymmetric dihydroxilation. A new ligand class and a process improvement. *J. Org. Chem.* **57**, 2768-2771

(43) Shi, X. W., Attygalle, A. B., Liwo, A., Hao, H.-H., Meinwald, J., Dharmaratne, H. R. W., Wanigasekera, and W. M. A. P. **1998**. Absolute stereochemistry of soulattrolide and its analogs. *J. Org. Chem.* **63**, 1233-1238

(44) Spring, O., Kupka, J., Maier, B., and Hager, A. **1982**. Biological-activities of sesquiterpene lactones from *Helianthus annuus*. Anti-microbial and cyto-toxic properties: influence on DNA, RNA, and protein-synthesis. *Z. Naturforsch.* **37c**, 1087-1091

(45) StatSoft Inc., Release 4.5, 1993.

(46) Stefinovic, M. and Snieckus, V. **1998**. Connecting directed ortho metalation and olefin metathesis strategies. Benzene-fused multiring-sized oxygen heterocycles. First syntheses of Radulanin A and Heliannane. *J. Org. Chem.* **63**, 2808-2809

(47) Takabatake, K., Nishi, I., Shindo, M., and Shishido, K. **2000**. Enantioselective total synthesis of heliannuols D and A. *J. Chem. Soc., Perkin Trans. 1*, 1807-1808

(48) Tuhina, K., Bhowmik, D. R., and Venkateswaran, R. V. **2002**. Formal syntheses of heliannuols A and D, allelochemicals from *Helianthus annuus*. *Chem. Commun.*, 634-635

(49) Varela, R. M. **1996**. Allelochemicals from the sunflower cultivars SH-222. Heliannuols and Heliespirones: two new families of sesquiterpenes. Ph. D. dissertation. University of Cadiz, Spain

(50) Vyvyan, J. R. and Looper, R. E. **2000**. Total synthesis of (±)-heliannuol D, an allelochemical from *Helianthus annuus*. *Tetrahedron Lett.* **41**, 1151-1154

(51) Wilson, R. E. and Rice, E. L. **1968**. Allelopathy as expressed by *Helianthus annuus* and its role in old-field succession. *Bull. Torrey Bot. Club* **95**, 432

(52) Wright, A. E., Pomponi, S. A., McConnel, O. J., Kohmoto, S., and McCarthy, P. J. **1987**. (+)-Curcuphenol and (+)-curcudiol, sesquiterpene phenols from shallow and deep-water collections of the marine sponge *Didiscus-flavus*. *J. Nat. Prod.* **50**, 976-978

(53) Zar, J. H. **1984**. *Statistical Analysis*, Prentice Hall, Inc., Englewood Cliffs, NJ

6 Chemistry of Host-Parasite Interactions

J. C. G. Galindo, F. A. Macías, M. D. García-Díaz, and J. Jorrín

CONTENT

Abstract 125
Introduction 126
Natural Germination Inductors 128
    Strigolactones 128
    Quinones 130
    Other Natural Nonhost Inductors 131
Synthetic Germination Inductors 134
    The GR Family of Compounds 134
    Sesquiterpene Lactones 137
Natural Haustorium Inductors 141
    Quinones 141
Methodology 142
References 143

ABSTRACT

Parasitic weeds represent an emerging branch of research in allelopathy. They constitute an economical threat for many important crops, and the tandem host-parasite is one of the systems where chemically mediated plant recognition is better proved. Germination in parasitic plants only takes place when the seed detects the presence in the soil of specific chemical signals from their hosts. Two families of parasitic weeds have received much attention due to economical reasons, witchweeds (*Striga* spp.) and broomrapes (*Orobanche* spp.). Up to date, only few compounds have been isolated and characterized as seed germination signals from natural hosts of these species. These compounds belong to two different chemical families, quinones (e.g., sorgoleone) and sesquiterpenes (e.g., sorgolactone). However, compounds isolated from nonhost plants and organisms, and synthetic derivatives as well, have also induced germination responses to different extents. A chemical model to explain the interaction between the inductor and the active site in the parasite has been proposed, but not all active compounds fit into this model, thus suggesting than more than one mechanism or site of action is involved.

INTRODUCTION

The study of host-parasite chemical interactions is a relatively new field of research in allelopathy that is receiving increasing attention for economical and scientific reasons. The existence of parasitic plants has been reported since ancient times. Dioscorides (s. I. a. C.) described plants belonging to the family Orobachaceae, the genera *Orobanche* being described by Linnaeus in 1793. Up to date, over 4000 species of parasitic plants grouped in 20 families have been described so far.[35] There are five families of special importance because of their adverse impact on different crops, namely Schrophulariaceae, Orobanchaceae, Cuscutaceae, Viscaceae and Loranthaceae. Among them, weeds belonging to the Schrophulariaceae and Orobanchaceae phyla are important economical threats on crops such as legumes, several Gramineae, tomato, sunflower, and tobacco (Table 6.1). However, parasitic weed control techniques have not been studied until recently, and proper control methods are not available yet.

Parasitic plants can be broadly divided into hemiparasites and holoparasites, according to the presence or absence of chlorophylls. The holoparasites depend on their hosts to get the nutrients and to complete their life cycle, as they are not able to fix carbon through photosynthesis. The hemiparasites take from their host just minerals and water, and their parasitism can be facultative.

Two well-differentiated phases can be established in the life cycle of most of the parasitic plants; the independent and the parasitic phases. The first one comprises seed dispersion, the latent phase, the seed-conditioning period, and germination. During this period, the plant does not need the presence of any host to survive. The parasitic phase includes the haustorium formation and penetration processes, the connection of the weed to the vascular system of the host, and the development and flowering of the parasite attached to its host. Among all of these different developmental stages, the germination and the formation and establishment of the haustorium are crucial for the survival of the plant.

Table 6.1

Some of the most important parasitic weeds according to the economical losses they cause.

Family	Genera	Species	Host Crop
Schrophulariaceae	*Striga*	*S. hermonthica*	sorghum, maize, millet
		S. asiatica	maize, sorghum
		S. gesnerioides	cowpea
Orobanchaceae	*Orobanche*	*O. cernua* (*O. cumana*)	sunflower, tomato, tobacco
		O. crenata	green pea, lentils, broadbean, chickpea, carrot, celery
		O. ramosa / *O. aegyptiaca*	onion, lettuce, sunflower, broadbean, greenpea, lentils, chickpea, tomato, tobacco, potato, carrot, celery, canola
		O. minor	lettuce, broadbean, tobacco, carrot, celery, red clover
	Agallinis	*A. purpurea*	
	Alectra	*A. vogelii*	cowpea
Convolvulaceae	*Cuscutaceae*	*C. campestris*	
Loranthaceae	*Amyema*	*A. sanguineum*	eucalyptus
	Dendropthoe	*D. curvata*	
	Tapinantus	*T. buchneri*	
Viscaceae	*Arceutobium*	*A. americanum*	pines
		A. abietinum	red fir
		A. pusillum	white spruce
		A. verticilliflorum	pines
	Phoradendron	*P. bolleanum*	western juniper
		P. juniperinum	western juniper
Cuscutaceae	*Cuscuta*	*Cuscuta campestris*	

The relationship between a parasite and its host is extremely specific: each species of parasite recognizes only its host(s). Host specificity depends upon such a diverse range of factors as the ability of the parasite to recognize and attack the host plant, to break down the defense responses of the host, and the existence of enough resources in the host to assure the growth development of the parasite. The interaction of the parasite and the host is chemically mediated and represents a clear example of allelopathy: the parasite recognizes certain chemicals exuded by the roots of their potential hosts. These chemical clues serve the parasite to "know" that there is a potential host in the vicinity to which to get attached. Depending on the parasite, the haustorium development can also be chemically mediated, as it will be noted later. However, the general process is not so easy. The germination conditions to break the dormancy of the seeds require a narrow range of temperatures and humidity before the inductor of the germination becomes effective. During this conditioning phase several changes occur inside the seed. The respiration changes, as does the protein synthesis, but the most important thing is the synthesis of high levels of gibberelins.[19] If there is not any germination inductor reaching the seed after this period, the seed can go into a second latent period. However, this second period of dormancy might affect adversely the capacity of the seed to germinate.

NATURAL GERMINATION INDUCTORS

STRIGOLACTONES

The denomination **strigolactones** comprises a relatively low number of structurally related compounds isolated from host and nonhost plants that induces the germination of *Striga* spp. and *Orobanche* spp. All of them share as common features a tricyclic degraded sesquiterpene skeleton (rings A, B, and C, Fig. 6.1) and a lactone-enol-γ-lactone moiety (rings C and D, Fig. 6.1).

Strigol (**1**) is the first natural germination inductor of a parasitic plant isolated from natural sources,[6] even though its structure was elucidated lately.[3] Strigol is obtained from the exudates of the nonhost cotton (*Gossypium hirsutum*), and it is able to induce the germination of *Striga asiatica* at extremely low doses. However, the first natural germination inductor of this family obtained from a host-plant is sorgolactone (**2**). This compound was isolated from the aqueous extracts of sorghum (*Shorgum bicolor*) roots,[15] and it is an even more potent inductor than strigol when assayed on *Striga asiatica* and *Striga hermonthica* seeds. Afterwards, Müller et al. isolated from cowpea (*Vigna unguiculata*) roots another germination stimulant agent belonging to this family, named alectrol (**3**).[34] However, the chemical structure of this compound is still controversial.[33] More recently, alectrol and orobanchol (**4**) were isolated as the first natural germination stimulants of broomrape (*Orobanche minor*), obtained from the root exudates of its natural host red clover (*Trifolium pratense*).[55] Going on with these studies, LC/MS/MS analysis of red clover root exudates discarded the presence of strigol and sorgolactone as other possible germination stimulants.[40]

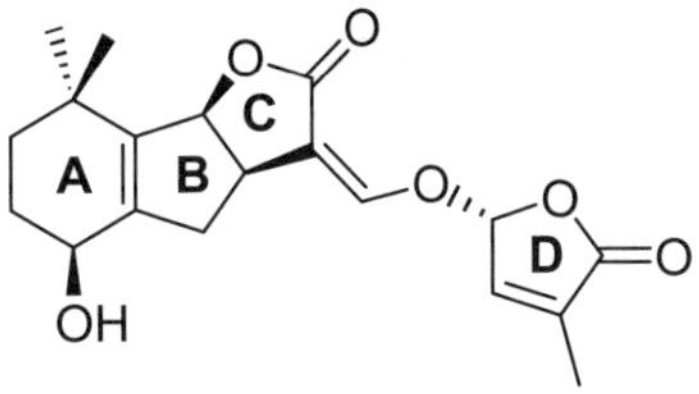

Strigol (1)

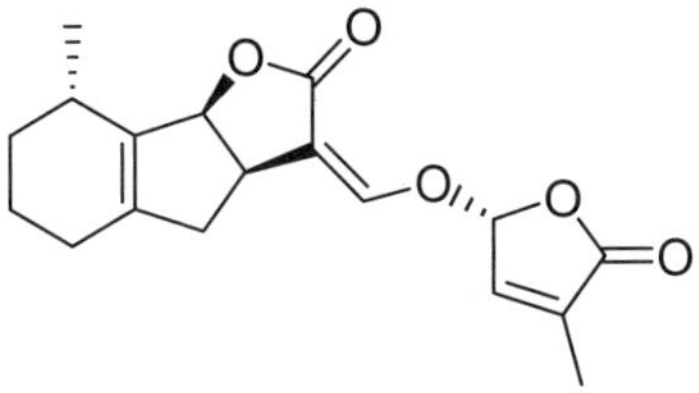

Sorgolactone (2)

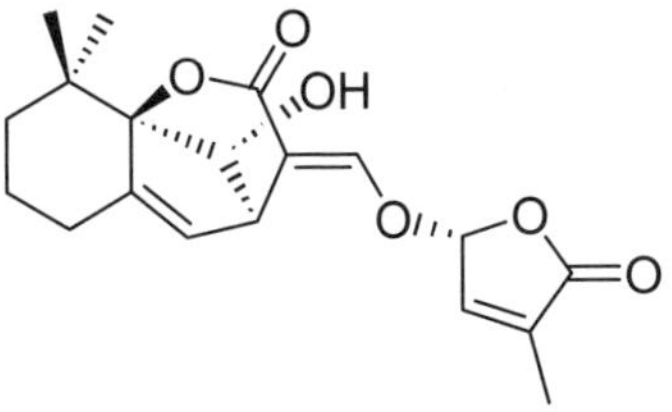

Alectrol, temptative (2)

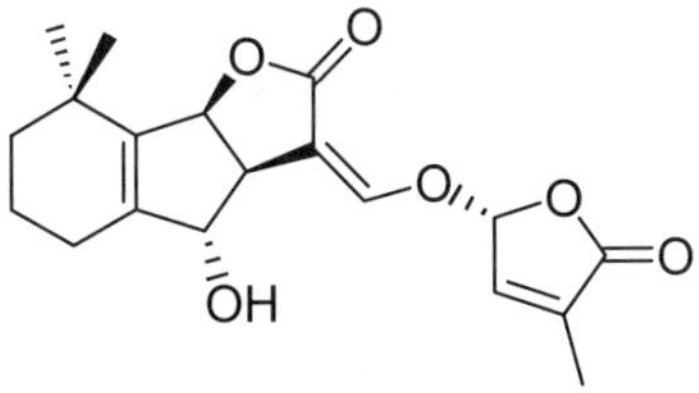

Orobanchol (4)

Figure 6.1
Structure of natural strigolactones: note that alectrol structure is tentative and not yet fully confirmed.

Strigol has also been reported recently in root cultures of the broad-leaved herbaceous weed *Menispermum dauricum*.[54] In this way, the use of LC/MS/MS techniques has revealed a powerful analysis tool for minor compounds. By using this methodology, a new isomer of strigol, the structure yet unknown, has been detected in sorghum root exudates, and the presence of alectrol or orobanchol in cotton root exudates has been discarded.[2] These results are summarized on Table 6.2.

All of these compounds are able to induce the germination of different *Striga* and *Orobanche* species, indistinctly. Structure-Activity Relationship (SAR) studies accomplished by Zwanenburg with natural and synthetic derivatives of strigolactones led to propose the common lactone-enol-γ-lactone moiety as the possible bioactiphore (Fig. 6.2).[31] In this model, a nucleophile present in the hypothetic receptors' cavity of the parasite (e.g., a sulfhydril or an amino group of a protein) should react in a Michael fashion with the enol-γ-lactone moiety to give **5**. Afterwards, a retro Michael reaction should liberate the D ring as the free 4-

carboxy-2-methyl-2-butenoic acid (**6**), thus leading to the nucleophile irreversibly attached to the germination signal (**7**).

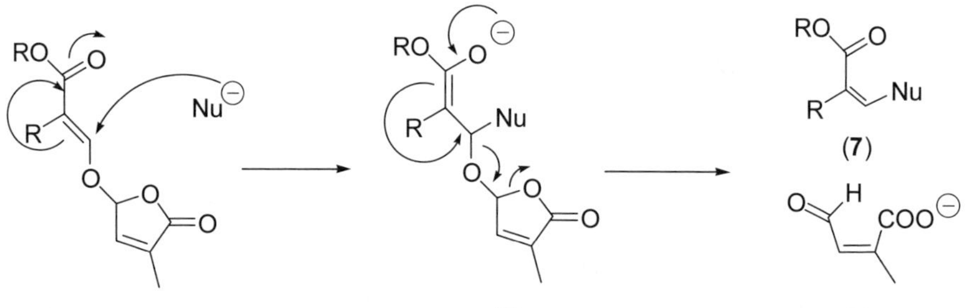

Figure 6.2

Parasite chemical recognition mechanism for strigolactones.

Table 6.2

Natural Scrophulariace seed germination inductors and their natural sources.

Compound	Source
Strigol	cotton,[2,3,6] sorghum,[11] *Menispermum dauricum*[54]
Sorgolactone	sorghum[15]
Alectrol	cowpea,[34] red clover[55]
Orobanchol	red clover[55]
New strigol isomer (unknown structure)	sorghum[2]

Ethylene is a potent inductor of *Striga* spp. germination.[26] It has been hypothesized as a possible mechanism for parasitic seed germination that attachment of the inductor to the receptor's site triggers a cascade of biosynthetic reactions leading to the synthesis of the enzymes necessary for ethylene production.[12]

QUINONES

The first germination stimulant isolated from a natural host was the hydroquinone SXSg (**8**), precursor of sorgoleone (**9**) (Fig. 6.3). This hydroquinone is exuded as oily droplets from the root hairs of *Sorghum bicolor* and readily oxidizes to the most stable quinone form (**9**).[5] The quinone sorgoleone is a highly phytotoxic product that has been proposed as one of the defense compounds responsible for the allelopathic activity of sorghum acting as a PS II inhibitor.[38] Hence, its storage under the form of a nonphytotoxic hydroquinone-like form should prevent the plant from autointoxication. The oxidation process from **8** to **9** is spontaneous and very fast. Consequently, the hydroquinone form is localized in a very narrow distance from the root, thus resulting in a "germination" zone close to the root that assures that the parasite will germinate within a reasonable distance from its host. This has been proposed as a mechanism to regulate chemically the germination security distance. However, this model is still controversial and remains under discussion since other germination inductors, like the sorgolactone, have been isolated from sorghum roots and do not fit this model: they are not able to undergo spontaneous oxidation, even though they easily degrade. The presence of the methoxy derivative of SXSg (**10**) has been also reported as an antioxidant enhancer of the SXSg (**8**) action.[9] Other quinones also play an important role in haustorium induction, as will be discussed later.

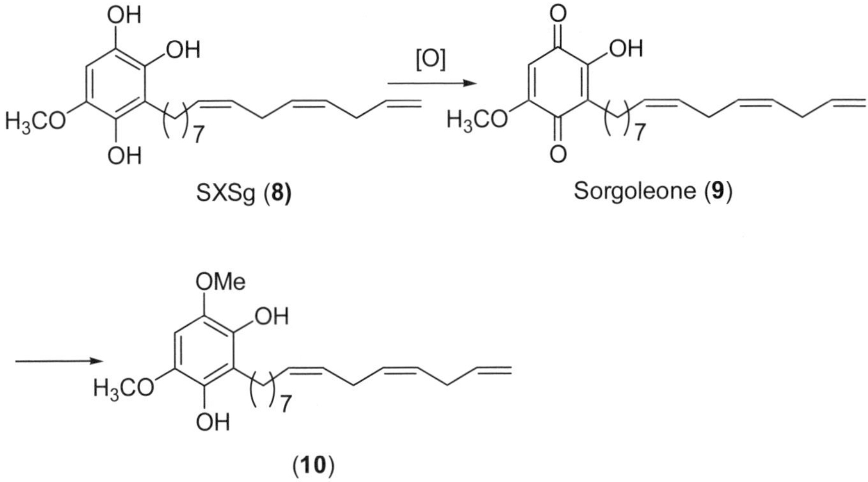

Figure 6.3

Quinones involved in germination and haustorium induction.

OTHER NATURAL NONHOST INDUCTORS

The continuous search for germination inductors to control parasitic weeds has led to testing a great variety of natural products. Some positive results have been obtained for *Striga* with compounds such as the coumarin scopoletin (**11**), the polyalcohol inositol (**12**), the aminoacid methionine (**13**), ethylene (**14**) and several citokynins (Fig. 6.4). However, they were not able to induce any positive response in germination of *Orobanche* seeds. This could be indicative that the structural requirements in *Orobanche* inductors for germination are different and more restrictive than those for *Striga* spp. This behavior will also be found for sesquiterpene lactones, as will be discussed later. Among these compounds, only the coumarin scopoletin presents an α,β-unsaturated carbonyl system, according to Zwanenburg's model. However, the high stability conferred by the conjugation of the double bond with the aromatic system makes alkylation very unlikely to occur.

Jasmonates constitute another family of compounds that show good results in the germination bioassays. Jasmonic acid (JA) (**15**) and the methyl jasmonate (MJA) (**16**) are recognized as endogenous phytohormones[18,43] that exhibit various biological activities in higher plants related to the regulation of senescence-promotion activities and growth-inhibition regulation. JA (**15**), MJA (**16**), the structurally related 6-*epi*-cucurbic acid (**17**) and methyl 6-*epi*-9,10-dihydro-cucurbate (**18**) (Fig. 6.4) induce high levels of germination in seeds of witchweed (*Striga hermonthica*) and clover broomrape (*Orobanche minor*),[56] *S. hermonthica* being the less sensitive species and MJA (**16**) the more active compound. It is also observed that the methyl esters present higher activities than the free acids, but the authors do not venture any explanation for this behavior. However, this could be related to an easier crossing of the membranes in the case of the more lipophilic ester derivatives.

The structure of jasmonates does not fit the requirements of the model proposed by Zwanenburg, as they do not present any moiety susceptible of giving Michael addition reactions. Whether it is possible that the data could be explained through the phytohormone properties of these compounds, such a hypothesis has not been yet explored although it constitutes another attractive line of research.

The enhancing effect of the brassinosteroids added during the pre-conditioning period on the germination rate of strigol-treated seeds could be related to this idea. The period of time previous to germination is a crucial step and it has been reported that all seeds present certain requirements of temperature and humidity during the period previous to germination.[23] During this time many changes in the metabolism of the seed can be observed, such as alteration in respiration, protein synthesis, and gibberellin synthesis.[19,20] In particular, exogenous gibberellin GA_3 is able to enhance seed germination in the presence of a proper germination inductor.[17,36] In the same way, the treatment with the brassinosteroids brassinolide (**19**) and castasterone (**20**) at the beginning of the conditioning period leads to a reduction in the preconditioning time needed. Moreover, when the brassinosteroid was added after the conditioning period, the seeds treated with a proper inductor (e.g., strigol or any other) germinate faster than without it.[46] They are also able to overcome to some extent the inhibitory effect that the light has on parasite seed germination.

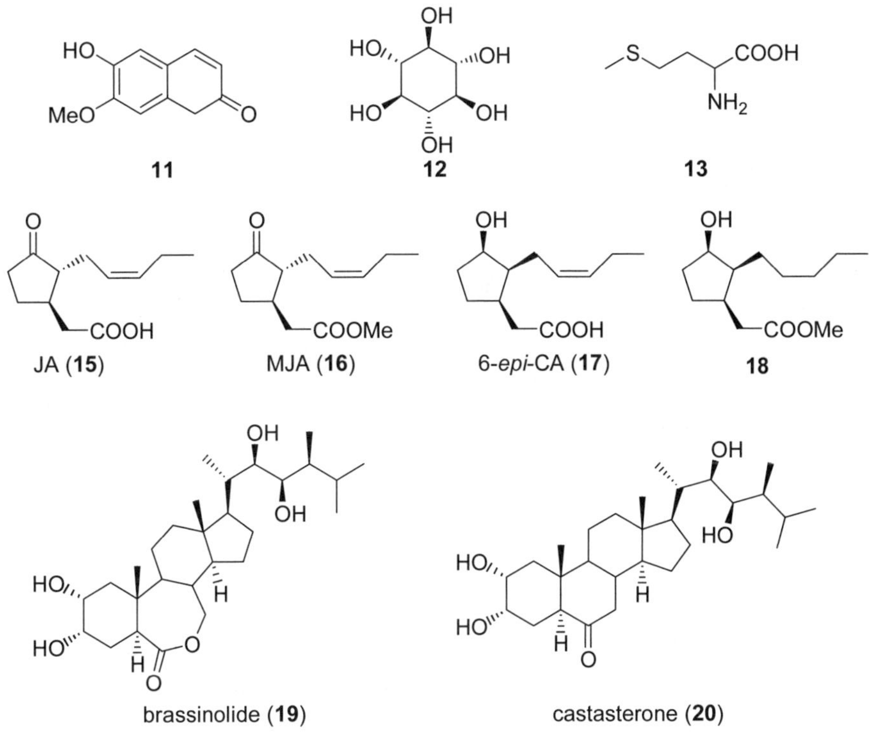

Figure 6.4

Nonhost inductors of the germination in parasitic plants.

Certain fungal metabolites have also been found to induce the germination of plant parasites effectively. Cotylenins **21** and **22** and fusicoccins **23** and **24** are effective inductors of *S. hermonthica* and *O. minor* seeds germination (Fig. 6.5). However, the activity of these compounds is lower than that of strigolactones (ED50 around 10 µM), the free alcohols **22** and **24** being the more active compounds.[57]

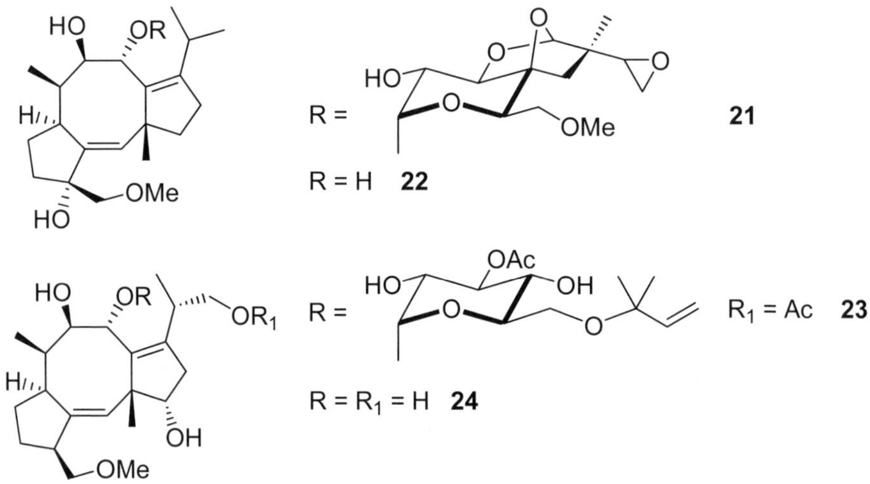

Figure 6.5

The fungal inductors cotylenins (**21, 22**) and fusiccocins (**23, 24**).

SYNTHETIC GERMINATION INDUCTORS

THE GR FAMILY OF COMPOUNDS

Based on the structure of the strigolactones, and looking for enough amounts of compounds with similar properties to perform parasitic weed control studies, a number of chemical analogues of sorgolactone have been synthesized (compounds **25-31**, Fig 6.6).[21,22] The most active compound of this family is commonly known as GR-24 (**28**), and these compounds are generally addressed as the GR family. Another promising lead compound is the phthaloylglycine derivative Nijmegen 1 (**29**); it has lower activity but is easier and cheaper to synthesize.[37] They have been developed in the course of Structure-Activity Relationship (SAR) studies, and several important conclusions could be established based on their activity as germination inductors:

- Compounds belonging to this structural type are equally recognized by seeds of *Striga* sp. and *Orobanche* sp. Comparative studies usually show higher levels of induction in witchweed (*Striga hermonthica*) than in broomrape (*Orobanche crenata*) seeds. Moreover, changes in the chemical structure induce bigger differences in the bioactivity in witchweed than in broomrape.[29,47,49,51]

- The ring D (the γ-butyrolactone moiety) is crucial for the activity. The change of this part of the molecule by any other leaving group greatly affects the germination activity.[31] However, it seems the better the leaving group, the lower the diminution in the activity is. Consequently, the introduction of a tosyl group as new D ring (compound **30**) results in a lower germination activity, while a 5,5-dimethyl-2-cyclohexenone D ring (compound **31**) yields a completely inactive compound (Fig. 6.6). Other modifications introduced in the D ring of the phthaloylglycine-derived Nijmegen 1 (**29**) greatly affect the activity, *O. crenata* being more sensitive to these changes than *S. hermonthica*.[47]

- The system enol-γ-lactone moiety characteristic of rings C and D is another crucial requirement for germination, and this fact led Zwanenburg to formulate a Michael addition/retro-Michael hypothesis for host recognition.[31] The introduction of changes such as the substitution of the oxygen atom by a methylene ($-CH_2-$) group led to inactive compounds.[49] However, if the enol ether moiety is preserved, changes in the stereochemistry of the double bond (from Z to E, or from exocyclic to endocyclic) or addition of a small substituent (e.g., a methyl group) do not adversely affect the activity.[51] This behavior can be explained since they do not complicate the course of the Michael addition reaction.

- The system formed by rings ABC seems not to be directly related to the activity. In fact, they are totally inactive by themselves. However, this modulates the activity by determining the spatial conformation of the molecule and thus the possibility of access to the reaction center. Compounds lacking ring C (**33**), ring B (**34**) or both (**35**) (Fig. 6.6) present a higher degree of rotation freedom and, consequently, exhibit lower activities.[29] Several analogues of the ring A (compounds **36-40**, Fig. 6.6) were also assayed on several *Orobanche* and *Striga* species and found inactive.[30]

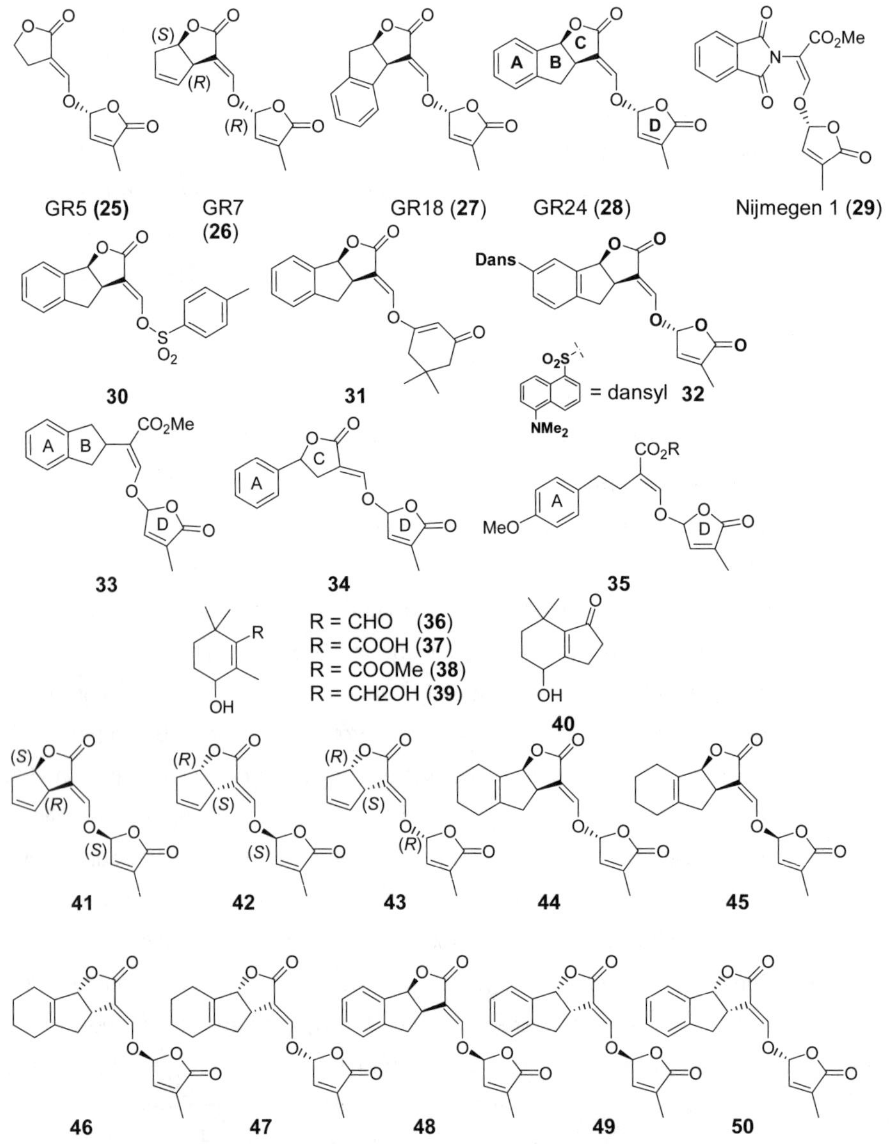

Figure 6.6

The GR family: representative chemical analogues synthesized for SAR studies in parasite seed germination.

- The influence in the stereochemistry of the chiral centers at rings C and D has also been explored. Asymmetric synthesis of all possible diastereoisomers of GR-7 (**26, 41-43**),[31] demethylsorgolactone (**44-47**),[48] and GR-24 (**28, 48-50**)[50] confirmed that among all possibilities, the isomers with the "natural" absolute configuration (**26, 28, 44**) were the most active, thus showing the importance of the stereochemistry in this part of the molecule. An adequate stereochemistry allows a correct binding at the receptor's site and also the subsequent reaction pathway. The isomers with the opposite configuration to the natural product in all chiral centers are the less active compounds (**42, 45, 49**). Finally, changes in the stereochemistry of only one lactone ring (ring D: **41, 48, 50**; ring C: **43, 47, 50**) lead to small changes in the activity (Fig. 6.6).

Current studies continue on developing new GR-derivatives, looking for better activities, higher chemical stability, and new starting materials.[7,52,53]

SESQUITERPENE LACTONES

Sesquiterpene lactones (SL) constitute a numerous group of compounds with several biological activities reported, usually related to the presence of an α,β-unsaturated carbonyl system in the lactone ring.[42] This is a very common chemical feature in these compounds, and their alkylating properties through Michael addition reactions are considered as responsible for such activity.[16]

The nucleophilic properties of groups such as –NHR and –SH have been reported in natural Michael acceptor systems present in many natural products. Typical examples are the addition of the sulfhydryl groups of glutathione to the exocyclic methylene of sesquiterpene lactones (Fig. 6.7).[14,41] Such interactions have been considered as responsible for the biological activity observed in many natural products, and this is again the case in Zwanenburg's hypothesis for strigolactones.

First studies with SL as witchweed (*S. asiatica*) germination agents used compounds with *trans,trans*-germacronolide (**51-59**), eudesmanolide (**60, 61**) and pseudoguaianolide (**62-67**) backbones (Fig. 6.8).[10,11] Most of these compounds induce germination responses similar to those of strigol at micro- and nanomolar levels; germacranolides and eudesmanolides resulted in by far the most potent compounds. The germacranolide dihydroparthenolide (**51**) and the eudesmanolides reynosin (**42**) and santamarin (**44**) were especially active, with nanomolar responses close to strigol. On the other hand, SL with a pseudoguaianolide backbone presented lower activity, being active usually at micromolar concentrations.

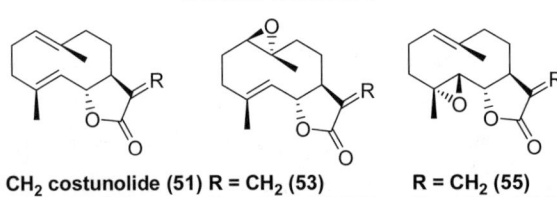

Figure 6.7

Michael addition of glutathione to the sesquiterpene lactones dehydrocostuslactone and helenalin.

Germacranolides

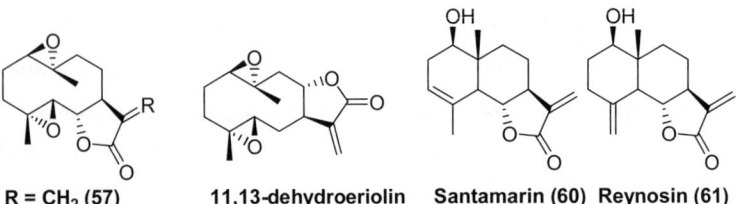

R = CH₂ costunolide (51)   R = CH₂ (53)   R = CH₂ (55)
R = αCH₃, βH (52)   R = αCH₃, βH (54)   R = αCH₃, βH (56)

Eudesmanolides

R = CH₂ (57)   11,13-dehydroeriolin   Santamarin (60) Reynosin (61)
R = αCH₃, βH (58)   (59)

Pseudoguaianolides

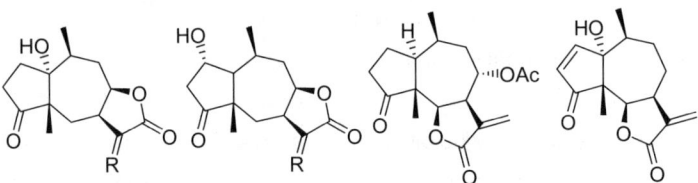

R = CH₂ Peruvin (62)   R = CH₂ Burrodin (64)   Parthenin (67)
R = αH, βCH₃ (63)   R = αH, βCH₃ (65)   Confertiflorin (66)

Figure 6.8

Some SL tested on *Striga asiatica* seed germination bioassays.

Interestingly, in many of these active compounds (**52, 54, 56, 58, 63**, and **65**) the double bond at the lactone ring is reduced, and thus it is not able to react in a Michael addition fashion. This is of special importance since such a reaction is proposed as the molecular recognition pathway in strigolactones and was thought to proceed also in SL.

There is also a clear differentiation of the activity depending on the type of backbone. The spatial disposition of the carbon skeleton in the pseudoguaianolides is different from those of the other two types. This fact has been correlated with the lower activities displayed by pseudoguaianolides, even though some of them present one or two α,β-unsaturated carbonyl systems (e.g. **67** in Fig. 6.8). The eudesmanolides reynosin (**60**) and santamarin (**61**) present a backbone resembling the spatial arrangement in germacranolides and also has an α,β-unsaturated carbonyl system. The authors hypothesized that compounds with what they call a "double crown" like spatial disposition (germacranolides and eudesmanolides) are similar to strigolactones and can fit into the receptor's cavity. However, the activity observed for those compounds without the unsaturated doubled bond in the lactone ring remains to be explained.

More recently, we have focused our interest in broomrapes as they constitute an important threat in sunflower, an important economic crop in many countries. *Orobanche cumana* is the specific parasite of sunflower and, to our knowledge, the chemical signals for germination response in this particular system have not been identified so far. On the other hand, sunflower constitutes a rich source of sesquiterpenoids, specially SL.[28] Considering the above mentioned results, we hypothesized that these compounds could act as the chemical signals that are recognized by broomrape seeds.[25] To test this hypothesis we have synthesized and assayed several SL with guaianolide (compounds **68-72**), eudesmanolide (compounds **60, 73-74**), *trans,trans*-germacranolide (compounds **53, 55, 56**), and melampolide (**75**) backbones on different broomrape species with interesting results (Fig. 6.9).[8,13]

- SL, when active, are specific of sunflower broomrape (*O. cumana*). Other broomrape species like tobacco (*O. ramosa*) or broadbean (*O. crenata*) broomrapes did not germinate in the presence of these compounds. Consequently, this behavior has been related with the high SL content in sunflower.
- Several SL present higher levels of activity than GR-24, tested in the same bioassay as internal standard. The most active compounds are

those with guaianolide and *trans,trans*-germacranolide backbones that also correspond to the most common structural types of SL in sunflower. Eudesmanolides resulted in being the less active compounds.

- Molecular modeling of compounds tested shows molecular volumes similar to those of strigolactones[25] and fits into the same hypothetic receptor's cavity. No correlation between the activity and changes in molecular properties such as dipole moments or heat of formation could be established.

- Differences in activity are observed when a double bond is present or not in the lactone ring.[13,25] However, several compounds without an α,β-unsaturated-γ-lactone system present higher levels of activity than GR-24. This is true for dihydroparthenolide (**56**) and the guaianolides **68** and **69**. As these compounds cannot react in a Michael addition fashion, other recognition mechanisms different from that proposed by Zwanenburg[31] could be operating in this case.

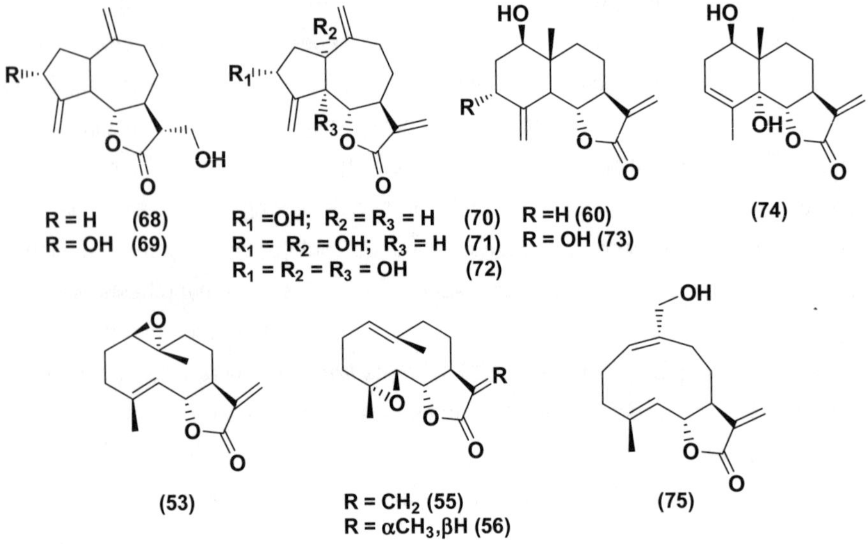

R = H   (68)
R = OH (69)

R_1 =OH; R_2 = R_3 = H   (70)
R_1 = R_2 = OH; R_3 = H   (71)
R_1 = R_2 = R_3 = OH   (72)

R = H (60)
R = OH (73)

(74)

(53)

R = CH_2 (55)
R = αCH_3,βH (56)

(75)

Figure 6.9
SL tested in broomrape germination studies.

Chemistry of Host-Parasite Interactions

NATURAL HAUSTORIUM INDUCTORS

QUINONES

Haustorium is the physiological organ of the parasite that effectively attaches to the host, allowing it to get into the tissue of the host, reach the phloem and uptake the nutrients. The physiological aspects of haustorium development have been extensively studied in Scrophulariaceae[25,39] and seem to be the result of the redirection of cellular expansion from longitudinal to radial, at least in *S. asiatica*.[24] However, the cascade of events that results in haustorium formation is still not well understood in most of parasitic plants. It is known that xenognosin A (**75**)[27] and B (**76**),[45] the quinones 2,6-dimethoxy-1,4-benzoquinone (**79**, DMBQ)[4] and 5,7-dihydroxynapthoquinone (**80**),[44] and, most recently, the flavonoid peonidin (**81**)[1] induce the formation of haustorium-like structures in the absence of any host (Fig. 6.10).

haustorial inducing compounds

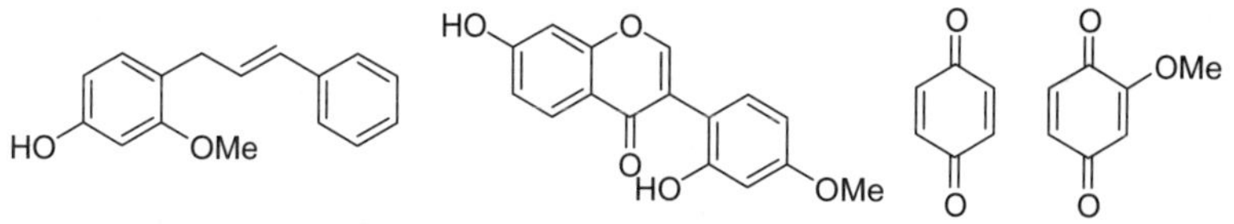

xenognosin A (**75**)   xenognosin B (**76**)   BQ (**77**)   MBQ (**78**)

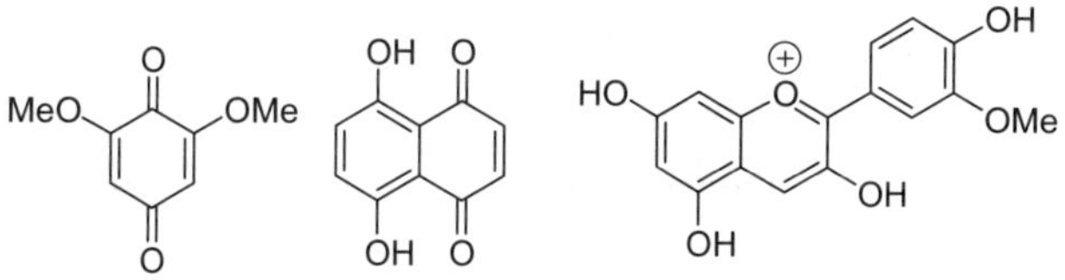

DMBQ (**79**)   **80**   peonidin (**81**)

haustorial inhibiting compounds

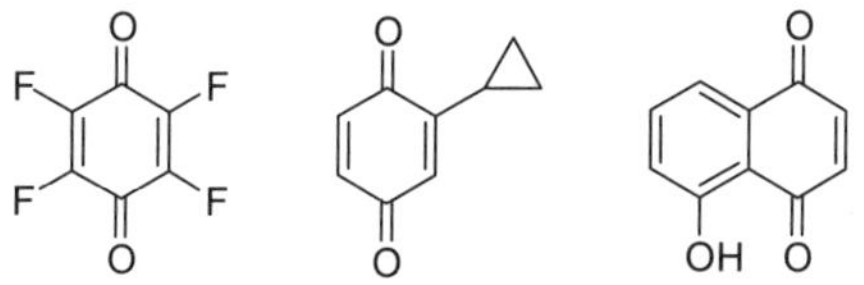

TFBQ (**82**)   CPBQ (**83**)   juglone (**84**)

Figure 6.10

Natural and synthetic haustorium inductors and inhibitors.

Recently, a chemical model to explain the formation of haustorium in the presence of benzoquinones has been reported, based on the Redox properties of these compounds.[24] This model explains the release of quinones from the host through oxidation of pectins, phenyl- propanoid esters decorating the pectins, and the syringic acid that are present in the surface of the root cell walls. The enzymes catalyzing this reaction are cell wall peroxidases from the host and need the presence of hydrogen peroxide (H_2O_2) as co-factor. This compound is widely present in nature (specially in plant tissues) and accumulates in the parasite root tip. The quinones generated from oxidation have been characterized as benzo-1,4-quinone (BQ, **77**), methoxybenzo-1,4-quinone (MBQ, **78**) and DMBQ (**79**). However, not all quinones are able to induce haustorium formation. In fact, and after assaying a wide variety of benzo- and naphtoquinones only those compounds, which can be easily reduced and oxidized are active. Those compounds having half-reduction redox potential out of the range –250 to 0 mV ("active window") are inactive.[44]

According to these findings, the model proposes as the active compound the semiquinone intermediate that reversibly binds the active site. Those electropositive compounds lying to the right of the "active window" (e.g., tetrafluorobenzo-1,4-quinone, TFBQ, **82**), will easily accept one electron. However, the next re-oxidizing step will be thermodynamically restricted, and the quinone will irreversibly remain bonded to the active site, thus inhibiting germination. In the opposite case, compounds in the left side of the window (mostly naftoquinones like juglone **84**) will be too difficult to half-reduce under the environmental conditions; semiquinones will not be produced, and the haustorial phase will not be induced. This is further proved when the cyclopropyl-*p*-benzoquinone (**83**) is assayed along with DMBQ. In this case, the semiquinone intermediate suffers a ring-opening rearrangement and the resulting carbon radical irreversibly binds the active site, thus inhibiting the haustorium formation induced by the DMBQ.

METHODOLOGY

Orobanche sp. germination bioassays

Seeds contained in the tip of a spatula were homogeneously dispersed in a Petri dish (55mm ⌀) on Whatman GF/A paper. For preconditioning, 1 mL of a solution of 0.3 mM of 2-[*N*-Morpholino]ethanesulfonic acid was added to the filter, and the Petri dishes were sealed to prevent drying and incubated in darkness at 20°C for 11 days.

Stock solutions were prepared with acetone and diluted with MES 0.3 mM to obtain a 10 µM (0.1 % acetone) solution; 0.1 and 1 µM solutions were prepared by diluting with MES 0.3 mM (0.1 % acetone aq. solns).

After the conditioning period, 500 µL of an aqueous solution of GR-24 (used as internal standard) or the test compounds were added to every Petri dish, sealed with parafilm and incubated for another 4 days in darkness at 20°C. Germination was observed under a microscope (30x) and the germinated seeds expressed as a percentage of the total seeds. Germination was considered when the radicle was at least 0.2 mm long.

ACKNOWLEDGMENTS

The authors acknowledge partial support from the Ministerio de Educación y Ciencia, CICYT, Project AGL2001-2420.

REFERENCES

(1) Albrecht, H., Yoder, J. I., and Phillips, D. A. **1999**. Flavonoids promote haustoria formation in the root parasite *Triphysaria*. *Plant Physiol.* **119**, 585-591

(2) Awad, A. A., Sato, D., Takeuchi, Y., Yokota, T. Sugimoto, Y., and Yoneyama, K. **2002**. Seed germination stimulants produced by cotton and sorghum. *Proceedings of the Third World Congress on Allelopathy. Challenge for the New Millennium.* Fujii, Y., Hiradate, S., Haraya, H., (Eds.) Tsukuba, Japan, 216

(3) Brooks, D. W., Bevinakatti, H. S., and Powell, D. R. **1985**. The absolute structure of (+) strigol. *J. Org. Chem.* **50**, 3779-3781

(4) Chang, M. and Lynn, D. G. **1986**. The haustorium and the chemistry of host recognition in parasitic angiosperms. *J. Chem. Ecol.* **12**, 561-579

(5) Chang, M., Netzly, D. H., Butler, L. G., and Lynn, D. G. **1986**. Chemical regulation of distance: characterization of the first natural host germination stimulant for *Striga asiatica*. *J. Am. Chem. Soc.* **108**, 7858-7860

(6) Cook, C. E., Whichard, L. P., Thurner, B., Wall, M. E., and Egley, G. H. **1966**. Germination of witchweed (*Striga lutea* Lour.): isolation and properties

of a potent stimulant. *Science* **154**, 1189-1190

(7) De Lima, M. E. F., Gabriel, A. J. A., and Castro, R. N. **2000**. Synthesis of new strigol analogue from natural safrol. *J. Braz. Chem. Soc.* **11**, 371-374

(8) de Luque, A. P., Galindo, J. C. G., Macías, F. A., and Jorrín, J. **2000**. Sunflower sesquiterpene lactone models induce *Oreobanche cumana* seed germination. *Phytochemistry* **53**, 45-50

(9) Fate, G., Chang, M., and Lynn, D. G. **1990**. Control of germination in *Striga asiatica*: chemistry of spatial definition. *Plant Physiol.* **93**, 207-207

(10) Fischer, N. H., Weidenhamer, J. D., and Bradow, J. M. **1989**. Dihydroparthenolide and other sesquiterpene lactones stimulate witchweed germination. *Phytochemistry* **28**, 2315-2317

(11) Fischer, N. H., Weidenhamer, J. D., Riopel, J. L., Quijano, L., and Menelaou, M. A. **1990**. Stimulation of witchweed germination by sesquiterpene lactones: a structure-activity study. *Phytochemistry* **29**, 2479-2483

(12) Gabbar, A., Babiker, T., Ejeta, G., Butler, L. G., and Woodson, W. **1993**. Ethylene biosynthesis and strigol-induced germination of *Striga asiatica*. *Physiol. Plant.* **88**, 359-365

(13) Galindo, J. C. G., de Luque, A. P., Jorrín, J., and Macías, F. A. **2001**. SAR studies of sesquiterpene lactones as *Orobanche cumana* seed germination stimulants. *J. Agric. Food. Chem.* **50**, 1911-1917

(14) Galindo, J. C. G., Hernández, A., Dayan, F. E., Téllez, M. R., Macías, F. A., Paul, R. N., and Duke, S. O. **1999**. Dehydrozaluzanin C, a natural sesquiterpenolide, causes rapid plasma membrane leakage. *Phytochemistry* **52**, 805-813

(15) Hauck, C., Müller, S., and Schildknecht, H. **1992**. A germination stimulant for parasitic flowering plats from *Sorghum bicolor*, a genuine host plant. *J. Plant Physiol.* **139**, 474-478

(16) Hejchman, E., Haugwithz, R. D. and Cushman, M. **1995**. Synthesis and cytotoxicity of water-soluble ambrosin prodrug candidates. *J. Med. Chem.* **38**, 3407-3410

(17) Hsiao, A. I., Warshem, A. D., and Moreland, D. E. **1988**. Effect of chemicals often regarded as germination stimulants on seed conditioning and germination of witchweed (*Striga asiatica*). *Ann. Bot.* **62**, 17-24

(18) Ichihara, A., and Toshima H. **1999**. Coronatine: chemistry and biological activities. In: Cutler, H.G. and Cutler, S. J. (Eds.). *Biologically Active Natural Products: Agrochemicals*. CRC Press, Boca Raton, FL, 93-105

(19) Joel, D. M., Back, A., Kleifeld, Y., Gepstein, S. **1991**. Seed conditioning and its role in *Orobanche* seed germination: inhibition by paclobutrazol. In: Wegmann, K. and Musselman, L. J. (Eds.). *Progress in Orobanche Research*. Proceedings of the International Workshop on *Orobanche* Research. *Obermarchtal,* FRG, 147-156

(20) Joel, D. M., Steffens, J. C., and Matthews, D. E. **1995**. Germination of weedy root parasites. In: Kiegel, J. and Galili, G. (Eds.). *Seed Development and Germination*. Marcel Dekker, Inc., New York, 567-598

(21) Johnson, A. W., Gowda, G., Hassanali, A., Knox, J., Monaco, S., Razavi, Z., Roseberry, G. **1981**. The preparation of synthetic analogs of strigol. *J. Chem. Soc., Perkin Trans. I,* 1734-1743

(22) Johnson, A. W., Roseberry, G., and Parker, C. **1976**. A novel approach to *Striga* and *Orobanche* control using synthetic germination stimulants. *Weed Res.* **16**, 223-227

(23) Kebreab, E. and Murdoch, A. J. **1999**. A model of the effects of a wide range of constant and alternating temperatures on seed germination of four *Orobanche* species. *Ann. Bot.* **84**, 549-557

(24) Keyes, W. J., O'Malley, R. C., Kim, D., and Lynn, D. G. **2000**. Signaling organogenesis in parasititc angiosperms: xenognosin generation, perception, and response. *J. Plant Growth Regul.* **19**, 217-231

(25) Kuijt, J. **1977**. Haustoria of phanerogamic parasites. *Annu. Rev. Phytopath.* **17**, 91-118

(26) Logan, D. C. and Stewart, G. R. **1991**. Role of ethylene in the germination of the hemiparasite *Striga hermonthica. Plant Physiol.* **97**, 1435-1438

(27) Lynn, D. G., Steffens, J. C., Kamat, V. S., Graden, D. W., Shabanowithz, J., and Riopel, J. L. **1981**. Isolation and characterization of the first host recognitions substance from parasitic angiosperms. *J. Am. Chem. Soc.* **103**, 1868-1870

(28) Macías, F. A., Molinillo, J. M. G., Varela, R. M., Torres, A., and Galindo, J. C. G. **1999**. Bioactive Compounds from the genus Helianthus. In: Macías, F. A., Galindo, J. C. G., Molinillo, J. M. G., and Cutler, H. G. (Eds.). *Recent Advances in Allelopathy. A Science for the Future*. Vol. 1. Servicio de Publicaciones de la UCA, Spain, 121-148

(29) Mangnus, E. M., van Vliet, L. A., Vandenput, D. A. L., and Zwanenburg, B. **1992** Structural modifications of strigol analogues. Influence of the B and C rings on the bioactivity of the germination stimulant GR24. *J. Agric. Food Chem.* **40**, 1222-1229

(30) Mangnus, E. M. and Zwanenburg, B. **1992**. Tentative molecular mechanism for germination stimulation of Striga and Orobanche seeds by strigol and its synthetic analogues. *J. Agric. Food Chem.* **40**, 1066-1070.

(31) Mangnus, E. M. and Zwanenburg, B. **1992**. Synthesis and biological evaluation of A-ring analogs of the natural germination stimulant strigol. *Recl. Trav. Chim. Pays-Bas.* **111**, 155-159

(32) Mangnus, E. M. and Zwanenburg, B. **1992**. Synthesis, structural characterization, and biological evaluation of all four enantiomers of strigol analogue GR7. *J. Agric. Food Chem.* **40**, 697-700

(33) Matsui, J., Bando, M., Kido, M., Takeuchi, Y., and Mori, K. **1999**. Plant Bioregulators, 3. Synthetic disproof of the structure proposed for Alectrol, the germination stimulant from Vigna unguiculata. *Eur. J. Org. Chem.* 9, 2195-2199

(34) Müller, S., Hauck, C., and Schildknecht, H. **1992**. Germination stimulants produced by Vigna unguiculata Walp cv. Saunders. *J. Plant Growth Reg.* **11**, 77-84

(35) Musselman, L. J. **1987**. Striga. In: Musselman, L. J., (Ed.). *Parasitic Weeds in Agriculture*. Vol. 1. CRC Press, Boca Raton, FL

(36) Nash, S. M. and Wilhelm, S. **1960**. Stimulation of broomrape seed germination. *Phytopathology* **50**, 772-774

(37) Nefkens, G. H. L., Thuring, J. F., Beenakkers, M. F. M., and Zwanenburg, B. **1997**. Synthesis of a phtaloylglycine-derived strigol analogue and its germination stimulatory activity toward seeds of the parasitic weeds Striga hermonthica and Orobanche crenata. *J. Agric. Food Chem.* **45**, 2273-2277

(38) Rimando, A. M., Dayan, F. E., Czarnota, M. A., Weston, L. A., and Duke, S. O. **1998**. A new photosystem II inhibitor from Sorghum bicolour. *J. Nat. Prod.* **61**, 927-930

(39) Riopel, J. L. and Baird, W. V. **1987**. Striga. In: Musselman, L. J., (Ed.). *Parasitic Weeds in Agriculture*. CRC Press, Boca Raton, FL, 107-126

(40) Sato, D., Awad, A. A., Takeuchi, Y., Yokota, T. Sugimoto, Y., and Yoneyama, K. **2002**. Analysis of germination stimulants for weedy root parasites by LC/MS/MS. *Proceedings of the Third World Congress on Allelopathy. Challenge for the New Millennium*. Fujii, Y., Hiradate, S., and Haraya, H., (Eds.). Tsukuba, Japan, 215.

(41) Schmidt, T. J. **1997**. Helenanolide-type sesquiterpene lactones-III. Rates and stereochemistry in the reaction of helenalin and related helenanolides with sulfhydryl containing biomolecules. *Bioorg. Med. Chem.* **5**, 645-653

(42) Schmidt, T. J. **1999**. Toxic activities of sesquiterpene lactones: structural and biochemical aspects. *Curr. Org. Chem.* **3**, 577-608

(43) Sembdner, G. and Parthier, G. **1993**. The biochemistry and the physiological and molecular actions of jasmonates. *Annu Rev. Plant Physiol. Plant Mol. Biol.*, **44**, 569-589

(44) Smith, C. E., Ruttledge, T., Zeng, Z., O'Malley, R. C., and Lynn, D. G. **1996**. A mechanism for inducing plant development – the genesis of a specific inhibitor. *Proc. Natl. Acad. Sci. USA* **93**, 6986-6991

(45) Steffens, J. C., Lynn, D. G., Kamat, V. S., and Riopel, J. L. **1982**. Molecular specificity of haustorial induction in *Agalinis purpurea* (L.) Raf. (Schrophulariaceae). *Ann. Bot.* **50**, 1-7

(46) Takeuchi, Y., Omigawa, Y., Ogasawara, M., Yoneyama, K., Konnai, M., and Worsham, A. D. **1995**. Effects of brassinosteroids on conditioning and germination of clover broomrape (*Orobanche minor*) seeds. *Plant Growth Reg.* **16**, 153-160

(47) Thuring, J. F., Bitter, H. H., de Kok, M. M., Nefkens, G. H. L., van Riel, A. M. D. A., and Zwanenburg, B. **1997**. N-phthaloylglycine-derived strigol analogues. Influence of the D-ring on seed germination activity of the parasitic weeds *Striga hermonthica* and *Orobanche crenata*. *J. Agric. Food Chem.* **45,** 2284-2290

(48) Thuring, J. F., Heinsman, N. W. J. T., Jacobs, R. W. A., Nefkens, G. H. L., and Zwanenburg, B. **1997**. Asymmetric synthesis of all stereoisomers of demethylsorgolactone. Dependence of the stimulatory activity of *Striga hermonthica* and *Orobanche crenata* seed germination on the absolute configuration. *J. Agric. Food Chem.* **45**, 507-513

(49) Thuring, J. F., Nefkens, G. H. L., and Zwanenburg, B. **1997**. Asymmetric synthesis of all stereoisomers of the strigol analogue GR24. Dependence of the absolute configuration on stimulatory activity of *Striga hermonthica* and *Orobanche crenata* seed germination. *J. Agric. Food Chem.* **45**, 2278-2283

(50) Thuring, J. F., Nefkens, G. H. L., and Zwanenburg, B. **1997**. Synthesis and biological evaluation of strigol analogue carba-GR-24. *J. Agric. Food Chem.* **45**, 1409-1414

(51) Thuring, J. F., van Gaal, A. M. A., Hornes, S. J., de Kok, M. M., Nefkens, G. H. L., and Zwanenburg, B. **1997**. Synthesis and biological evaluation of strigol analogues modified in the enol ether part. *J. Chem. Soc., Perkin Trans. I,* 767-774

(52) Wigchert, S. C. M., Kuiper, E. Boelhouwer, G. R., Nefkens, G. H. L.,

Verkleif, J. A. C., and Zwanenburg, B. **1999**. Dose-response of seeds of the parasitic weeds *Striga* and *Orobanche* toward the synthetic germination stimulants GR24 and Nijmegen 1. *J. Agric. Food Chem.* **47**, 1705-1710

(53) Wigchert, S. C. M. and Zwanenburg, B. **1999**. An expeditious preparation of all enantiopure diastereoisomers of aromatic A-ring analogues of strigolactones, germination stimulants for seeds of the parasitic weeds *Striga* and *Orobanche. J. Chem. Soc., Perkin Trans. I*, 2617-2623

(54) Yasuda, N., Sugimoto, Y., Kato, M., Inanaga, S., and Yoneyama, K. **2002**. Isolation of (+)-strigol, a germination stimulant of root parasitic weeds, from *Menispermum dauricum* root culture. *Proceedings of the Third World Congress on Allelopathy. Challenge for the New Millennium.* Fujii, Y., Hiradate, S., Haraya, H., (Eds.) Tsukuba, Japan, 206.

(55) Yokota, T., Sakai, H., Okuno, K., Yoneyama, K., and Takeuchi, Y. **1998**. Alectrol and Orobanchol, germination stimulants for *Orobanche minor*, from its host red clover. *Phytochemistry* **49**, 1967-1973

(56) Yoneyama, K. Ogasawara, M., Takeuchi, Y., Konnai, M., Sugimoto, Y., Seto, H., and Yoshida, S. **1998**. Effect of jasmonates and related compounds on seed germination of *Orobanche minor* Smith and *Striga hermonthica* (Del.) Benth. *Biosci. Biotechnol. Biochem.* **62**, 1448-1450

(57) Yoneyama, K., Takeuchi, Y., Ogasawara, M., Konnai, M., Sugimoto, Y., and Sassa, T. **1998**. Cotylenins and Fusicoccins stimulate seed germination of *Striga hermonthica* (Del.) Benth and *Orobanche minor* Smith. *J. Agric. Food. Chem.* **46**, 1583-1586

7 Application of Analytical Techniques to the Determination of Allelopathic Agents in Wheat Root Exudates – Practical Case Study

T. Haig

CONTENT

Abstract 149
Introduction 150
Results and Discussion 152
Methodology 157
References 160

ABSTRACT

An example of the analytic power of highly selective, multi-stage, GC-MS/MS, quantitatively targeting seven phenolic acids and DIMBOA as allelochemicals in 17-day old wheat root exudates, is described. The enhanced analyte signal-to-noise ratio offered by this type of instrumentation readily overcomes the analytical problems arising from the search for active, trace, allelopathic agents within a complex matrix. Results from the analyses of the allelochemical contents from a worldwide selection of 58 wheat (*Triticum aestivum* L.) cultivar exudates, and their corresponding allelopathic inhibitions in bioassay upon the root growth of test weed *Lolium rigidum* G., imply a direct relationship between degree of inhibition and the total molar concentration of allelochemicals in the exudate.

INTRODUCTION

That different species of wheat plants possess an allelopathic capability toward other plants (weeds being of the greatest interest) is now a widely recognized phenomenon.[6,14,15] Investigations into this area, and into plant allelopathy generally, are commendably moving from those of a qualitative nature onto a more quantitative basis as scientific enquiry seeks deeper levels of causative explanation.[4,10] With the availability of accurate quantitative data, progress in the correlation between the number, class, and amounts of allelochemicals exuded by plants and their observed phytotoxic effect upon nearby receiver plants, will be greatly enhanced.

At Charles Sturt University (CSU), Australia, a multidisciplinary research group has been interested in the chemical basis for plant allelopathy for a number of years. One major impediment to progress has been the difficulty in tracking trace amounts of biologically active substances within complex plant and soil matrices. Recent improvements in bench-top chromatography-mass spectrometry have, however, greatly improved this situation.[3,5]

Thus, in our laboratories the use of ion-trap GC-MS, particularly in the MS/MS mode, has led to faster results, especially in studies on wheat (*Triticum aestivum*). The analytic power of "hyphenated" chromatography-mass spectrometry systems combining a high-resolution physical mixture separation with a strongly selective, highly sensitive, structure-identifying detector can avoid laborious individual allelochemical isolations. The power to deal with interfering background noise from the analyte's matrix using hyphenated systems arises from the enhancement of signal-to-noise ratio which occurs at each new hyphenation stage. Therefore, a GC-MS/MS or LC-MS/MS instrument is capable of remarkable analyte selectivity via its three-stage analysis.[7] The MS/MS mode has proven very powerful in LC work. Some sample clean-up steps can be avoided, method development expedited, limits of detection improved, and powerful structural analysis techniques applied to improve upon the limited molecular weight information normally available on unknown compounds with LC-MS.[2] The CSU group has employed GC-MS procedures in wheat studies because the analytes were of the lighter, stable, derivatisable type from within two distinct molecular categories (phenolics and benzoxazinoids), where the powerful resolving ability of long capillary GC columns gave an edge over LC in reducing co-elution, and fingerprint identifications against large commercial libraries of standard EI mass spectra were also available.[5]

Because this present study has been focused on contributing to the proof of a plant-plant allelopathic effect along the donor exudation – soil transportation – receiver uptake pathway (*T. aestivum / L. rigidum*), our attention has been given to the analysis of donor wheat exudate compounds, but which ones to measure has been a vexing question, as there are several hundred from which to choose, and many of them are unknown. Partial answers to this question may be found by searching the literature, or extending the range of candidates by further experimentation, but with the present incomplete state of knowledge surrounding the chemical basis for plant allelopathy, only an incomplete selection can be made. Indeed, in this matter, compound mixture synergism undoubtedly plays a significant role, making any choice of analytes even more uncertain. Nevertheless, in the face of these difficulties, the analyst must draw up a list of analytes which are as relevant as current information allows. For commercial wheat exudates we have chosen the benzoxazinoid DIMBOA, together with the seven phenolic acids: *p*-hydroxybenzoic, vanillic, syringic, *cis* & *trans*-coumaric, and *cis* & *trans*-ferulic. In very recent work (as yet unpublished) we have extended the list of benzoxazinoids to include HBOA; DIBOA; HMBOA and DHBOA (Fig. 7.1).

As this project's main concern has been with the chemical form of the benzoxazinoids which would be present in a transportation soil medium where microbial enzymatic hydrolysis would occur, these compounds have been determined as their aglucones. The acidity of their analysis sample medium has also to be controlled, as it is known that certain of these compounds have very short half-lives in neutral to alkaline solution, and such aglycones are best generated and stored within acidic solution (pH 3).[11] The exudate studies performed in this project have been carried out under ECAM conditions,[14] where wheat seedlings have been grown in a neutral 0.3% agar/water medium (without nutrients) and the exudates examined after seedling removal and agar acidification. Furthermore, the active constituents of exudate media are usually very dilute (typically μmol/L) and, therefore, prior concentration is required before determination. The literature records a variety of methods for this, including circulating exudate trapping on XAD-4 resin; adsorption on charcoal; C_{18} SPE, SDVB disks, and diethyl ether solvent extraction, but a comparison among these methods by Mattice et al. 1998[8] showed ether extraction to give the most reliable results. The CSU group has also adopted this practice.

Finally, consideration must be given to the availability of pure reference compounds from which to construct analytical calibration curves. Some compounds (such as phenolic acids) may be purchased commercially, but for others (such as benzoxazinoids) it will be necessary to first isolate and purify them from recognized natural sources described in the literature[12] or to synthesise them.[11]

The above analytical considerations have been combined here to present a GC-MS/MS study carried out on the exudates of 17-day old wheat seedlings from 58 worldwide accessions of varied allelopathic activity, as ascertained in a previous screening.[14] The target allelopathic agents chosen are the seven previously mentioned phenolic acids and DIMBOA.

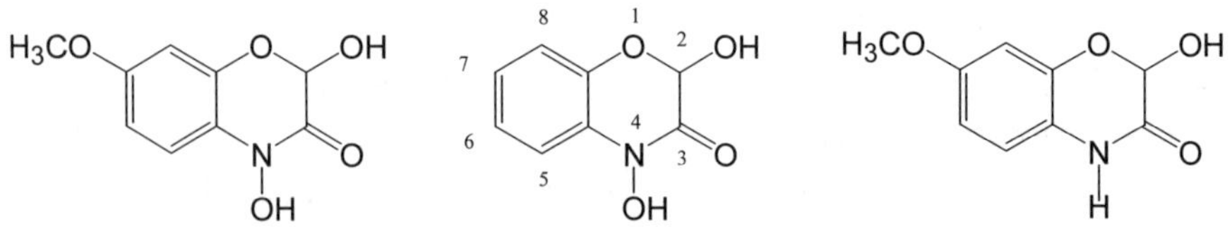

DIMBOA [1] DIBOA [2] HMBOA [3]

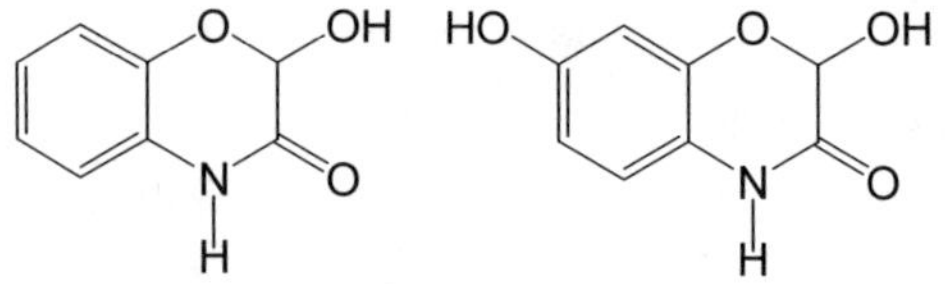

HBOA [4] DHBOA [5]

[1] = 2,4-dihydroxy-7-methoxy-2*H*-1,4-benzoxazin-3(4*H*)-one

[2] = 2,4-dihydroxy-2*H*-1,4-benzoxazin-3(4*H*)-one

[3] = 2-hydroxy-7-methoxy-2*H*-1,4-benzoxazin-3(4*H*)-one

[4] = 2-hydroxy-2*H*-1,4-benzoxazin-3(4*H*)-one

[5] = 2,7-dihydroxy-2H-1,4-benzoxazin-3(4H)-one

Figure 7.1

Five bioactive benzoxazinoid allelochemicals of wheat.

RESULTS AND DISCUSSION

Analytical results for the agar/water root exudate concentrations of DIMBOA across all 58 wheat accessions for 17-day old seedlings are listed in Table 7.1.

Table 7.1

DIMBOA contents in the agar root exudates of 58 wheat accessions.

Country of Origin	Wheat Accession	DIMBOA Conc.[a]	Country of Origin	Wheat Accession	DIMBOA Conc.[a]
Australia	Hartog	nd	Canada	Canada 51	nd
Australia	RAC 710	nd	Canada	Canada 4125	nd
Australia	Sunstate	nd	Canada	Canada 3740	nd
Australia	Currawong	nd	Canada	Jing Hong	nd
Australia	Triller	nd	Canada	Canada 56	nd
Australia	Excalibur	30.7 ± 2.5	Greece	Eretria	13.3 ± 0.2
Australia	Robin	nd	Germany	Wattines	nd
Australia	Sunco	11.6 ± 0.5	Italy	Virgilio	10.3 ± 2.4
Australia	Janz	nd	India	HY-65	nd
Australia	Trident	nd	India	WG-357	nd
Australia	Meering	nd	India	Khapli	69.8 ± 4.9
Australia	Baroota wonder	nd	Mexico	AUS#18056	nd
Australia	Cranbrook	15.7 ± 1.4	Mexico	Opata	nd
Australia	Batavia	nd	Mexico	AUS#18060	nd
Australia	Halberd	nd	Mexico	Altar 84	nd
Australia	Cadoux	nd	New Zealand	Wakanui	nd
Australia	Dollarbird	nd	New Zealand	Batten	nd
Australia	Matong	nd	Poland	Emika	nd
Australia	Sunstar	nd	Peru	AUS#12627	nd
Australia	Federation	nd	Sudan	Sudan 8	nd
Australia	Kite	nd	South Africa	SST 6	8.6 ± 0.6
Australia	RAC 820	nd	South Africa	SST 16	nd
Australia	CD 87	nd	South Africa	AUS#18364	nd
Australia	Insignia	nd	Switzerland	Bernina	nd
Australia	Egret	nd	Tunisia	Tunis 2	14.9 ± 1.0
Australia	Tasman	34.2 ± 4.9	USA	AUS#12788	nd
Afghanistan	Afghanistan 19	nd	USA	Lamar	nd
Afghanistan	Afghanistan 9	nd	Un. Arab Rep.	L1512-2721	79.1 ± 5.9
Brazil	PF 8716	42.0 ± 1.7	Yugoslavia	Studena	nd

[a]: concentrations in µg/L agar; values expressed as mean ± S.D; nd: not detected.

There were notable differences in the amounts of DIMBOA exuded by this worldwide selection of wheat accessions. Forty-seven out of the 58 (81%) did not exude any detectable amounts of DIMBOA into their agar growth medium, and only 11 accessions (19%) were found capable of exuding this hydroxamic acid, with amounts ranging from 8.6 ± 0.6 µg/L of agar/water for SST 6 (South Africa) to 79.1 ± 5.9 µg/L for L 1512-2721 (United Arab Republic). The release of strongly bioactive DIMBOA appears accession-specific, suggesting the influence of genetic factors governing the exudation process.

On the other hand, the less bioactive phenolic acids were distributed across the 58 accession exudates more evenly. Vanillic and p-hydroxybenzoic acids were normally distributed within the group and appeared in every exudate. The average levels of these two phenolic acids in the 17-day old wheat exudates were 7.33 ± 3.38 µg/L for vanillic acid, and 7.15 ± 3.50 µg/L for p-hydroxybenzoic acid, with a range for each from 0.6 to 17.5 µg/L, and from 2.3 to 18.6 µg/L, respectively. The remaining five phenolic acids were distributed differently among the 58 accessions and had average agar/water concentrations (µg/L) as follows: syringic acid 21.09 ± 8.29 (range 0.0 to 52.7); *trans*-ferulic acid 9.87 ± 5.58 (range 1.6 to 23.4); *trans-p-*coumaric acid 6.22 ± 3.84 (range 1.5 to 20.5); *cis*-ferulic acid 2.79 ± 2.57 (range 0.33 to 12.7); and *cis-p*-coumaric acid 1.07 ± 1.11 (range 0.1 to 4.9). Specific accessions containing the highest and lowest concentrations of each of these seven phenolics are described in Table 7.2.

Five of the accessions in Table 7.2 displaying the highest contents of phenolic acids had been shown in a previous study[14] to be strongly allelopathic to the growth of annual ryegrass (*Lolium rigidum* Gaud.). These accessions, which gave an average ryegrass root length of 7.0 mm compared to a nil-wheat control of 55.0 mm, were Wattines, Meering, AUS12627, AUS18056, and Khapli. Similarly, the four accessions Sunstate, Eretria, Sudan 8, and Afghanistan 19, which are included among the six wheat accessions exuding the lowest levels of phenolic acids (Table 7.2), had also been previously found[14] to be weakly allelopathic towards the growth of annual ryegrass, giving an average root length of 31.0 mm.

Such results indicate that for ryegrass growth there is a chemical basis for the differential allelopathy exhibited by the wheat accessions studied. Similar findings were made by Baghestani et al.,[1] who studied the phenolic contents of root exudates from four crop types, including wheat, but did not determine any exuded benzoxazinoids. These workers analyzed for 16 common phenolic compounds (including the group in the present study) using HPLC, and applied wheat root exudates to *Brassica kaber* (DC.) L. C. Wheeler as their bioassay test plant. They found a similar allelopathic response, with *B. kaber* root and shoot length decreasing with increasing concentration of wheat exudate. The Baghestani group stated that vanillic acid, *o*-coumaric acid, and scopoletin (7-hydroxy-6-methoxycoumarin) may be responsible for the allelopathic effect in wheat but do not appear to have considered the exudation of the more biologically active hydroxamic acids and the contribution they would make to the overall effect.

Table 7.2

Highest and lowest concentrations of phenolic acids in 17-day old wheat root exudates.

Phenolic Acid	Accession with Highest Contents	
syringic acid	Dollarbird	52.72 ± 0.61[a]
vanillic acid	AUS 12627	17.53 ± 0.12
p-hydroxybenzoic acid	Meering	18.57 ± 0.82
cis-p-coumaric acid	Wattines	4.89 ± 1.39
trans-p-coumaric acid	Canada 51	20.46 ± 2.85
cis-ferulic acid	Khapli	12.69 ± 0.25
trans-ferulic acid	AUS 18056	23.41 ± 0.15
	Accession with Lowest Contents	
syringic acid	AUS 12627	nd
vanillic acid	Eretria	0.61 ± 1.15
p-hydroxybenzoic acid	Sudan 8	2.32 ± 0.29
cis-p-coumaric acid	Sunstate	0.07 ± 0.07
trans-p-coumaric acid	Sudan 8	1.50 ± 1.00
cis-ferulic acid	AUS 18364	0.33 ± 0.17
trans-ferulic acid	Afghanistan 19	1.60 ± 0.90

[a] Mean values ± SD in µg/L of agar/water; nd = not detected.

The exudation of allelochemicals by living plants is one of the basic tenets of allelopathy theory and would be an indispensable component of any harnessed crop capacity against weeds.[9] The exudation of relatively large amounts of key phenolics and benzoxazinoids from wheat (*Triticum aestivum*) cultivars (and other *Triticum* species) may well serve as an easily measured indicator of valuable genetic material which could be used to breed for wheat cultivars of elevated allelopathic activity and weed control. In this respect, we have recently taken the present study a further step ahead, based upon a series of measurements taken across a 15-day Khapli wheat growth period (unpublished results). Analytical attention was focused upon this strongly allelopathic Indian wheat because it not only exuded high levels of phenolic acids (Table 7.2) but also was second-highest in the levels of exuded DIMBOA (69.8 µg/L agar/water, Table 7.1). When a plot of allelopathic effect (measured as ratio of nil-wheat control root length divided by root length of test plant *Lolium rigidum* after addition of Khapli exudate) against log [sum of molar concentrations] of the seven phenolics plus DIMBOA and DIBOA was drawn, the co-efficient of determination (r^2) on this linear model dose-response curve was 0.914, demonstrating that for Khapli wheat under ECAM[14] conditions (non-nutrient agar/water growth medium) most of the molecular explanation of its allelopathic activity on ryegrass is tied up in those nine active compounds.

As expected in the GC-MS analysis procedure, despite the high resolving power of a 30 meter GC capillary column (0.25 mm ID), much co-elution of background peaks still occurred with the chosen analytes, and great difficulty was experienced in locating any unique, characteristic ions for quantification by extracted ion chromatograms.[13] However, the third-stage of the MS/MS technique successfully removed most of the background interfering ions and provided clear extracted ion MS/MS chromatograms for quantifying each of the eight analytes. The signal-to-noise ratio in these types of plant-product MS/MS analyses is typically a 5-fold improvement for DIMBOA, a 10-fold improvement for *trans-p*-coumaric acid, and a 27-fold improvement for *p*-hydroxybenzoic acid over the normal GC-MS response. Naturally, if the need were evident, a further stage of product ion formation using GC-MS^3 could be easily introduced by the (time-based) ion trap mass spectrometer.

In summary, it is expected that, for allelopathy studies, the analytic power of multi-stage chromatography-mass spectrometry techniques will generate much useful qualitative and quantitative data, thereby helping to unravel some of the remaining mysteries of the allelopathy phenomenon.

METHODOLOGY

Collection of 17-day old wheat root exudates from an agar/water growth medium

Samples from a worldwide collection of 58 wheat accessions with varied allelopathic activities were grown according to the following procedure.

Sterilization. Seeds of each wheat accession were surface-sterilised by soaking in 70% ethanol for 2.5 min, followed by four rinses in sterilised distilled water. They were then soaked in 2.5% sodium hypochlorite solution for 15 min, followed by five rinses in sterilized distilled water.

Pre-germination. Surface-sterilized seeds of the wheat genotypes were each soaked in sterilised water for imbibition of water in light at 25°C for 24 h and then rinsed with fresh sterilised water. The wheat seeds were then incubated in light at 25°C for another 24 h.

Collection of exudates. Twelve pre-germinated, surface-sterilized wheat seeds of each accession were uniformly selected and aseptically sown (embryo up) in an autoclaved 500 mL glass beaker pre-filled with 30 mL of 0.3% agar/water. Each beaker was wrapped with parafilm to prevent contamination and evaporation from the agar surface and then placed into a controlled growth cabinet with a daily light/dark cycle of 13 h/11 h and a temperature cycle at 25°C/13°C. The fluorescent light intensity in the cabinet was $3.56 \pm 0.16 \times 10^3$ lux (Quantum/Radiometer/Photometer, Model LI-185B, LI-COR Inc.).

After the growth of seedlings for 17 days, all twelve were gently uprooted from their soft agar medium (nutrient-free) and the roots rinsed twice with 5 mL portions of distilled water to remove the residual agar. The washings were pooled into the agar medium. For better averaging, this seedling growth process was repeated three times, so that three lots of agar medium were collected, pooled together, and stirred thoroughly. One third of the pooled agar was then adjusted to pH 3.0 by dropwise addition of 0.06M HCl, stirred, and sonicated at 5°C for 15 min. The agar medium was extracted three times with 60 mL portions of diethyl ether. The combined ether layers were then evaporated in a rotary evaporator under reduced pressure at 35°C, in readiness for derivatization.

Derivatization, quantification and GC-MS/MS analysis of wheat exudates

Preparation of calibration standards: Seven standard solutions each containing all the target compounds (phenolics plus DIMBOA) were first prepared in HPLC-grade methanol at concentrations of 0.05, 0.1, 0.5, 1.0, 5.0, 10.0, and 20.0 µg/mL respectively. In separate analyses, 1.0 mL from each of the methanolic standards was pipetted into a 2 mL minivial and dried by gentle nitrogen gas blow-down. A 1.0 mL volume of internal standard p-chlorobenzoic acid at a concentration of 1.0 µg/mL in methanol was then pipetted into the dry minivial and dried again by blow-down prior to formation of silylated derivatives.

GC-MS/MS instrumentation and conditions: GC-MS/MS analysis was carried out on a Varian 3400 CX gas chromatograph coupled with a Varian Saturn 2000 ion trap mass spectrometer. Samples were introduced via a DB-5 MSITD fused-silica, capillary column of 30 m × 0.25 mm ID × 0.25 µm dimensions. The column temperature program was 1 min at 80°C, then to 160°C at 10°C/min, from 160°C to 235°C at 5°C/min, from 235°C to 280°C at 50°C/min, with a final hold time of 5 min (total run time 29.9 min). Helium (99.9999%) at 34 cm^3/s was used as carrier gas. The injector temperature was 280°C and injection volume 1 µL in the splitless mode.

For MS work, the electron impact (EI) mode with automatic gain control (AGC) was used. The electron multiplier voltage for MS/MS was 1450 V, AGC target was 10,000 counts, and filament emission current was 60 µA with the axial modulation amplitude at 4.0 V. The ion trap was held at 200°C and the transfer line at 250°C. The manifold temperature was set at 60°C and the mass spectral scan time across 50-450 m/z was 1.0 s (using 3 microscans). Nonresonant, collision-induced dissociation (CID) was used for MS/MS. The associated parameters for this method were optimized for each individual compound (Table 7.3). The method was divided into ten acquisition time segments so that different ion preparation files could be used to optimize the conditions for the TMS derivatives of the chemically distinct internal standard, phenolic acids, and DIMBOA. Standard samples of both p-coumaric and ferulic acids consisted of *trans* and *cis* isomers so that four segments were required to characterize these two acids. The first time segment was a 9 min solvent delay used to protect the electron multiplier from the solvent peak.

Table 7.3

GC-MS/MS ion trap method parameters.

Parameters	MS/MS Acquisition Segment Number (No. 1 = solvent delay period)				
	2	3	4	5	6
Compound (as TMS-deriv)	CBA	HBA	VAN	c-COU	SYR
Retention time (min)	9.78	13.3	15.74	16.23	18.23
Precursor ion selected (m/z)	213	267	297	293	297
Excitation amplitude (V)	54	64	43.6	45	43.6
Excitation storage level (m/z)	60	80	65	65	65
Quantifying product ion (m/z)	169	223	267	249	253
Segment duration (min)	4.1	0.4	2.6	2	0.3

Parameters	MS/MS Acquisition Segment Number (No. 1 = solvent delay period)			
	7	8	9	10
Compound (as TMS-deriv.)	c-FER	t-COU	DIMBOA	t-FER
Retention time (min)	18.63	19.06	20.16	21.19
Precursor ion selected (m/z)	323	293	340	323
Excitation amplitude (V)	41	45	46.5	41.2
Excitation storage level (m/z)	65	65	75	65
Quantifying product ion (m/z)	293	249	194	293
Segment duration (min)	0.5	1.1	1.8	8.1

(For all analyses, the mass isolation window was set at 3 m/z and the excitation time at 20 milliseconds.)

CBA = *p*-chlorobenzoic acid (internal standard) HBA = *p*-hydroxybenzoic acid
VAN = vanillic acid   c-COU = *cis*-coumaric acid
SYR = syringic acid   c-FER = *cis*-ferulic acid
t-COU = *trans*-coumaric acid   t-FER = *trans*-ferulic acid

<u>Identification and quantification</u>: Two spectral user-libraries (MS and MS/MS) were developed using injections of TMS-derivatized authentic reference compounds. The MS library recorded the retention times and normal EI mass spectra of trimethylsilyl (TMS) derivatives of authentic standards under the chosen chromatographic conditions, while the MS/MS library recorded retention times and product ion spectra derived from the specifically chosen precursor ions (Table 7.3) of TMS

derivatives of each authentic standard after CID with helium gas inside the ion trap. The allelopathic analytes were then identified by comparing retention times and mass spectral data with those in the respective user-library. Quantifications were achieved via the internal standard method using *p*-chlorobenzoic acid as the standard at a concentration of 1.0 μg/mL of injection solution. The quantification of allelochemicals was based on the chromatographic peak area of the selected product ions listed in Table 7.3, and is reported in units of μg/L of agar/water. All calibration standards and wheat samples are reported as the averages of triplicate analyses.

ACKNOWLEDGMENTS

Grateful acknowledgement of financial assistance and support is extended to Charles Sturt University, Australia; the Australian Grains Research and Development Corporation; and the Australian Co-operative Research Centre for Weed Management Systems. Generous provision of wheat accessions was made by the Australian Winter Cereals Collection, and of reference samples of benzoxazinoids by Prof. Dieter Sicker, University of Leipzig, Germany. The author thanks Prof. James Pratley (CSU) and Dr. Deidre Lemerle (NSW, ARI) for helpful consultations during the project, and extends appreciation to Dr. Hanwen Wu and Dr. Min An for their invaluable experimental contributions.

REFERENCES

(1) Baghestani, A., Lemieax, C., Leroux, G. D., Baziramakenga, R. and Simard, R. R. **1999**. Determination of allelochemicals in spring cereal cultivars of different competitiveness. *Weed Sci.* **47**, 498-504

(2) Cambier, V. and de Hoffmann, E. **1999**. Noninjured maize contains several 1,4-benzoxazin-3-one related compounds but only as glucoconjugates. *Phytochem. Anal.* **10**, 119-126

(3) Eljarrat, E. and Barcelo, D. **2001**. Sample handling and analysis of allelochemical compounds in plants. *Trends in Anal. Chem.* **20**, 584-590.

(4) Friebe, A., Roth, U., Kuck, P., Schnabl, H. and Schulz, M. **1997**. Effects of 2,4-dihydroxy-1,4-benzoxazin-3-ones on the activity of plasma membrane H^+-ATPase. *Phytochemistry* **44**, 979-983

(5) Haig, T. **2001**. Application of hyphenated chromatography-mass

(6)   spectrometry techniques to plant allelopathy research. *J. Chem. Ecol.* **27**, 2363-2396

(6)   Hashem, A. and Adkins, S. W. **1998**. Allelopathic effects of *Triticum speltoides* on two important weeds of wheat. *Plant Prot. Quart.* **13**, 33-35

(7)   Lee, C. W., Yoneyama, K., Takeuchi, Y., Konnai, M., Tamogami, S., and Kodama, O. **1999**. Momilactones A and B in rice straw harvested at different growth stages. *Biosci. Biotechnol. Biochem.* **63**, 1318-1320

(8)   Mattice, R., Lavy, T., Skulman, B. and Dilday, R. **1998**. Searching for allelochemicals in rice that control ducksalad. Ch. 8 in: Olofsdotter, M. (Ed.), *Allelopathy in Rice*. International Rice Research Institute, Los Baños, Philippines

(9)   Rice, E. **1984**. *Allelopathy*, 2nd ed. Academic Press, Orlando, FL

(10)   Seigler, D. S. **1996**. Chemistry and mechanisms of allelopathic interactions. *Agron. J.* **88**, 876-885

(11)   Sicker, D., Hartenstein, H. and Kluge, M. **1997**. Natural benzoxazinoids – synthesis of acetal glucosides, aglucones, and analogues. *Recent Res. Devel. Phytochem.* **1**, 203-223

(12)   Woodward, M. D., Corcuera, L. J., and Helgeson, C. D. **1978**. Decomposition of 2,4-dihydroxy-7-methoxy-2H-1,4-benzoxazin-3(4H)-one in aqueous solutions. *Plant Physiol.* **61**, 796-802

(13)   Wu, H., Haig, T., Pratley, J., Lemerle, D., and An, M. **1999**. Simultaneous determination of phenolic acids and 2,4-dihydroxy-7-methoxy-1,4-benzoxazin-3-one in wheat (*Triticum aestivum* L.) by gas chromatography-tandem mass spectrometry. *J. Chromat. A*, **864**, 315-321

(14)   Wu, H., Pratley, J., Lemerle, D. and Haig, T. **2000**. Laboratory screening for allelopathic potential of wheat (*Triticum aestivum*) accessions against annual ryegrass (*Lolium rigidum*). *Aust. J. Agric. Res.* **51**, 259-266

(15)   Wu, H., Pratley, J., Lemerle, D. and Haig, T. **2001**. Allelopathy in wheat. *Ann. Appl. Biol.* **139**, 1-9

8 The Importance of Alkaloidal Functions

M. S. Blum

CONTENT

Abstract 163
Introduction 164
Chemical Defense Based on Alkaloids 164
Alkaloids for Defense and Exploitation 166
    Sequestration for Defense and Communication 169
    Alkaloid Sequestration in Vertebrates and Invertebrates 169
    Sequestration of Alkaloids: Defense, Metabolism, and
    Communication 170
PAs Synthesized by Animals and their Functions 173
Alkaloid Parsimony 175
Physiological and Biochemical Targets for Alkaloids 176
References 178

ABSTRACT

A large variety of alkaloids is produced by microorganisms, plants, and animals, and some of these natural products have been demonstrated to cause severe physiological malfunctions that can eventuate in fatalities. Multifarious physiological systems constitute targets for these commonly encountered compounds, but the specific manifestations of intoxication may reflect the mode of administration of these alkaloidal intrusions.

Alkaloids are generally bitter, which suggests that these compounds could be utilized as either animal deterrents or in intra- or interspecific competition of plant species. In some cases repellent alkaloids (e.g., pyrrolizidine alkaloids) are sequestered by herbivores and converted to compounds that function as sex pheromones while still possessing deterrent activities. The insect derivation of sex pheromones from known repellent alkaloids that accompany the ingested nutrients illustrates the common principle of alkaloidal parsimony in which these natural products can possess unrelated functions, in this case being both a predator deterrent and a sexual pheromone.

INTRODUCTION

The modes of action of alkaloids are often studied in pharmacological terms that reflect the perturbations of physiological and biochemical systems after exposure to these natural products. However, although it is convenient to obtain toxicological data on alkaloids after parenteral administration, most of these compounds probably gain access to the the bodies of herbivores by oral application following the ingestion of alkaloid-fortified food. Furthermore, herbivores may readily expectorate plant materials from the mouth, based on unpalatable gustatory reactions to alkaloids in the food. These considerations emphasize the importance of treating alkaloidal toxins in terms of their biological (ecological) activities in natural systems *in vivo*. As a consequence, the following brief analysis of the modi operandi of alkaloids will attempt to treat these compounds as versatile natural products that have been evolved to subserve a surprising diversity of functions that are eminently adaptive.

CHEMICAL DEFENSE BASED ON ALKALOIDS

Overall, plants are quite capable of defending themselves from diverse microorganisms, herbivores, and other plants. Candidates for this protective role are their allelochemicals, which may possess deterrent/repellent or toxic properties for a wide range of competitors.[37] Included in the allelochemicals are the alkaloids, which are generally considered to be bitter (e.g., quinine, brucine) and in some cases very toxic (e.g., tetrodotoxin, batrachotoxin). Although at a glance the alkaloids appeared to be outstanding candidates to play defensive roles for their producers, they were initially described as "secondary metabolites", as if their functions were not at all adaptive. Furthermore, plant-derived natural products were regarded as waste products that were equivalent to functionless molecules of little selective value.[30] However, a concatenation of physiological, ecological, and chemical investigations has unambiguously established the critical role of alkaloids, for both survival and fitness, of a variety of plant species.

The importance of studying biosynthetic pathways in terms of alkaloidal functions was stressed more than forty years ago in an investigation of the biogenesis of nicotine.[15] Recognizing that alkaloids had been customarily regarded as wastes of metabolism, it was concluded that detailed investigations of their syntheses may "provide us with a back door into the intermediary metabolism of growth," and in addition, *"an entirely new concept of their biological significance."*

The evolution of complex metabolic pathways for the synthesis of alkaloids is simply not consonant with the conclusion that this biosynthetic elegance has been evolved in order to synthesize nitrogenous "waste" products for elimination. Plant ecologists emphasize that nitrogen is the most limiting nutrient for plants, being required for key processes such as reproduction, growth, and development. Consequently, it would hardly be adaptive to jettison the element that has been incorporated into novel ring structures and which possesses great biological activities, if for no other reason than the plant is also eliminating an essential component of the peptides, proteins, and other nitrogen-containing compounds that are of such great metabolic significance for all organisms. It may well be that the excretion of uric acid or urea by carnivorous animals, which ingest an excess of proteins and nucleic acids, is responsible for the belief that organisms eliminate large amounts of nitrogenous compounds when this element is readily available.

On the other hand, a diversity of complex alkaloids is synthesized *de novo* by animals, particularly species of amphibians.[14] These compounds, which are often novel natural products, constitute some of the most toxic alkaloids that have been identified and emphasize their potential as agents of deterrency. For example, batrachotoxin, a steroidal alkaloid of five species of dendrobatid frogs in the genus *Phyllobates*, has an LD50 of 40 ng. by subcutaneous injection in mice. Indians in South America utilize skin secretions of these frogs in order to poison blow darts that are used against a variety of prey with great effectiveness. Similarly, parotid skin gland secretions of salamanders (*Salamandra* spp.) are fortified with steroidal alkaloids such as samandarine which are very toxic to mammals.[22] A variety of invertebrates is also identified with the utilization of alkaloids as toxins.

Tetrodotoxin, probably of bacterial origin, is present in the venom gland of the octopus *Hapalochlaena maculosa*, clearly functioning to augment the toxicity of the *de novo*-synthesized poison gland constituents.[34] Venoms, fortified with alkaloids, characterize the poison gland secretions of ants in the genera *Solenopsis* and *Monomorium*.[27] The venoms, 2,5-dialkylpyrrolidines and 2,6-

dialkylpiperidines, can be applied topically to the cuticle of insects to function as repellents or insecticides.[4] In addition, fire ant species in the genus *Solenopsis* can introduce the venoms by injection, unleashing a series of pharmacological lesions that emphasize the versatile defensive properties of these alkaloidal venoms.[27] Severe anaphylactic reactions can result in death in individuals that manifest extreme hypersensitivity to the venomous constituents. Nonvenomous alkaloids are employed by other species of insects as defensive secretions but may possess other functions as well.

At this juncture, alkaloids would appear to constitute defensive compounds that are highly adaptive for their producers against a variety of predatory organisms with whom they share their world. Examining some of the defensive devices employed, *in vivo*, by these plant and animal synthesizers of alkaloids, can help to illuminate the versatility of these nitrogen-containing compounds as agents of deterrence. In addition, by focusing on these alkaloidal chemical defenses, it will also be possible to examine the successful offensive strategies of specialist herbivores that exploit host plants fortified with alkaloidal "forbidden fruits."

ALKALOIDS FOR DEFENSE AND EXPLOITATION

Plant alkaloids often demonstrate defensive activity against a wide variety of predators and competitors among which are microorganisms, fungi, viruses, invertebrate and vertebrate herbivores, and plants, including immatures of their own species. Since these alkaloids, termed allelochemicals, have been evolved as deterrents for very different classes of organisms, in plants their production and storage often result in mixtures of structurally diverse alkaloids that are stored at specific sites. Furthermore, these alkaloidal accumulations may change both quantitatively and qualitatively in response to factors that optimize reproduction. Indeed, it has been suggested that "the breadth of activities of alkaloids in ecological interactions parallels that seen in pharmacological tests."[10]

In general, plants do very well in their environments, notwithstanding the omnipresence of a multitude of potential insect herbivores and a number of vertebrate herbivores as well. Some plant species are very insecticidal, as a consequence of their producing a variety of alkaloids including nicotine, piperine, lupine alkaloids, steroidal alkaloids, ephedrine, berberine, strychnine, gramine, and caffeine.[37] These biologically active alkaloids also function as deterrents. In terms of families, caffeine is the the most widely distributed alkaloid, a fact that may be significant because this compound is very insecticidal for a large number of species. Nicotine is also very deterrent and toxic to insect herbivores and has been utilized as a commercial insecticide for many years. On the other hand, plants in the genus *Nicotiana* (Solanaceae) are fed upon readily by larvae of the tobacco hornworm, *Manduca sexta*. This insect is a specialist on nicotine-fortified plants, tolerating diets containing up to 1% nicotine. Larvae either degrade this alkaloid or eliminate it directly through the excretory system. The ability of *M. sexta* larvae to tolerate nicotine, which binds to the acetylcholine (ACH) receptor, is probably possible because the receptor of the larva may have been modified but still can bind ACH, but not nicotine.[10]

Manduca sexta emphasizes that alkaloidal defense against herbivores is not absolute. Plant armies are no different than human armies—no defense is perfect! A number of unrelated insect herbivores specialize on plants fortified with toxic alkaloids (e.g., cocaine, pyrrolizidine alkaloids, quinolizidine alkaloids), and often these alkaloidal plants are fed upon exclusively by these specialists. However, while some insects have "broken through" the alkaloidal defenses of selected plant species, plant alkaloids in general are very active in deterring a wide range of potential enemies.[24] Alkaloids clearly emerge as a demonstrated *sine qua non* for protecting plants from hungry herbivores.

In some cases, symbiotic relationships between plants and alkaloid-producing fungi provide the plants with protection from herbivores, probably at some cost, but without compromising reproduction.[12] For example, fungi infecting many species of grasses (e.g., rye plants) produce mixtures of ergot alkaloids, some of which are very toxic to human beings. Fungi such as *Claviceps purpurea* synthesize relatively high concentrations of ergothioneine in the sclerotia, and this compound endows the grasses with a strong vertebrate toxin that mimics the activity of neurotransmitters such as dopamine, serotonin, and noradrenaline. Indeed, grasses that were infected by ergot-producing fungi suffered less damage from herbivores than grasses that were free of fungi. The fungi appear to utilize nutrients from the grasses while supplying alkaloidal deterrents for their hosts. This type of symbiosis appears to be highly adaptive and will probably be encountered with some frequency when the so-called "parasitic" relationships of fungi and their plant hosts are examined in some detail.

Demonstrating unambiguously that alkaloids are solely responsible for the deterrency of plants to an herbivore has been difficult to establish except in selected instances. This has been possible by examining several species of *Lupinus* (Leguminosae) that synthesize quinolizidine alkaloids (QAs) which may be concentrated in the seeds as an obvious means of promoting reproductive fitness.[37] These seeds, which can contain 2-8% alkaloids, have long been utilized as human food, but only after they have been boiled to leach out the bitter and toxic QAs. However, this problem of purification was apparently solved by plant breeders who successfully produced sweet lupines, containing an alkaloid concentration of less than 0.01%, about 100 years ago. But although sweet lupines were indeed very palatable for humans, they were also a gustatory delight to a variety of herbivores. In the virtual absence of QAs, the sweet lupines were readily selected by rabbits and hares, as well as aphids, beetles, and thrips. On the other hand, the QA-fortified plants were almost untouched, clearly demonstrating that these alkaloids were highly active as deterrents for both vertebrate and invertebrate herbivores.

It has been suggested that sometimes plant species must make a "choice" between two herbivorous specialists, each of which is tolerant of the alkaloidal products synthesized by their host plant. This appears to be the case for *Senecio jacobaea* (Asteraceae), a species which is rich in PAs. Two PA-tolerant herbivores, the aphid *Aphis jacobaeae,* and larvae of the tiger moth, *Tyria jacobaeae* (Arctiidae), are commonly associated with *S. jacobaea* and if possible, readily utilize it as a host plant. However, whereas the aphids are sap suckers and generally constitute a feeding load that the plant can tolerate, the moth larvae are leaf chewers and can rapidly defoliate the plant. Thus, the plant must make an adaptive "choice" in terms of herbivores, by selecting the phytophage that will have less impact on the plant's fitness.[35] The voracious herbivory of the moth larvae clearly eliminates them from consideration, and it is therefore necessary for the plant to obtain effective help in the form of antilarval defenses. The plant does this through "choosing" aphids which attract aggressive ants that attack tiger moth larvae and, in exchange, provide honeydew, a sweet excretory product, rich in PAs. However, although the PA-fortified honeydew is clearly an intestinal excretory product of aphids that is derived from the carbohydrate-rich phloem in *S. jacobaea,* it is significant that the aphids have also sequestered repellent pyrrolizidine alkaloids in nonexcretory tissues and have thus appropriated the chemical defenses of their plant hosts. Analyzing the sequestration of alkaloids by herbivores can provide surprising insights into how these plant-derived compounds, in two-trophic systems, are manipulated by either animal sequestrators or ingestive predators.

SEQUESTRATION FOR DEFENSE AND COMMUNICATION

Specialists feeding on either plants or animals that are fortified with toxic alkaloids are unpredictable in terms of how these natural products are processed. For example, nicotine, as previously mentioned, is degraded or rapidly eliminated by larvae of the tobacco hornworm, *Manduca sexta*.[37] Nicotine is not known to be stored by any tobacco-feeding insects and may simply be too reactive to be easily sequestered in tissues. Of course, the inability of these larvae to accumulate nicotine internally deprives them of a powerful defense, a development that must reflect the nonadaptive nature of this alkaloid for sequestration. Similarly, a tropane alkaloid, cocaine, is only sequestered in trace amounts after moth larvae, *Eloria noysei*, have fed on coca plants.[5] The larvae of *E. noysei* excrete the alkaloid very rapidly, and the feces contains most of the ingested compound. Significantly, the adults are cryptically colored, which is consistent with a species that lacks protective defensive compounds and maintains an inconspicuous posture in the environment. Species that contain defensive compounds are generally aposematic (brightly colored) and virtually advertise their unpalatability/toxicity. In essence, these colorful sequestrators have evolved a lifestyle predicated on a close association with toxic plants whose alkaloids are forcefully utilized to deliver an unambiguous message of deterrence.

ALKALOID SEQUESTRATION IN VERTEBRATES AND INVERTEBRATES

In some cases very toxic alkaloids are not produced *de novo* by their hosts but rather represent sequestered products that have been synthesized by microorganisms present in the host animals. This is the case for tetrodotoxin, a steroidal alkaloid with great mammalian toxicity whether administered intraperitoneally (LD_{50} in mice about 0.2 ug.) or orally in the case of human beings (Japan) who have died after eating a portion of the "specialty of the house," the tetrodotoxin-containing Japanese puffer fish, *Fuga rubipes*.[14] This compound has also been identified in the venom gland of the octopus *Hapalochlaena maculosa* where it presumably augments the toxicity of the other venomous constituents.[34] This guanidinium alkaloid has a very wide distribution, especially among marine organisms. This compound has been identified in genera of newts and salamanders in the family Salamandridae as well as frogs and toads (Anura) in the families Bufonidae, Hylidae, Ranidae, and Pipidae.[8] In addition to the puffer fish, tetrodotoxin has been identfied in goby fish in the family Gobiidae. Significantly, this steroidal alkaloid is synthesized by bacteria in the genera *Alteromonas, Vibrio,* and *Pseudomonas*, all of which had been isolated from a red alga in the genus *Jania*.[2] Although complementary *in vivo* studies are generally lacking, it appears that a remarkable diversity of animals has been able to safely sequester tetrodotoxin-producing microorganisms and, in so doing, possess a powerful defensive system. Other highly toxic microbial toxins (e.g., saxitoxin) mirror tetrodotoxin in being present in a variety of marine animals[25] and in some cases these alkaloids, after sequestration from prey, have been demonstrated to be potent deterrents.

Sea slugs (nudibranchs), *Tambja abere* and *T. eliora,* feed on the bryozoan ectoproct *Sessibugula translucens*, and sequester several tambjamines (bipyrrolic alkaloids), probably of bacterial origin, from their nudibranch prey.[11] A more ferocious nudibranch predator, *Roboastra tigris, is* deterred by the sequestered bipyrroles in the the mucus eliminated by the *Tambja* species. While low concentrations of tambjamines are used by the sea slugs such as *R. tigris* to locate prey, higher concentrations of the pyrroles are toxic to *R. tigris* and inhibit feeding by the spotted kelpfish *Gibbonsia elegans.* Similarly, three tambjamines and a blue tetrapyrrole have been demonstrated to be feeding deterrents for seven genera of coral reef fishes.[11]

Not surprisingly, although the evolution of sequestered microbial toxins appears to be rather widespread in marine environments, sequestration of defensive alkaloids in the *apparent* absence of microorganisms may generally characterize the chemical defenses of terrestrial animals. Careful searches for possible microbial syntheses of defensive compounds (allomones) have not been generally implemented, but recent studies in a few laboratories raise the possibility that microbial endosymbionts may be of major importance in the biogenesis of selected insect deterrents.

SEQUESTRATION OF ALKALOIDS: DEFENSE, METABOLISM, AND COMMUNICATION

In a sense, alkaloidal sequestration constitutes the transfer of compounds from the producer to the receiver (predator). In effect, this transfer effectively arms the predator with at least some of the alkaloidal defenses of the ingested plant or animal, probably at little energetic expense to the receiver. Significantly, if a specialist herbivore feeds on an alkaloid-rich plant, it is adapted to tolerate these nitrogen heterocycles which usually render it unpalatable to diverse predators. On the other hand, nonadapted herbivores, constituting the vast majority of plant feeders, reject the specialist's host plant immediately. Clearly these alkaloids are outstanding feeding deterrents for their plant producers, notwithstanding the fact that relatively few herbivores have been able to specialize on this plant species.[6] Indeed, the protective role of alkaloids against nonspecialized herbivorous species is easily demonstrated if alkaloid-fortified plants are offered to herbivores not adapted to feed on these compounds. Alkaloid-intolerant herbivores will either starve to death in the presence of plants containing these nitrogen heterocycles or, if they reluctantly feed, manifest symptoms of intoxication.

The association of herbivores with toxic plants clearly provides the plant feeders with adaptive characteristics that are very beneficial. Butterfly and moth larvae deposit pyrrolizidine alkaloids (PAs) in the cuticle (skin) and are thus armored with a PA shield that renders them unpalatable and toxic to predators. Adults of these species, having ingested PAs as larvae, deposit these compounds in the wings, usually the most accessible target for flying predatory birds. These very unpalatable compounds are known to have deterred avian predators after they had "tasted" them on wing fragments that they had ingested.

Ingested alkaloids can also be utilized for defense even if these compounds are still present in the intestine. For example, grasshoppers and the larvae of many species of butterflies and moths regurgitate when tactually disturbed. The enteric fluid, fortified with alkaloids such as PAs, can function as a highly effective topical deterrent against small invertebrate predators.[33] Thus the plant has adapted its host plant's allelochemicals to function as a secretion that can be accurately applied in aggressive confrontations.

Insects have also adapted some plant alkaloids to function as toxicants for pathogenic fungi and larval parasites as well. A pathogenic fungus, *Nomuraea rileyi*, attacks larvae of the corn earworm, *Helicoverpa zea*, with great pathogenicity. However, if these larvae ingest -tomatine, an alkaloid produced by tomato plants, fungal toxicity can be reduced substantially.[21] Since -tomatine is also toxic to larval parasites, it is obvious that these moth larvae have exploited effectively the defense of the tomato plant.[17]

Sequestration of toxic alkaloids also has another benefit for the insects that feed on these "forbidden fruits." In a very real sense these herbivores are conspicuously marked with colorful patterns that designate them as toxic animals. They are described as aposematic or warningly colored organisms, in much the same way as a Gila monster or a coral snake. Many of the aposematic larvae are colored red, yellow, and black, being easily recognized and rejected by experienced predators.[33] The same is true of adults of many of these species which are eminently conspicuous and can be recognized as unpalatable whether on a plant or in the air. Bright colors signify danger in the animal kingdom and these sequestrators of plant alkaloids have evolved aposematic "coats" made possible by their association with the toxins in their diet. The aposematism–toxicity correlation has recently been extended to a bird, *Pitohui dichrous*, a brilliant orange and blue species that contains a toxic alkaloid, homobatrachotoxin, in its skin and feathers.[17]

Skin toxins, characteristic of many species of amphibians, are generally synthesized *de novo,* but in some cases defensive compounds are sequestered from a variety of animal sources. The alkaloid precoccinelline, a tricyclic defensive compound secreted by adults of ladybird beetles (Coccinellidae),[26] along with related compounds, has been identified in skin extracts of dendrobatid frogs and bufonid toads.[14] Compounds identical to the 2,6-dialkylpiperidines identified in the venoms of fire ants (*Solenopsis* species) and 2,5-dialkylpyrrolidines in thief ants in a subgenus of *Solenopsis,* have a limited distribution in dendrobatid frogs in the genera *Dendrobates* and *Epipedobates*. These possible sequestrations of ant-derived alkaloids, which can persist for months, may result in enhancing the toxicities of the frog skin secretions. On the other hand, the sequestration of the alkaloid morphine in the skin of the large toad *Bufo marinus* is unexpected unless this insectivorous amphibian ate animals that had ingested this alkaloid.[31] However, these sequestrations of animal toxins by amphibians may result in producing skin toxins with a wider range of deterrent activities, although it is hard to imagine improving on the toxicities of compounds like batrachotoxins, samandrines, and tetrodotoxin.

Sequestration of plant natural products by herbivorous insects is widespread. This is not surprising, since most insects are herbivorous and it is estimated that there are between two and six million species. While these sequestrators obviously exploit their alkaloid-rich host plants, such specialists represent a very minor component of the total herbivorous population. In essence, these nitrogenous compounds are highly effective deterrents for most of the herbivorous species with which they share their world. On the other hand, a brief discussion of some of the alkaloids and their specialist herbivores illustrates the versatiliity of these insects as exploiters of toxic compounds This alkaloidal treatment has been derived from the tables of Brown and Trigo[10] and emphasizes the fact that even adapted species can be deterred by high concentrations of their host-plant alkaloids.

Species in a relatively small number of herbivorous families dominate the list of plant feeders associated with alkaloid-rich foods. Lepidoptera (butterflies and moths) have catholic tastes when it comes to alkaloid-fortified plants, being represented by the families Nymphalidae (calystegine A-3, pyrrolizidine alkaloids: lycopsamine type, harman), Arctiidae (senecionine type), Papilionidae (synephrine, isoquinolines), and Pterophoridae (monoterpene alkaloid rhexifoline).[6] In addition, beetles (Chrysomelidae) sequester PAs (senecionine), grasshoppers (Acrididae) store senecionine, and aphids (Aphididae) sequester QAs (sparteine and diterpene alkaloids).

Clearly these insects are capable of sequestering a variety of alkaloids, belonging to different classes, from their host plants, but these animals do not necessarily store alkaloids with the same pattern as they occur in their hosts. For example, the African tiger moth, *Amphicallia bellatrix*, when reared on PA-rich *Crotolaria semperflorans*, sequesters two *Crotalaria* alkaloids, crispatine and trichodesmine. Although the major alkaloid in the plant is crosemperine, an octonecine ester, adults of *A. bellatrix* contain primarily trichodesmine in their bodies.[33]

Similarly, the cinnabar moth, *Tyria jacobaeae*, is a very efficient sequestrator of PAs, storing 8-12 times (dry weight) more alkaloids than are present in the leaves of the ragwort *Senecio jacobaea*. The great efficiency of *T. jacobaeae* as a sequestrator is demonstrated by the fact that the larval feces contain only minor amounts of *Senecio* alkaloids in contrast to the concentration of PAs that are stored in the body. For example, senecionine, which is a trace constituent in plant leaves, is the major compound stored in pupae of *T. jacobaeae*.[1]

Furthermore, insects are not necessarily passive sequestrators of pyrrolizidine alkaloids, but rather can convert these secondary compounds into novel structures that are utilized adaptively both for defense and chemical communication.

PAs SYNTHESIZED BY ANIMALS AND THEIR FUNCTIONS

Although PAs are responsible for poisonous plant toxicoses in livestock throughout the world,[32] these compounds are, in general, outstanding deterrents for a large variety of invertebrate and vertebrate herbivores. Pyrrolozidine alkaloids are mutagenic for insects and hepatotoxic and pneumotoxic for mammals, usually exhibiting a delayed toxicity that is characterized by lesions on the liver and other organs. In livestock pyrrolizidine poisoning is asymptomatic at the time of ingestion and is only detected months or years later. Human poisoning is manifested by similar symptoms in response to ingestion of herbal preparations from PA-producing plants (e.g., borage and comfrey) or honey contaminated with high concentrations of these alkaloids.[19] PAs clearly constitute alkaloidal defensive compounds *par excellence*.

PAs are found widely in the families Asteraceae, Boraginaceae, Fabaceae, and Apocynaceae. About 360 diverse structures have been characterized,[23] and these widespread alkaloids are produced by at least 3% of all flowering plants. It is remarkable that one group of animals--the insects–have broken through the formidable alkaloidal defenses represented by the highly toxic pyrrolizidine alkaloids. This evolutionary adaptation has provided a large resource for these insects that is not available to PA-intolerant species, which constitute most of the herbivores.

Hydroxydanaidal, the male courtship pheromone of tiger (Arctiidae) moths and danaidone, a PA metabolite utilized as a sex pheromone by male danaine and ithomiine butterflies, are derived from PA esters produced by the plants.[24] Moth larvae (*Creatonotus transiens*) synthesize of hydroxydanaidal from ingested pyrrolizidines after ester hydrolysis and oxidation of the primary alcohol into the aldehyde. Danaidone, the related male sex pheromone, is not produced from ingested PAs but rather from PA-containing plant tissues which the adult male butterflies scrape from dead or injured leaves.[29] This phenomenon is called pharmacophagy and refers to the selective acquisition of toxic compounds for specific utilization.[7]

Males of the moth *C. transiens* possess huge eversible (abdominal) scent organs (coremata) that disseminate the sex pheromone hydroxydanaidal. The great importance of PAs to the males is illustrated by the fact that the degree of development of the coremata is dependent upon their concentration in the larval diet.[24] Small coremata liberate considerably less sex pheromone than large organs. In addition, the concentration of sex pheromone reflects the concentration of PAs in the diet. Reduced PA levels result in males that are at a competitive disadvantage.[7]

Arctiid moths are also capable of synthesizing their own PAs from a necine base of plant origin and a necic acid synthesized from isoleucine in their own metabolism. Callimorphine, a PA not produced by plants, is synthesized by many arctiid moths and is sequestered as a defensive compound after re-esterification of retronecine of plant origin. Creatonotine, another insect PA, is produced by adults of *C. transiens* by esterifying ingested retroncine with a distinctive necic acid. These moths are unique in adding *de novo*-synthesized PAs to plant-derived alkaloids as part of their defensive arsenal.

In addition, copulating male and female arctiid moths provide PAs that are mainly tranferred to the eggs to provide protection against predators.[24] Males provide females with a copulatory "bonus" in the form of a seminal ejaculate fortified with pyrrolizidine alkaloids. The PAs contributed by the male are allocated to both the female and the eggs, supplementing the female's own PA contribution. Obviously the alkaloids play a pivotal role in ensuring reproductive success.

A lifestyle in intimate association with compounds as pernicious as pyrrolizidine alkaloids requires physiological and biochemical adaptations that are capable of blunting the pronounced toxicities of these diverse nitrogen-containing compounds. This is possible because PAs can exist in two interchangeable forms, a nontoxic *N*-oxide and its conversion to a toxic tertiary alkaloid. The pronounced toxicity of the tertiary alkaloid is realized if a herbivore ingests a plant rich in PA *N*-oxides or a predator swallows adapted insects containing PA *N*-oxides.[24] Subsequently, reduction in the gut (abnormal detoxication) of the herbivore/predator generates the highly toxic tertiary alkaloids which are absorbed passively. The tertiary PA, unlike the PA *N*-oxide, is highly toxic to organisms with an accessible microsomal multisubstrate cytochrome P-450.

ALKALOID PARSIMONY

Ecological theory would predict that the natural products produced by plants and animals would be multifunctional so as to be maximally adaptive for the producers. This phenomenon, described as alkaloid parsimony, has been detected with some frequency in spite of the fact that more appropriate *in vivo* ecological studies utilizing realistic bioassays are still required.

At least 50 alkaloids have been demonstrated to inhibit the germination or the growth of seedlings. This phenomenon, referred to as allelopathy, is important in competition between plants and is frequently directed against seedlings by mature plants of the same species.[37] A diversity of alkaloids, including quinine, cinchonine, ergotamine, harmaline, strychnine, berberine, colchicine, morphine, cocaine, caffeine, coniine, and nicotine, have been demonstrated to possess allelopathic activity, but many of the compounds possess additional ecological roles. In essence, these usually bitter and toxic alkaloids are also herbivore deterrents, and in many cases they possess antibacterial and antifungal activities.[3] These examples of alkaloid parsimony are illustrative of the mutifunctionality of these compounds and serve to focus on the great adaptiveness of these nitrogen heterocycles.

Alkaloid parsimony has also been detected in animals that produce alkaloids. Fire ants in the genus *Solenopsis* produce venoms dominated by novel 2,6-dialkylpiperidines that are delivered into vertebrates subcutaneously. The venoms constitute outstanding examples of alkaloid parsimony as they provide their ant producers with a diversity of deterrents. The alkaloids cause dermal necrosis in humans and are very algogenic in addition to being very lytic. The piperidines also perturb enzymatic pathways and block neuromuscular junctions. Fire ants have obviously adapted their piperidines to function as extreme exanples of alkaloid parsimony.

Some low molecular-weight alkaloids subserve the role of pheromones but in addition they possess at least one other function of great ecological significance. Ant workers in a variety of genera synthesize trisubstituted alkylpyrazines in their mandibular glands and these compounds generally function as alarm or alerting pheromones.[36] Workers of *Odontomachus* species frequently generate a very effective alarm signal with 2,5-dimethyl-3-isopentylpyrazine, but they also use this compound parsimoniously as an effective repellent against other ants in physical confrontations. It has also been suggested that these alkaloids may be utilized as antimicrobials in the moist environment of the nest, as is the case for the fire ant alkaloids.

PHYSIOLOGICAL AND BIOCHEMICAL TARGETS FOR ALKALOIDS

Since some alkaloids exhibit great mammalian toxicity (e.g., batrachotoxin, samandarine), it is obvious that these compounds are capable of initiating severe biochemical and physiological lesions. However. it should not be overlooked that toxicological evaluations of most alkaloids have been implemented by *in vitro* studies in which the alkaloids have been introduced subcutaneously. On the other hand, in nature, in the absence of a venom apparatus, these compounds would be introduced into the body of a vertebrate through oral administration. Thus, rejection. can reflect noxious or toxic effects on nerves and muscles of buccal tissue before alkaloids reach the intestine as a prelude to absorption into the circulatory system. It can be anticipated that toxicological differences may reflect the mode of introduction of toxic alkaloids into the body of a vertebrate. For example, the response to caffeine in laboratory studies with rats demonstrated that toxicity patterns were pronounced with bolus oral doses by intubation, whereas administration of the alkaloid in the drinking water did not result in any of the expressions of toxicity encountered with intubation.[13] However, this consideration notwithstanding, diverse studies with a variety of alkaloids have demonstrated unambiguously that these compounds perturb many physiological and biochemical systems as agents of pronounced toxicity. A brief examination of alkaloids as selective toxins emphasizes the many roles that these compounds play as deterrents.

Wink[37] has identified a considerable diversity of cellular targets that are vulnerable to the cytotoxic effects of various alkaloids. Although some of these data have been obtained. by screening these natural products as anticancer agents, there are no grounds for believing that the susceptibility of a cancer cell and a "normal" cell will be somewhat similar. Since herbivores would be characterized by "normal" cells, the production of allelochemical alkaloids by plants can be regarded as a powerful deterrent for herbivores.

A variety of alkaloids bind to or intercalate with DNA or DNA/RNA processing enzymes and affect either transcription or replication (quinine, harmane alkaloids, melinone, berberine), act at the level of DNA and RNA polymerases (vinblastine, coralyne, avicine), inhibit protein synthesis (sparteine, tubulosine, vincrastine, lupanine), attack electron chains (pseudane, capsaicin, solenopsine), disrupt biomembranes and transport processes (berbamine, ellipticine, tetrandrine), and inhibit ion channels and pumps (nitidine, caffeine, saxitoxin). In addition, these natural products attack a variety of other systems that can result in serious biochemical destabilization

Some alkaloids interfere with the assembly of microtubules (taxol, colchicine, maytansine), inhibit key enzymes such as adenylate cyclase (papaverine, theophylline, theobromine), activate neuromuscular systems involving ACH (physostigmine, coniine, nicotine), inhibit digestive processes (emetine, lobeline, morphine), modulate liver and kidney function (pyrrolizidine alkaloids, amanitine), and destabilize the blood and circulatory system (vinblastine, colchicine).

These results, while requiring considerable *in vitro* extension, document the fact that a variety of alkaloids are capable of inhibiting central processes at different levels (organismic, organ, cellular), thus causing serious biochemical and physiological lesions. Although the chemistry of these compounds and their detailed modes of action are far from complete, the results as presented clearly show that these nitrogen heterocycles are structural and pharmacological "gold mines" that ultimately affect animals significantly and often deleteriously. It would be no exaggeration to state for alkaloid researchers, "The best is yet to come."

ACKNOWLEDGMENTS

The author wishes to thank Dr. Thomas Spande and N. A. Blum for providing important technical help in the preparation of this manuscript.

REFERENCES

(1) Aplin, R.T., Benn, M. H. and Rothschild, M. **1968**. Poisonous alkaloids in the body tissues of the cinnibar moth (*Callimorpha jacobeae*). *Nature* **219**, 747-748

(2) Atta-ur-Rahman and Choudhary, M .I. **1996**. Toxic alkaloids and other nitrogenous compounds from marine plants. In: Blum, M.S., (Ed.), *Chemistry and Toxicology of Diverse Classes of Alkaloids*. Alaken Inc., Fort Collins, CO, 121-143

(3) Blum, M. S. **1996**. Semiochemical parsimony in the Arthropoda. *Annu. Rev. Ent.* **41**, 353-374

(4) Blum, M. S., Everett, D. M., Jones, T. H. and Fales, H. M. **1991**. Arthropod natural products as insect repellents. In: Hedin, P. (Ed.), *Naturally Occurring Pest Regulators*. ACS Symposium Series No. 149. American Chemical Society, Washington D.C., 14-26

(5) Blum, M. S., Rivier, L. and Plowman, T. **1981**. Fate of cocaine in the lymantriid *Eloria noyesi*, a predator of *Erythoxylum coca*. *Phytochemistry* **20**, 2499-2500

(6) Boppre, M. **1990**. Lepidoptera and pyrrolizidine alkaloids. Exemplification of complexity in chemical ecology. *J. Chem. Ecol.* **16**, 165-185

(7) Boppre, M. and Schneider, D. **1985**. Pyrrolizidine alkaloids quantitatively regulate both scent organomorphogenesis and pheromone biosynthesis in

male *Creatonotus* moths (Lepidoptera: Arctiidae). *J. Comp. Physiol. A* **157,** 569-577.

(8) Brodie, E. D., Hensel, J. L. and Johnson, D. A. **1974.** Toxicity of the urodele amphibians *Taricha, Notophthalmus, Cynops,* and *Paramesotriton*. *Copeia* 506-511

(9) Brown, K. S. **1984.** Adult obtained pyrrolizidine alkaloids defend ithomiine butterflies against a spider predator. *Nature* **309,** 707-709

(10) Brown, K. S. and Trigo, J. R, **1995.** The ecological activity of alkaloids. In: Cordell, G.A. (Ed.), *The Alkaloids: Chemistry and Pharmacology*. Academic Press, San Diego, CA, 227-354

(11) Carte, B. and Faulkner, D. J. **1986.** Role of secondary metabolites in feeding associations between a predatory nudibranch, two grazing nudibranchs, and a bryzoan. *J. Chem. Ecol.* **12,** 795-804

(12) Clay, K. **1990.** Fungal endophytes of grasses. *Annu. Rev. Ecol. Syst.* **21,** 275-290

(13) Collins, T., Welsh, J., Black, T., Whitby, K. and O'Donnell, M. **1987.** Potential reversibility of skeletal effects in rats exposed *in utero* to caffeine. *Food Chem. Toxicol.* **25,** 647-663

(14) Daly, J. W., Garraffo, H. M. and Spande, T. F. **1993.** Amphibian alkaloids. In: Cordell, G.A. (Ed.), *The Alkaloids: Chemistry and Pharmacology*. Vol. 43. Academic Press, San Diego, CA, 185-288

(15) Dawson, R. F., Christman, D. R., D'adamo, A., Solt, M. L. and Wolf, A. P. **1960.** Biosynthesis of nicotine from nicotinic acid: chemistry and radiochemical yield. *Arch. Biochem. Biophys.* **9,** 144-150

(16) Duffey, S. S., Bloem, K. A. and Campbell, B. C. **1986.** Consequences of sequestration of plant natural products in plant-insect interactions. In: Boethel, D. F. and Eikenbary, R. D. (Eds.), *Interactions of Plant Resistance and Parasitoids and Predators of Insects*. Horwood, Chichester, England. 31-60

(17) Dumbacher, J. A., Beeler, B., Garraffo, H. M., Spande, T. F. and Daly, J. W. **1993.** Homobrachotoxin in the genus *Pitohui*: chemical defense in birds? *Science* **258,** 799-801

(18) Dussourd, D. E., Ubik, K., Harvis, C., Resch, J., Meinwald, J. and Eisner, T. **1988.** Biparental defensive endowment of eggs with acquired plant alkaloid in the moth *Utethesa ornatrix*. *Proc. Natl. Acad. Sci.* **85,** 5992-5996

(19) Edgar, J. A. **2002.** Honey from plants containing pyrrolizidine alkaloids. *J. Agr. Food Chem.* **50,** 2719-2730

(20) Escoubas, P. and Blum, M. S. **1990**. The biological activities of ant-derived alkaloids. In: Vander Meer, R. K., and Cedeno, A. (Eds.), *Applied Myrmecology*. Westview Press, Boulder, CO, 482-489

(21) Gallardo, F., Boethel, D. J., Fuxa, J. R. and Richter, A. **1990**. Susceptibility of *Heliothis zea* larvae *Nomuraea rileyi* (Farlow) Samson. Effects of tomatine at the third trophic level. *J. Chem. Ecol.* **16,** 1751-1759

(22) Habermehl, G. **1966**. Chemistry and toxicology of salamander alkaloids. *Naturwissenschaften* **53,** 123-128

(23) Hartmann, T. **1996**. Diversity and variability of plant secondary metabolism: a mechanistic view. *Ent. Exper. Appl.* **80,** 177-188

(24) Hartmann, T. **1999**. Chemical ecology of pyrrolizidine alkaloids. *Planta* **207,** 483-495

(25) Hayes, E. S., Guppy, L. J. and Walker, M. J. A. **1995**. Toxic alkaloids from marine invertebrates. In: Blum, M. S. (Ed.), *The Toxic Action of Marine and Terrestrial Invertebrates*. Alaken, Inc., Fort Collins, CO, 1-52

(26) Holloway, G. J., de Jong, P. W. and Oltenheim, M. **1993**. The genetics and cost of chemical defense in the two-spot ladybird (*Adalia bipunctata*). *Evolution* **47,** 1229-1239

(27) Jones, T. H. and Blum, M. S. **1983**. Arthropod alkaloids: Distribution, functions and chemistry. In: Pelletier, S .W. (Ed.), *Alkaloids: Chemical and Biological Perspectives.* Vol. 1. John Wiley and Sons, New York, 33-84

(28) Lindigkeit, R., Biller, A., Buchus, M., Scheibel, H. M., Boppre, M. and Hartmann, T. **1997.** The two faces of pyrrolizidine alkaloids: the role of the tertiary amine and its *N*-oxide in chemical defense of insects with acquired plant alkaloids. *Eur. J. Biochem.* **245,** 626-636

(29) Meinwald, J., Meinwald, Y. C. and Mazzochi, P. H. **1969**. Sex pheromone of the queen butterfly: chemistry. *Science* **164,** 1174-1175

(30) Mothes, K., Schutte, H. R. and Luckner, M. **1976**. Secondary plant substances as materials for chemical breeding in higher plants. *Recent Adv. Phytochem.* **10,** 385-405

(31) Oka, K., Kantrowitz, J. D. and Spector, S. **1985**. Isolation of morphine from toad skin. *Proc. Natl. Acad. Sci.* **82,** 1852-1854

(32) Roitman, J. N. and Panter, K. E. **1995**. Livestock poisoning caused by plant alkaloids. In: Blum, M. S. (Ed.), *The Toxic Action of Marine and Terrestrial Alkaloids*. Alaken Inc., Fort Collins, CO, 53-124

(33) Rothschild, M., Aplin, R. T., Cockrum, P. A., Edgar, J. A., Fairweather, P.,

and Lees, R., **1979**. Pyrrolizidine alkaloids in arctiid moths (Lep.) with a discussion on host plant relationships and the role of these secondary plant substances in the Arctiidae. *Biol. J. Linn. Soc.* **12,** 305-326

(34) Sheumack, D. D., Howden, M. E. H., Spence, I. and Quinn, R. J. **1978**. Maculotoxin: a neurotoxin from the venom glands of the octopus *Hapalochlaena maculosa* identified as tetrodotoxin. *Science* **199,** 188-189

(35) Vrieling, K., Smit, W. and Van der Meijden, E. **1991**. Three-trophic interactions with pyrrolizidine alkaloids lead to general variation in PA concentrations between aphid species (*Aphis jacobaea*) and *Tyria jacobaeae*. *Oecologia* **86,** 177-182

(36) Wheeler, J. W. and Blum, M. S. **1973**. Alkylpyrazine alarm pheromones in ponerine ants. *Science* **182,** 501-503

(37) Wink, M. **1993**. Allelochemical properties or the *raison d'etre* of alkaloids. In: Cordell, G. A. (Ed.), *The Alkaloids: Chemistry and Pharmacology*. Vol. 43. Academic Press, New York, 1-118

9 Allelochemical Properties of Quinolizidine Alkaloids

M. Wink

CONTENT

Abstract 183
Introduction 184
    Evolution and Function of Secondary Metabolites 184
Methods to Study Allelochemical Interactions at the Molecular Level 188
    Interaction of Alkaloids with DNA 188
    Inhibition of Protein Biosynthesis 190
    Influence of Alkaloids on Membrane Permeability 190
    Membrane Preparation for Receptor Binding Studies 190
    Radio Receptor Binding Assays 191
Functions and Molecular Modes of Actions of QA 192
    Pharmacological Properties of QA 194
Conclusions 195
References 195

ABSTRACT

For quinolizidine alkaloids (QA) the major molecular targets and the likely modes of action were identified. These alkaloids modulate nicotinic and muscarinic acetylcholine receptors and inhibit Na^+/K^+ channels. These findings can explain the toxic and pharmacological effects of QA in insects, vertebrates and other organisms. The experimental data provide a solid base to infer the function of QA as defense compounds against herbivores, microorganisms and competing plants. Field data with alkaloid-rich and alkaloid-free lupins clearly show that quinolizidine alkaloids are, indeed, the crucial factor controlling herbivory under natural conditions.

INTRODUCTION

Secondary metabolites (SM) show large structural diversity and more than 80.000 structures have been described from plants, over 20.000 from microorganisms and fungi, more than 20.000 from amphibia, reptiles, arthropods, and marine organisms. These compounds have been isolated and their structures determined by mass spectrometry (EI-MS, CI-MS, FAB-MS), nuclear magnetic resonance (^{1}H-NMR, ^{13}C-NMR) or X-ray diffraction.[9,50,51] It is likely that many more compounds will be found in the future, as only a small number of existing organisms have been thoroughly analyzed.

EVOLUTION AND FUNCTION OF SECONDARY METABOLITES

Several SM have been used by mankind for thousands of years[22,27] as dyes (e.g., indigo, shikonine), flavors (e.g., vanillin, capsaicin, mustard oils), fragrances (e.g., rose oil, lavender oil and other essential oils), stimulants (e.g., caffeine, nicotine, ephedrine), hallucinogens (e.g., morphine, cocaine, mescaline, hyoscyamine, scopolamine, tetrahydrocannabinol), insecticides (e.g., nicotine, piperine, pyrethrin), vertebrate and human poisons (e.g., coniine, strychnine, aconitine) and even therapeutic agents (e.g., atropine, quinine, cardenolides, codeine, etc.).

The putative functions of SM have been discussed controversially. For many years, SM were considered to be waste products or otherwise functionless metabolites. Alternatively, it was argued 110 years ago by E. Stahl in Jena (Germany) that secondary metabolites serve as defense compounds against herbivores. This hypothesis has been elaborated during the last decades[7,8,20,37] and a large body of experimental evidence supports the following concept.[2,9,46,47,49] Functions of SM include

- Defense against herbivores (insects, vertebrates)
- Defense against fungi and bacteria, and viruses
- Defense against other plants competing for light, water, and nutrients
- Signal compounds to attract pollinating and seed dispersing animals
- Signals for communication between plants and symbiotic microorganisms (N-fixing Rhizobia or mycorrhizal fungi)
- Protection against UV-light or other physical stress factors

Because of the large number of SM, determining the functions of most compounds is rather speculative and needs experimental support. Among the better studied groups are pyrrolizidine[3,11] and the quinolizidine alkaloids (QA). This article will mainly deal with our own studies that were carried out to understand the physiology of QA formation in legumes and their ecological functions.

Occurrence of quinolizidine alkaloids in legumes

Legumes are able to fix atmospheric nitrogen *via* symbiotic *Rhizobia* in root nodules. Thus, nitrogen is easily available for secondary metabolism, and it is probably not surprising that nitrogen-containing SM (alkaloids, nonprotein amino acids [NPAAs], cyanogens, protease inhibitors, lectins) are a common theme in legumes.[10,36]

Quinolizidine alkaloids figure as the most prominent group of alkaloids in legumes, present in members of "genistoid alliance s.l." of the subfamily Papilionoideae including the tribes Genisteae, Crotalarieae, Podalyrieae/Liparieae, Thermopsideae, Euchresteae, Brongniartiaeae, and Sophoreae.[15,48] Also ,dipiperidine alkaloids of the ammodendrine type, which also derive from lysine as a precursor, exhibit a comparable distribution pattern.

As can be seen from Figure 9.1, most taxa of the genistoid alliance accumulate QAs. An obvious exception are members of the large tribe Crotalarieae, that either sequester PAs and/or NPAAs. In *Lotonis* some taxa produce QAs, others PAs. Since *Crotalaria* and *Lotonis* derive from ancestors which definitely produced QAs but not PAs, we suggest that the genes encoding biosynthetic enzymes of QA formation must still be present. It is unlikely that corresponding genes have been lost. More likely the QA genes have been turned off in *Crotalaria* and partially in *Lotononis*. The formation of PAs instead appears to be a new acquisition for chemical defense, which probably evolved independently. In a few other taxonomic groups which cluster within QA accumulating genera, QAs are hardly detectable or levels are very low, such as in *Ulex, Calicotome*, or *Spartocytisus*. These taxa have extensive spines in common that apparently have substituted chemical defense; in this case, the presence or absence of QAs is clearly a trait reflecting rather different ecological strategies than taxonomic relationships.

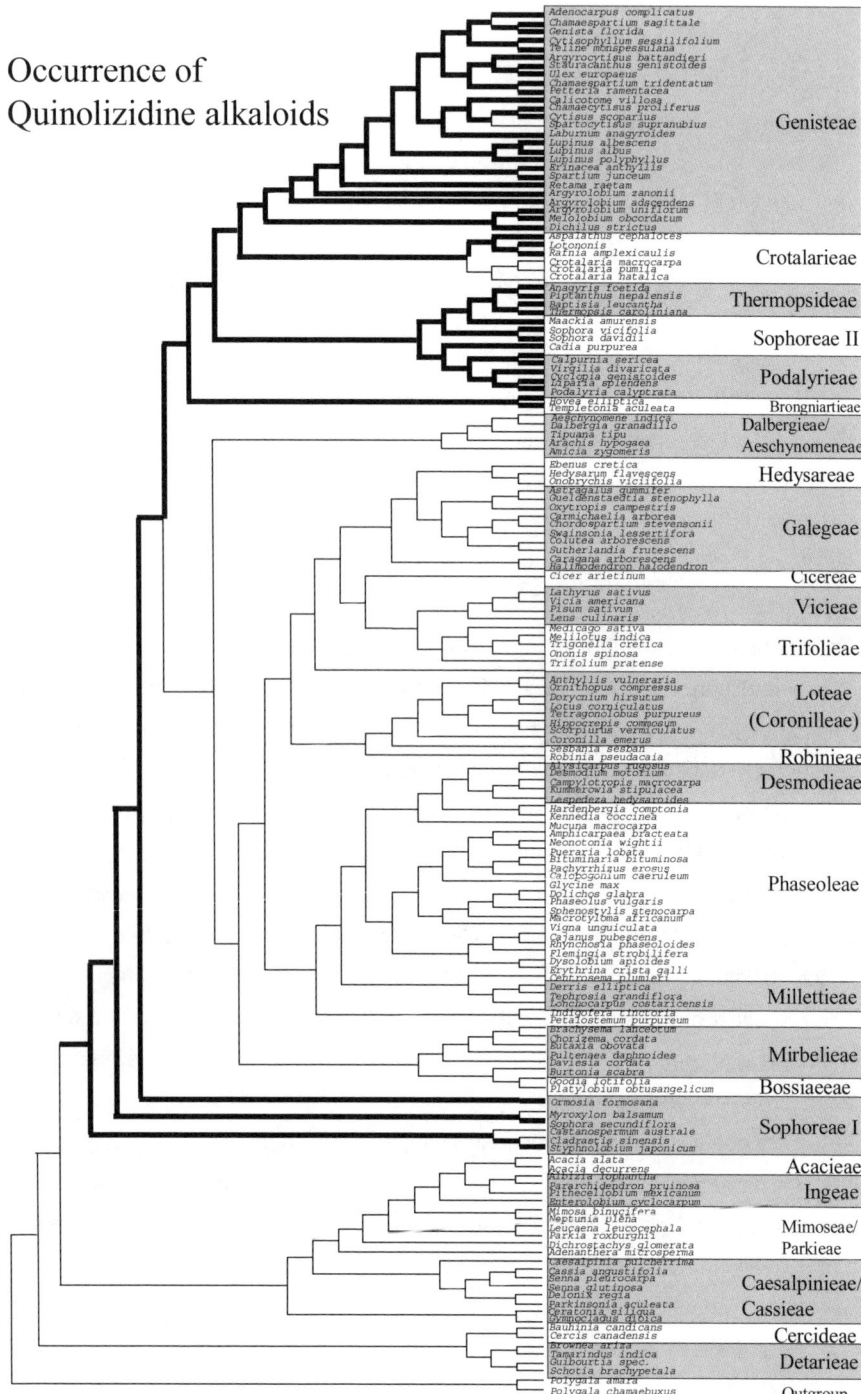

Figure 9.1

Distribution of alkaloids in legumes. Branches leading to taxa that accumulate QA are printed in bold.

Within the genistoid alliance, all taxa (except the few examples mentioned before) produce alkaloids of the sparteine/lupanine type, at least as minor alkaloids. α-Pyridone alkaloids, such as anagyrine and cytosine, are apparently already present in the more ancestral tribes of the Papilionoideae, but also in the more advanced *Cytisus/Genista* complex of the Genisteae, suggesting that already the ancestor of genistoids must have possessed the biosynthetic capacity to produce these alkaloids. This suggests also that the pathway leading to α-pyridone alkaloids is present at the genomic level in the early stages of legume phylogeny, but not expressed in most advanced taxa.[59]

QA formation in plants

QA biosynthesis from lysine *via* cadaverine to sparteine/lupanine takes place in the aerial green parts of legumes. Its intracellular site is the chloroplast.[53] Biosynthesis is regulated by light; thus QA concentrations display a diurnal rhythm, with a stimulated production period during the day.[60] QA are translocated by the phloem all over the plant,[60,62] and they accumulate predominantly in epidermal and subepidermal cell layers (Fig. 9.2).[44,53] The subcellular site of QA storage is the vacuole, into which the alkaloids are pumped by a selective carrier system.[24,56] Especially rich in alkaloids are the seeds, which can store up to 8% of their dry weight as alkaloids. During germination the alkaloids are translocated from the cotyledons to the newly formed tissues, where the alkaloids are partly degraded, obviously serving as nitrogen storage compounds.[61] QA also disappear from senescing leaves during the vegetation period, indicating that they are not end products but metabolically mobile compounds.[61]

Function of QA

If we want to understand the function of a particular group of compounds, we must study their interactions with potential molecular targets. Some useful methods are described in the following paragraph.

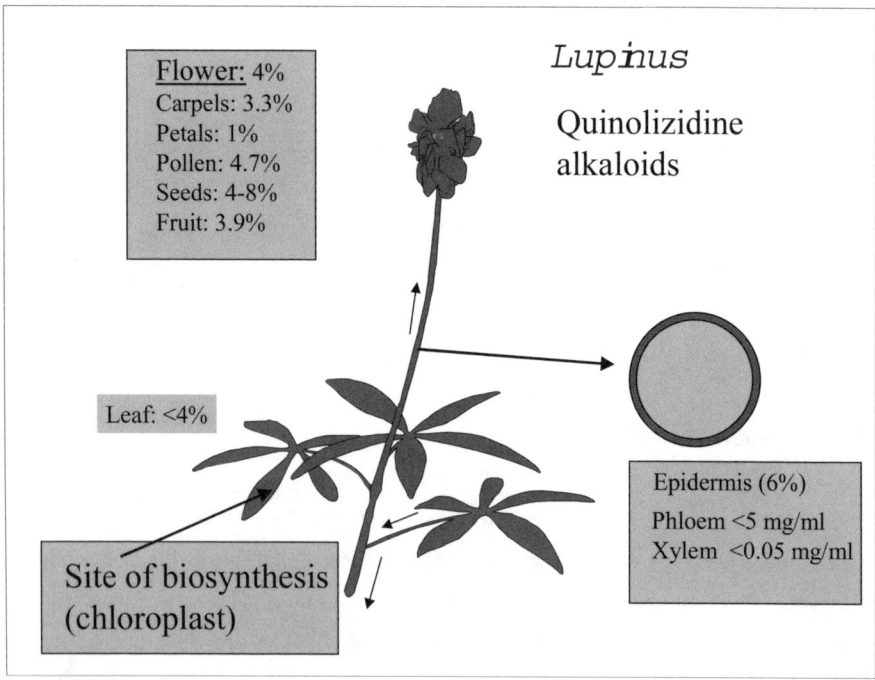

Figure 9.2

Biosynthesis, transport and storage of quinolizidine alkaloids in lupins. Alkaloid concentrations are given as % dry weight.

METHODS TO STUDY ALLELOCHEMICAL INTERACTIONS AT THE MOLECULAR LEVEL

A number of assays were established in our laboratory to determine the interaction of alkaloids with DNA and related enzymes, with biomembranes, protein biosynthesis, and neuroreceptors. These assays were all optimized and standardized in terms of linearity, reproducibility, sensitivity and specificity.[19,35,57]

INTERACTION OF ALKALOIDS WITH DNA

l. Melting point determination. If compounds intercalate with DNA, then the melting point is shifted to higher temperatures.[21,25] 70 µM *Sinapis* DNA was incubated in TE-buffer (pH 7.4) with 70 µM alkaloids for 30 min at 22°C. Then the temperature was increased by 1°C/min to 90 °C and the absorption was continuously determined in a spectrometer at 256 nm. Differences between two consecutive measurements were plotted, to determine the kinetics of the process.[19]

II. Methylgreen assay. Methylgreen (MG) binds to DNA and bound MG displays an absorption maximum at 642 nm, whereas free MG shows no absorption at this wavelength.[4,18] When alkaloids bind or intercalate with DNA, then MG is released, which can be measured as a decrease of optical density at 642 nm. 70 µM DNA-methylgreen (Sigma, USA) was incubated in the dark in 20 mM Tris-HCl (pH 7.4) together with up to 5 mM alkaloids. After 24 h the OD_{642} of untreated controls and treated samples was determined.[19]

Inhibition of DNA-Polymerase I. To determine the activity of DNA-polymerase I, we modified a "nick translation assay".[30] The assay buffer contained 50 mM Tris-HCl (pH 7.5), 10 mM $MgSO_4$, 0.1 mM DDT, 500 ng of a linearized plasmid (pUC19), 625 µM dNTPs, 0.01 µCi α^{32}P-dCTP, 1 U DNA-polymerase I, 25 pg DNAse I and up to 10 mM alkaloids. The reaction was started by adding DNAse I; after 15 min at 37°C, the reaction was terminated by adding 100 mM EDTA (pH 8.0). Two variations were carried out: 1) a preincubation of DNA polymerase I with alkaloids for 15 min, prior to adding plasmid DNA, and 2) a preincubation of DNA and alkaloids for 15 min before adding the enzymes in order to differentiate between alkaloidal effects on DNA polymerase I and on DNA. The incorporated radioactivity was removed from the nonincorporated $\alpha[^{32}P]$-dCTP by gelfiltration on Sephadex G 50 (Pharmacia) and measured in a liquid scintillation counter.[19]

Inhibition of reverse transcriptase (RT). To measure the activity of reverse transcriptase, a protocol for the synthesis of cDNA was modified,[30] and mRNA was isolated from rat liver according to standard protocol.[30] mRNA (500 ng) and 500 ng random primer (Boehringer Mannheim) were denatured at 70°C for 5 min and immediately cooled afterwards in ice-water. Then 0.3 mM dNTPs, 0.01 µCi $\alpha[^{32}P]$-dCTP, 6 U AMV reverse transcriptase (Promega) in RT buffer (50 mM Tris-HCl, pH. 7.8, 10 mM $MgCl_2$, 80 mM KCl, 10 mM DTT) were added and incubated 30 min at 42°C. The reaction was terminated by adding 100 mM EDTA; the incorporation of $\alpha[^{32}P]$-dCTP was measured as described before in the DNA-polymerase assay. Again, two preincubation strategies as previously described were employed.[19]

INHIBITION OF PROTEIN BIOSYNTHESIS

An *in vitro* reticulocyte translation assay (Boehringer Mannheim) was modified to determine an inhibition of translation by alkaloids. An assay (total volume 25 µl) contained 2 µl 12.5 x translation mix (Boehringer), 10 µl reticulocyte lysate, 200 mM K-acetate, 1.5 mM Mg-acetate, 0.25 µCi L-[4,5-^{3}H(N)]-leucine, 0.5 µg TMV-RNA (Boehringer) and up to 5 mM alkaloids (buffered to pH 7). The mixture was incubated at 30 °C; reactions were terminated after 0, 10, 20, 30 and 40 min. The radiolabeled protein was precipitated by adding 200 µl ice-cold trichloroacetic acid (TCA) (50%; w/v) and, after 30 min, filtered through GF 34 filters (Schleicher-Schüll), which binds proteins. After washing the filters three times with 50% TCA, they were dried at 85 °C. Radioactivity of the filters was determined in a liquid scintillation counter.[19]

INFLUENCE OF ALKALOIDS ON MEMBRANE PERMEABILITY

Sheep erythrocytes were purified and incubated in 50 µl PBS (8 g NaCl, 0.2 g KCl, 1.44 g Na_2HPO_4, 0.24 g KH_2PO_4 in 1 l H_2O). Erythrocytes were incubated for 15 min at 10°C together with up to 5 mM alkaloids. Then erythrocytes were precipitated by centrifugation (4 min at 2000 g), and the hemoglobin released from erythrocytes was determined photometrically at 543 nm.[19]

MEMBRANE PREPARATION FOR RECEPTOR BINDING STUDIES

Porcine brains, which were obtained within 30 min after death of the animals from a local slaughterhouse, were used to prepare receptor-rich membranes. The brains were immediately frozen in liquid N_2; 50 g brain per 200 mL ice-cold buffer (0.32 M sucrose, 10 mM potassium phosphate buffer, pH 7.0, 1 mM EDTA) were homogenized twice for 15 sec in a blender and then for 1 min with an ultraturrax. The homogenate was centrifuged three times for 15 min at 1.400 g and 4°C to separate cellular debris. The supernatant was spun down at 100.000 g for 60 min. The resulting pellet was resuspended in buffer (as above but without sucrose). Aliquots were stored frozen at -80°C. Protein content was determined by the Lowry method, using bovine serum albumin as a standard.[32-35]

RADIO RECEPTOR BINDING ASSAYS

Binding assays (in triplicates) were performed using a rapid filtration technique.[32-35]

Muscarinic receptor (mAChR). Membrane preparations adjusted to 500 µg protein in a final volume of 500 µL buffer were incubated with [^{3}H]-quinuclidinyl benzilate (QNB) (52.3 Ci/mmol; Dupont NEN) for 1 h at 20° in the absence and presence of alkaloids, employing 20 µM atropine as a positive control substance. The incubation was stopped with 3 mL ice-cold 0.9% NaCl-solution and filtered (by suction) through Whatman GF/C glass fiber filters. The filters were washed three times with 3 ml 0.9% NaCl, placed in vials, and dried for 30 min at 60°C. Their radioactivity was measured in a liquid scintillation counter (RackBeta, Pharmacia) using "Ultima-Gold" (Packard) as the scintillation cocktail.

Nicotinic receptor. [^{3}H]-Nicotine (85 Ci/mmol; Amersham) was used to assay specific binding of alkaloids to the nicotinic ACh receptor (nAChR). The membrane preparation was incubated for 40 min with differing concentrations of alkaloids or 1 mM nicotine as a positive control. The GF/C filters were presoaked with polyethylene glycol 8000 (5% in water) for 3 h to reduce nonspecific binding of [^{3}H]-nicotine. Further procedures were the same as described above for mAChR.

Alpha$_1$ receptor. [^{3}H]-Prazosine (78 Ci/mmol; DuPont NEN) was used to assay specific binding of alkaloids to the alpha$_1$ receptor. The membrane preparation was adjusted to 400 µg in a final volume of 500 µL and incubated for 45 min at 20°C with differing concentrations of alkaloids or 400 µM phentolamine as a positive control. Further procedures were the same as described above for mAChR.

Alpha$_2$ receptor. [^{3}H]-yohimbine (81 Ci/mmol; DuPont NEN) was used instead of [^{3}H]-prazosine; other conditions were the same as in the alpha$_1$ receptor assay.

Serotonin$_2$ receptor. [^{3}H]-ketanserine (85,1 Ci/mmol; DuPont NEN) was used to assay specific binding of alkaloids to the serotonin$_2$ receptor (5-HT$_2$). The membrane preparation was adjusted to 400 µg in a final volume of 500 µL and incubated for 40 min at 20°C with differing concentrations of alkaloids or 100 µM mianserine as a positive control. Further procedures were the same as described above for mAChR.

FUNCTIONS AND MOLECULAR MODES OF ACTIONS OF QA

It could be shown experimentally that pure QA or QA mixtures are biologically active.[15] They inhibit the multiplication of potato-X virus,[44] the growth of gram-negative and gram-positive bacteria,[38,41] and of certain fungi.[41,43,63] QA deter or repel feeding of a number of herbivores (nematodes, bees, caterpillars, beetles, aphids, locusts, snails, rabbits, cows) or are directly toxic (Fig. 9.3) or mutagenic to them.[1,6,13,14,23,42-44,46,47,58] In addition, QA shows a certain degree of allelopathic (i.e. phytotoxic properties) and can inhibit germination and growth of other plants.[43]

Having performed the above-mentioned bioassays with molecular targets, we assume that the toxic effects seen in animals are probably due to inhibition of Na^+ and K^+ ion channels,[17,26] to interactions with the nicotinic and muscarinic acetylcholine receptor[34,57] (Fig. 9.4) and to inhibition of protein synthesis.[16,58] Minor targets are dopamine receptor, GABA receptor, NMDA receptor, alpha 2 receptor, membrane permeability,[56,57] and DNA (anagyrine, ammodendrine).[14] The effects seen in microorganisms and plants are likely due to interactions with protein biosynthesis and membrane integrity. Considering the targets and organisms which are affected, QA can be considered as broad range allelochemical defense compounds.

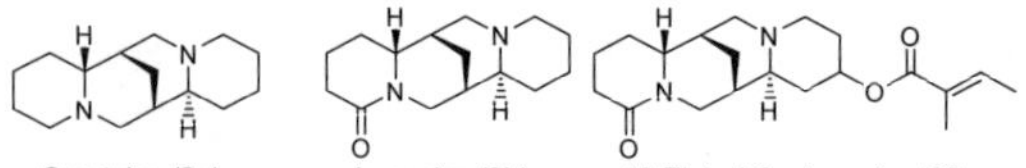

	Sparteine (3a)	Lupanine (3b)	13-Tigloyl-Oxylupanine (3f)
Vertebrate toxicity (LD50)			
(i.v., i.p.)	20-70 mg/kg	20-200 mg/kg	n.d.
(p.o.)	300-500 mg/kg	410-1464 mg/kg	n.d.
Insect toxicity (LD100)			
Ceratitis	0.2%	0.07%	0.2%
Phaedon	1%	0.3%	0.2%
Plutella	1%	0.3%	0.2%
Dysdercus	0.9%	0.3%	0.2%

- Toxicity also found in worms, molluscs and other animals
- QA exhibit some
  - antimicrobial
  - antiviral
  - phytotoxic properties

Figure 9.3

Overview over toxic properties of quinolizidine alkaloids.

Allelochemical Properties of Quinolizidine Alkaloids

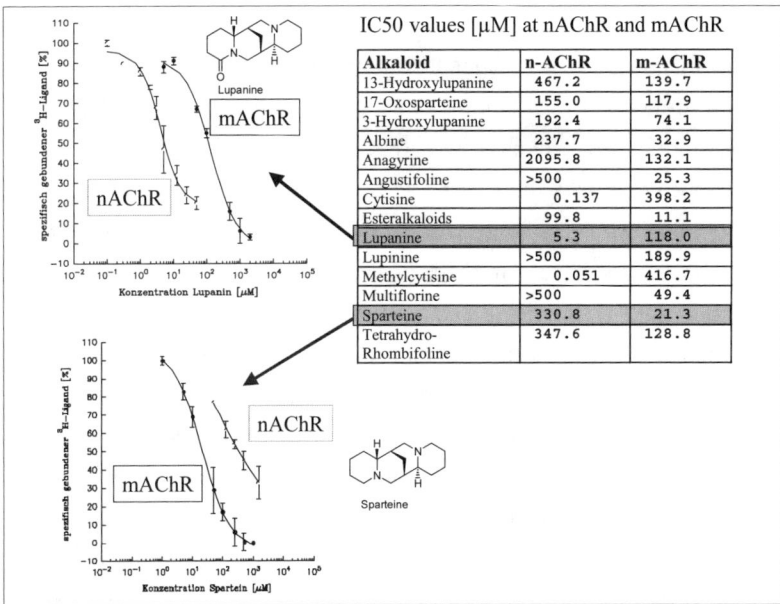

Figure 9.4

Binding of quinolizidine alkaloids to acetylcholine receptors. Binding was analyzed using radio receptor assays. (From Schmeller, T., et al. **1994**. *J. Nat. Prod.* **57**, 1316-1319; with permission).

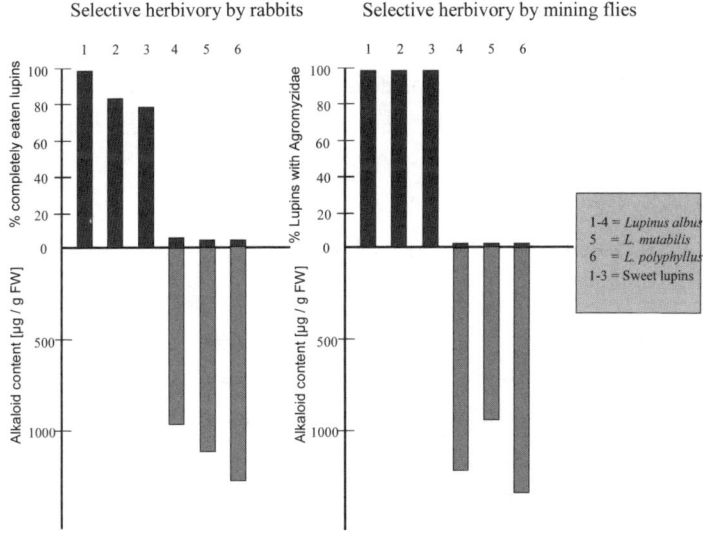

Figure 9.5

Selective advantage of lupins with alkaloids as compared to "sweet" lupins.

The concentration of QAs in the plant is sufficiently high to perform their inhibitory effects observed *in vitro*. In addition QA contents can be increased by wounding: this effect was highest under greenhouse conditions, but also measurable in the field.[12,40,43] In this context, QA localization in epidermal tissues can be interpreted as a strategically important adaptation (Fig. 9.2), since this tissue has to ward off small herbivores and pathogens in the first instance.[54]

In general, we find a series of related compounds in each plant; often a few major metabolites and several minor components which differ in the position of their substituents (Fig. 9.4). Sparteine and lupanine only differ by a single keto function, but their modes of action differ substantially (Fig. 9.2). The profile usually varies between plant organs, within developmental periods, and sometimes even diurnally (e.g., in lupin alkaloids[60]). Also marked differences can usually be seen between individual plants of a single population, even more so between members of different populations. This variation, that is part of the apparent evolutionary arms race between plants and herbivores, makes adaptations by herbivores more difficult, since even small changes in chemistry can be the base for new pharmacological activities.

Plant breeders have selected lupin mutants which produce only minute amounts of alkaloids. These so-called "sweet" lupins have a dramatically reduced fitness under natural conditions as compared to their "bitter" wild forms. They are preferentially eaten by rabbits, leaf miners (Fig. 9.5), aphids or beetles and are vulnerable to other pathogens.[5,43,44,46,47] Thus, it seems well established that the ecological function of QA is that of defense against herbivores but also against microorganisms. Some evidence suggests activity against other competing plants.[39,58]

PHARMACOLOGICAL PROPERTIES OF QA

In addition to toxic and repellent properties, lupin alkaloids have a number of pharmacological activities.[15,47,52]

Sparteine, lupanine, and other QA exhibit antiarrhythmic properties. Since only sparteine can be isolated easily from broom (*Cytisus scoparius*) it is the only lupin alkaloid that is commercially available and exploited in medicine as an antiarrhythmic drug. However, about 10% of all patients are unable to metabolize sparteine and suffer from sparteine intoxication. Because of these side effects and the availability of more reliable synthetic heart drugs, the use of sparteine in modern medicine is declining and restricted. The utilization of sparteine as a uterus contracting drug has been abandoned for the same reasons. Sparteine, lupanine and 13-hydroxylupanine have hypotensive and CNS-depressant properties and furthermore, are hypoglycemic and, thus, might be interesting antidiabetic drugs. In addition to QA, the alkaloid-free fraction of *L. albus* seeds seems to have antidiabetic activities. Matrine and cytisine are amoebicidal; matrine and 17-oxolupanine are effective inflammatory compounds. Some of these pharmacological properties can be explained through activation of the acetylcholine receptors and inhibition of K^+ and Na^+-channels.[48,52]

CONCLUSIONS

For quinolizidine alkaloids, we have identified the major molecular targets and the likely mode of action. These findings can explain the toxic and pharmacological effects of QA in insects, vertebrates and other organisms. The experimental data provide a solid base to infer the function of QA as defense compounds against herbivores, microorganisms and competing plants. Field data with alkaloid-rich and alkaloid-free lupins clearly show that our conclusions appear to be correct (Fig. 9.5). Using this information we can extrapolate, to a certain degree, the evolution and function of other groups of SM, even if particular experimental data are missing. However, it should be the aim of science to unravel the intricate evolution and interactions of all groups of secondary metabolites. Although this demands a tremendous amount of work, such an effort is worthwhile.

REFERENCES

(1) Bentley, M. D., Leonard, D. E., Reynolds, E. K., Leach, S., Beck, A. B. and Murakoshi, I. **1984** Lupine alkaloids as larval feeding deterrents for spruce budworm, *Choristoneura fumiferana* (Lepidoptera: Tortricidae). *Ann. Entomol. Soc. Am.* **77**, 398-400

(2) Bernays, E. A. and Chapman, R. F. **1994**. In: Host-plant Selection by Phytophagous Insects. Chapman & Hall, New York. 312 pp

(3) Brown, K. and Trigo, J. R. **1995**. The ecological activity of alkaloids. In: Cordell, G. A. (Ed.), *The Alkaloids*, Vol 47, 227-354

(4) Burres, N. S., Frigo, A., Rasmussen, R. R., and McAlpine, J. B. **1992**. A colorimetric microassay for the detection of agents that interact with DNA. *J.*

Nat. Prod. **55**, 1582-1587

(5) Cantot, P. and Papineau, J. **1983**. Discrimination des lupins basse teneur en alcaloides par les adultes de *Sitonia lineatus* L. (Col.; Curculionidae). *Agronomie* **3**, 937-940

(6) Dreyer, D., Jones, K. C. and Molyneux, R. J. **1985**. Feeding deterrency of some pyrrolizidine, indolizidine, and quinolizidine alkaloids towards pea aphid (*Acyrthosiphon pisum*) and evidence for phloem transport of the indolizidine alkaloid swainsonine. *J. Chem. Ecol.* **11**, 1045-1051

(7) Ehrlich, P. R. and Raven, P. H. **1964**. Butterflies and plants: a study of coevolution. *Evolution* **18**, 586-608

(8) Fraenkel, G. **1959**. The raison d'être of secondary substances. *Science* **129**, 1466-1470

(9) Harborne, J. B. **1993**. In: *Introduction to Ecological Biochemistry*. 4th. Ed., Academic Press, London

(10) Harborne, J. B., Boulter, D., and Turner, B. L. **1971**. In: *Chemotaxonomy of the Leguminosae*. Academic Press, London

(11) Hartmann, T. and Witte, L. **1995**. Chemistry, biology and chemoecology of the pyrrolizidine alkaloids. In: Pelletier, S. W. (Ed.), *Alkaloids: Chemical and Biological Perspectives*, Vol. 9., Pergamon, Oxford, 155-233

(12) Johnson, N. D., Rigney, L., and Bentley, B. L. **1989**. Short-term induction of alkaloid production in lupines. Differences between N2-fixing and nitrogen-limited plants. *J. Chem. Ecol.* **15**, 2425-2434

(13) Johnson, N. D. and Bentley, B. B. **1988**. Effects of dietary protein and lupine alkaloids on growth and survivorship of *Spodoptera eridania*. *J. Chem. Ecol.* **14**, 1391-1403

(14) Keeler, R. F. **1976**. Lupin alkaloids from teratogenic and nonteratogenic lupins: III. Identification of anagyrine as the probable teratogen by feeding trials. *J. Toxicol. Envir. Health* **1**, 887-898

(15) Kinghorn, A. D. and Balandrin, M. F. **1984**. Quinolizidine alkaloids of the Leguminosae: Structural types, analysis, chemotaxonomy, and biological activities. In: Pelletier, S. W. (Ed.), *Alkaloids: Chemical and Biological Perspectives*, Wiley, New York, 105-148

(16) Korcz, A., Markiewicz, M., Pulikowska, J., and Twardowski, T. **1987**. Species-specific inhibitory effects of lupine alkaloids on translation in plants. *J. Plant Physiol.* **128**, 433-442

(17) Körper, S., Wink, M., and Fink, R. A. **1998**. Differential effects of alkaloids on

sodium currents of isolated single skeletal muscle fibres. *FEBS Lett.* **436**, 251-255

(18) Krey, A. K. and Hahn, F. E. **1969**. Berberine: complex with DNA. *Science* **166**, 755-757

(19) Latz-Brüning, B. **1994**. Molekulare Wirkmechanismen von Alkaloiden. PhD dissertation, University of Heidelberg

(20) Levin, D. A. **1976**. The chemical defences of plants to pathogens and herbivores. *Annu. Rev. Ecol. Syst.* **7**, 121-159

(21) Maiti, M., Nandi, R., and Chaudhuri, K. **1982**. Sanguinarine: a monofunctional intercalating alkaloid. *FEBS Lett.* **142**, 280-284

(22) Mann, J. **1992**. In: *Murder, Magic and Medicine*. Oxford University Press, London

(23) Matsuda, K., Kimura, M., Komai, K., and Hamada, M. **1989**. Nematicidal activities of (-)-N-methylcytisine and (-)-anagyrine from *Sophora flavescens* against Pine wood nematodes. *Agric. Biol. Chem.* **53**, 2287-2288

(24) Mende, P. and Wink, M. **1987**. Uptake of the quinolizidine alkaloid lupanine by protoplasts and isolated vacuoles of suspension-cultured Lupinus polyphyllus cells. Diffusion or carrier-mediated transport? *J. Plant Physiol.* **129**, 229-242

(25) Nandi, R. and Maiti, M. **1985**. Binding of sanguinarine to desoxyribonucleic acid of differing base composition. *Biochem. Pharmacol.* **34**, 321-324

(26) Paolisso, G., Nenquin, M., Schmeer, W., Mathot, F., Meissner, H. P., and Henquin, J. C. **1985**. Sparteine increases insulin release by decreasing K^+ permeability of the B-cell membrane. *Biochem. Pharmacol.* **34**, 2355-2361

(27) Roberts, M. F. and Wink, M. **1998**. In: *Alkaloids-Biochemistry, Ecological Functions and Medical Applications*. Plenum, New York

(28) Rosenthal, G. A. and Berenbaum, M. R. **1991** In: *Herbivores: their Interactions with Secondary Plant Metabolites. Vol. **1**. The Chemical Perticipants*. Academic Press, San Diego

(29) Rosenthal, G. A. and Berenbaum, M. R. **1991** In: *Herbivores: their Interactions with Secondary Plant Metabolites. Vol. **2**. Ecological and Evolutionary Processes*. Academic Press, San Diego

(30) Sambrook, J., Fritsch, E. F., and Maniatis, T. **1989**. In: *Molecular Cloning: a Laboratory Manual*. Cold Spring Harbour Labs: New York, NY

(31) Schmeller, T. **1995**. Interaktionen von Alkaloiden mit Neurotransmitter-Rezeptoren. PhD. Dissertation, University of Heidelberg

(32) Schmeller, T., El-Shazly, A., and Wink, M. **1997**. Allelochemical activities of

pyrrolizidine alkaloids: interactions with neuroreceptors and acetylcholine related enzymes. *J. Chem. Ecol.* **23**, 399-416

(33) Schmeller, T., Latz-Brüning, B., and Wink, M. **1997**. Biochemical activities of berberine, palmatine and sanguinarine mediating chemical defence against microorganisms and herbivores. *Phytochemistry* **44**, 257-266

(34) Schmeller, T., Sauerwein, M., Sporer, F., Müller, W. E., and Wink, M. **1994**. Binding of quinolizidine alkaloids to nicotinic and muscarinic receptors. *J. Nat. Prod.* **57**, 1316-1319

(35) Schmeller, T., Sporer, F., Sauerwein, M., and Wink, M. **1995**. Binding of tropane alkaloids to nicotinic and muscarinic receptors. *Pharmazie* **50**, 493-495

(36) Southon, I. W. **1994**. *Phytochemical Dictionary of the Leguminosae.* Chapman & Hall, London

(37) Swain, T. **1974**. Secondary compounds as protective agents. *Ann. Rev. Plant Physiol.* **28**, 479-501

(38) Tyski, S., Markiewicz, M., Gulewicz, K. and Twardowski, T. **1988**. The effect of lupin alkaloids and ethanol extracts from seeds of *Lupinus angustifolius* on selected bacterial strains. *J. Plant Physiol.* **133**, 240-242

(39) Wink, M. **1983**. Inhibition of seed germination by quinolizidine alkaloids. Aspects of allelopathy in *Lupinus albus* L. *Planta* **158**, 365-368

(40) Wink, M. **1983**. Wounding-induced increase of quinolizidine alkaloid accumulation in lupin leaves. *Z. Naturforsch.* **38c**, 905-909

(41) Wink, M. **1984**. Chemical defence of Leguminosae. Are quinolizidine alkaloids part of the antimicrobial defence system of lupins? *Z. Naturforsch.* **39c**, 548-552

(42) Wink, M. **1984**. Chemical defense of lupins. Mollusc-repellent properties of quinolizidine alkaloids. *Z. Naturforsch.* **39c**, 553-558

(43) Wink, M. **1985**. Chemische Verteidigung der Lupinen: Zur biologischen Bedeutung der Chinolizidinalkaloide. *Plant Syst. Evol.* **150**, 65-81

(44) Wink, M. **1987**. Chemical ecology of quinolizidine alkaloids. In: Waller, G.R., (Ed.), *Allelochemicals. Role in Agriculture and Forestry*, ACS Symposium Series 330; American Chemical Society, Washington, DC, 524-533

(45) Wink, M. **1987**. Site of lupanine and sparteine biosynthesis in intact plants and in vitro organ cultures. *Z. Naturforsch.* **42**, 868-872

(46) Wink, M. **1988**. Plant breeding: importance of plant secondary metabolites for protection against pathogens and herbivores. *Theor. Appl. Gen.* **75**, 225-

233
(47) Wink, M. **1992**. The role of quinolizidine alkaloids in plant insect interactions. In: Bernays, E. A. (Ed.), *Insect-Plant Interactions*, Vol. IV. 133-169

(48) Wink, M. **1993**. Allelochemical properties and the raison d'être of alkaloids. In: Cordell, G. (Ed.), *The Alkaloids*, Vol. **43,** Academic Press, Orlando, 1-118

(49) Wink, M. **1993**. Quinolizidine alkaloids. In: Waterman, P. G. (Ed.), *Methods in Plant Biochemistry*. Academic Press, London, 197-239

(50) Wink, M. **1999**. *Biochemistry of Plant Secondary Metabolism. Annual Plant Reviews* Vol. 2. Sheffield Academic Press and CRC Press, 358 pp.

(51) Wink, M. **1999**. *Function of Plant Secondary Metabolites and their Exploitation in Biotechnology. Annual Plant Reviews*. Vol. 3. Sheffield Academic Press and CRC Press, 362 pp.

(52) Wink, M. **2000**. Interference of alkaloids with neuroreceptors and ion channels. In: Atta-Ur-Rahman (Ed.). *Bioactive Natural Products*. Vol 11. Elsevier, pp. 3-129

(53) Wink, M. and Hartmann, T. **1982**. Localization of the enzymes of quinolizidine alkaloid biosynthesis in leaf chloroplast of *Lupinus polyphyllus*. *Plant Physiol*. **70,** 74-77

(54) Wink, M., Heinen, H. J., Vogt, H. and Schiebel, H. M. **1984**. Cellular localization of quinolizidine alkaloids by laser desorption mass spectrometry (LAMMA 1000). *Plant Cell Rep*. **3,** 230-233

(55) Wink, M. and Latz-Brüning, B. **1995**. Allelopathic properties of alkaloids and other natural products: Possible modes of action. In: Inderjit, K.M.M. Dakshini and F.A. Einhellig, (Eds), *Allelopathy. Organisms, Processes and Applications*. ACS Symposium Series 582, American Chemical Society, Washington, DC, 117-126

(56) Wink, M. and Mende, P. **1987**. Uptake of lupanine by alkaloid-storing epidermal cells of Lupinus polyphyllus. *Planta Medica* **53**, 465-469

(57) Wink, M. and Schmeller, T., and Latz-Brüning, B. **1998**. Modes of action of allelochemical alkaloids: interaction with neuroreceptors, DNA and other molecular targets. *J. Chem. Ecol.* **24**, 1881-1937

(58) Wink, M. and Twardowski, T. **1992**. Allelochemical properties of alkaloids. Effects on plants, bacteria and protein biosynthesis. In: Rizvi, S. J. H. and Rizvi, V. (Eds), *Allelopathy: Basic and Applied Asspects*. Chapman & Hall, London, 129-150.

(59) Wink, M. and Waterman, P. **1999**. Chemotaxonomy in relation to molecular phylogeny of plants. In: Wink, M. (Ed.), *Biochemistry of Plant Secondary*

Metabolism. *Annual Plant Reviews*. Vol. 2. Sheffield Academic Press and CRC Press, 300-341

(60) Wink, M. and Witte, L. **1984**. Turnover and transport of quinolizidine alkaloids: diurnal variation of lupanine in the phloem sap, leaves and fruits of *Lupinus albus* L. *Planta* **161**, 519-524

(61) Wink, M. and Witte, L. **1985**. Quinolizidine alkaloids as nitrogen source for lupin seedlings and cell suspension cultures. *Z. Naturforsch.* **40c**,767-775.

(62) Wink, M. and Witte, L. **1991**. Storage of quinolizidine alkaloids in *Macrosiphum albifrons* and *Aphis genistae* (Homoptera: Aphididae). *Entomol. Gener.* **15**, 237-254

(63) Wippich, C. and Wink, M. **1985**. Biological properties of alkaloids. Influence of quinolizidine alkaloids and gramine on the germination and development of powdery mildew, *Erysiphe graminis* f. sp. hordei. *Experientia* **41**,1477-1478

10 Mode of Action of Phytotoxic Terpenoids

S. O. Duke and A. Oliva

CONTENT

Abstract 201
Introduction 202
Results and Discussion 202
    Monoterpenes 202
    Sesquiterpenes 205
    Diterpenes 208
    Triterpenes and their Derivatives 210
Methodology 211
References 212

ABSTRACT

The actual molecular site of action of few phytotoxic terpenoids is known. General physiological effects of representatives of various terpenoid categories and methods used to focus mode of action studies on their potential molecular targets are discussed. In particular, methods such as complementation studies, determining effects on mitosis, and measuring effects on plasma membrane integrity are described. The mode of action of 1,4-cineole and how it was discovered are provided in detail. This monoterpene and a metabolic conversion product of its synthetic herbicide analogue, cinmethylin, are potent inhibitors of asparagine synthetase. The sesquiterpene lactone, artemisinin, has an apparently unique yet still unknown mode of action. The sesquiterpene lactone, dehydrozaluzanin C, appears to be generally cytotoxic, causing rapid plasma membrane disruption, perhaps through interaction with sulfhydryl groups of membrane proteins. Quassinoids appear to have a different but still unknown mode of action in plants than in mammalian cells. Saponins may be phytotoxic through their effects on membrane lipids or by effects on specific enzymes. In summary, much more work needs to be done on the mode of action of phytotoxic terpenoids.

INTRODUCTION

Natural products, through evolution of the producing organisms, have many different biological activities that are manifested through an array of different molecular target sites. Thus, phytotoxic naturally occurring compounds provide opportunities for the discovery of new and useful herbicide target sites.[15,17] So far, there is relatively little overlap between the molecular target sites of commercial herbicides and those known for natural phytotoxins.[14,17] Although a relatively large number of highly phytotoxic allelochemicals are derived from the terpenoid pathway,[31,32] the mode of action of few of these phytotoxins is well understood.

In this review, we provide what we consider the most important information on the mode of action of phytotoxic terpenoids. Information on the methods used in some of these studies is provided, excluding details of simple growth studies. There are thousands of potential molecular target sites of phytotoxins, so scientists must have a strategy to reduce the number of possible target sites that must be examined. A recent review contains the general approach used by our laboratory and a few others to accomplish this.[7] By the time one identifies the actual molecular target site the methods needed are often esoteric and quite specific for the particular molecular site, and are unlikely to be useful in other mode of action studies. Therefore, we will not provide details on methods at this level. One must go to the literature to find the appropriate methods for the molecular site indicated by more general experiments.

RESULTS AND DISCUSSION

MONOTERPENES

A large number of monoterpenes have been reported to be phytotoxic, and they have been proposed as potential starting structures for herbicides.[44] Nevertheless, of the monoterpenes, we only know anything of significance about the mode of action of the cineoles. Early work showed that relataively high concentrations of 1,8-cineole (Fig. 10.1) inhibit mitochondrial respiration of isolated organelles.[28] The high concentration used makes this mechanism of action suspect as a primary site of action. Both 1,8-cineole and 1,4-cineole (Fig. 10.1) are strong growth inhibitors, but only 1,8-cineole strongly inhibits all stages of mitosis (Fig. 10.2).[39] Swollen root tips have also been reported in 1,8-cineole-treated onion root tips.[27] Also, 1,4-cineole causes growth abnormalities in shoots, such as helical growth, that 1,8-cineole does not cause. Thus, despite the similar structures, the two compounds apparently have different modes of action. Camphor (Fig. 10.1) has effects on mitosis and respiration that are similar to those of 1,8-cineole.[27, 28]

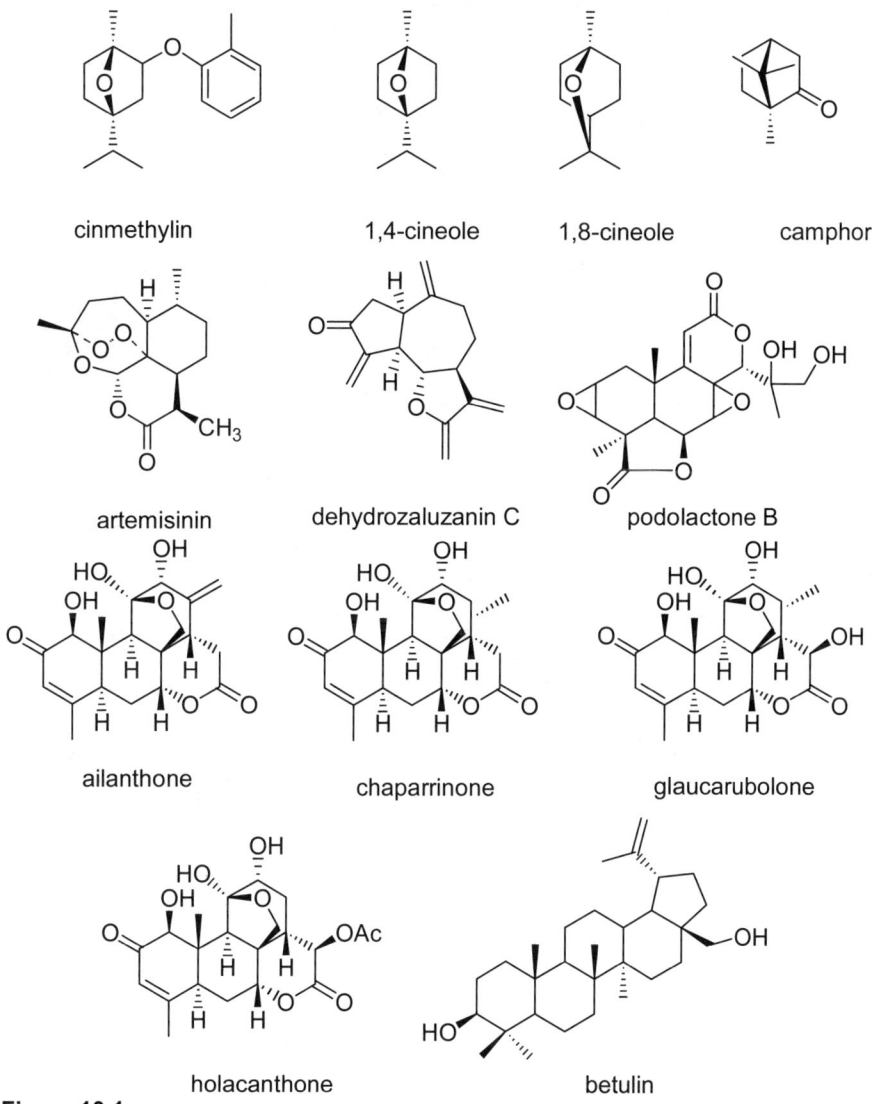

Figure 10.1
Structures of some of the compounds mentioned in the text.

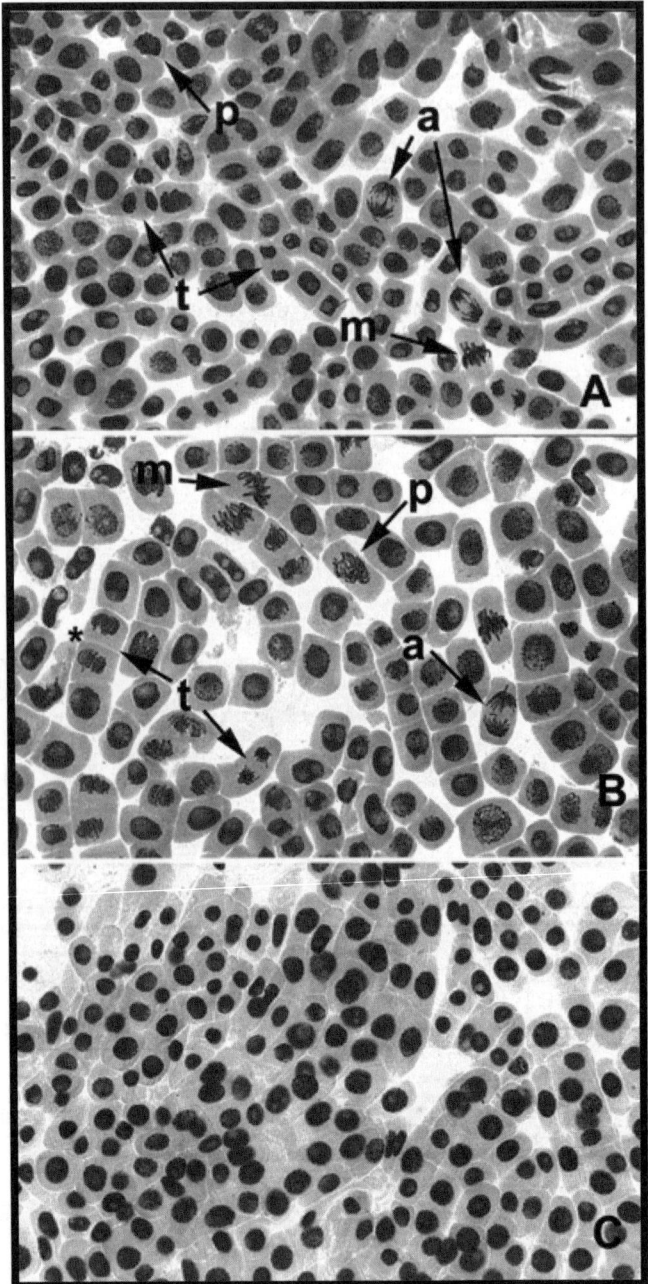

Figure 10.2

Micrographs of (A) control, (B) 1,4-cineole-, and (C) 1,8-cineole-treated onion root tip cells. p = prophase, m = metaphase, a = anaphase, and t = telophase. Asterisk indicates formation of a cell plate. (From Romagni, J. G. et al. **2000**, *J. Chem. Ecol.* **23**, 303-313. With permission).

The mode of action of 1,4-cineole is perhaps the most clearly understood of a terperoid as a phytotoxin. Using the procedure for determination of a mode of action described in Dayan et al.,[7] the molecular target site of 1,4-cineole was determined to be asparagine synthetase.[40] The critical experiment that pointed to this critical enzyme was a series of growth reversal studies in which the effects of different amino acids on phytotoxicity of 1,4-cineole and its herbicidal analogue, cinmethylin (Fig. 10.1), were determined (Fig. 10.3). The physicochemical properties of cinmethylin make it much less volatile than 1,4-cineole, allowing it to be used as a commercial herbicide.

The symptoms caused by 1,4-cineole and cinmethylin are identical. The rationale behind the approach of supplementing the plant with various metabolites is that the toxic effects of inhibitors or metabolic pathways can sometimes be ameliorated by supplying the products of those pathways. This approach has been successful with other phytotoxins, such as cornexistin,[1] an apparent inhibitor of aspartate amino transferase. Asparagine provided substantial reversal of growth inhibition by both 1,4-cineole and cinmethylin (Fig. 10.3A, B). Furthermore, 1,4-cineole strongly inhibited asparagine synthetase *in vitro*, although cinmethylin had no effect at the enzyme level (Fig. 10.3C). A potential metabolite of cinmethylin, cis-2-hydroxy-1,4-cineole, is an even better inhibitor of asparagine synthetase than 1,4-cineole. Thus, we hypothesized that cinmethylin is a proherbicide that must be metabolically activated. Metabolic activation is quite common for natural phytotoxins. Other examples are hydantocidin,[18] bialaphos,[29] and 2,5-anhydro-*D*-glucitol.[8]

The modes of action of other phytotoxic monoterpenes are relatively unexplored, providing future opportunities.

SESQUITERPENES

A great deal has been published on the phytotoxicity of sesquiterpenoids,[e.g., 10,12,15,31,32] but there is relatively little in the literature that identifies specific molecular target sites. We will discuss two of the better studied sesquiterpenes.

Artemisinin (Fig. 10.1) is a highly phytotoxic sesquiterpene endoperoxide,[12] found only in the glandular trichomes of *Artemisia annua*.[11] Duke et al.[13] first tested it for phytotoxicity because of its high level of activity against malarial parasites (*Plasmodium* spp).[23] Others later verified that it is strongly phytotoxic.[3,4,30] These studies showed little more than that the compound is highly phytotoxic. Chen and Leather[4] reported that artemisinin caused duckweed to release proteins into the

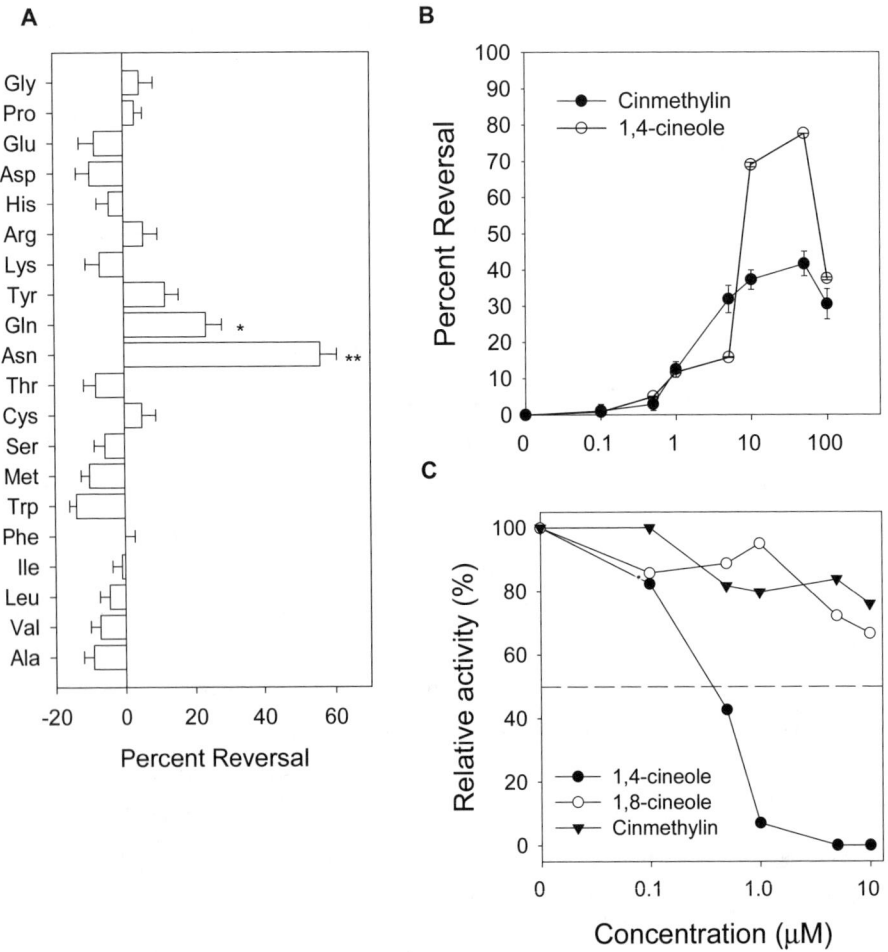

Figure 10.3

    A. Reversal of cinmethylin-inhibited lettuce seedling root growth by exogenous supplies of amino acids. All 20 essential amino acids were tested individually at 100 µM (except for Try, Trp, and Met at 50 µM). Asterisks indicate significance at the $P \leq 0.0001^{**}$ and 0.01^{*} levels. **B.** Reversal effect of exogenous asparagine on the phytotoxic effect of 1 µM cinmethylin and 1,4-cineole on lettuce seedling growth. **C.** Relative activity (percent of control) of asparagine synthetase at different concentrations of 1,8-cineole, 1,4-cineole, and cinmethylin. (From Romagni, G. J. et al. **2000**, *Plant Physiol.* **123**, 725-732. With permission.)

medium. Based on this, they speculated that the site of action is the plasma membrane. Chen et al.[5] later reported that there was no effect of artemisinin on peroxidase secretion (leakage?) from mung bean. Furthermore, Dayan et al.[6] found no effect on electrolyte leakage from plant tissues. Chen et al.[5] reported that artemisinin inhibited peroxidase synthesis; however, this is an unlikely site of action of a phytotoxin.

Stiles et al.[43] found artemisinin to inhibit respiration in Lemna minor, but the compound strongly stimulates oxygen uptake by lettuce root tips.[6] All mitotic phases of onion root tips are inhibited by the compound, and it induces a low level of abnormal mitotic figures.[6] Dayan et al.[6] conducted a battery of simple physiological tests with several artemisinin analogues. All phytotoxic compounds had similar effects, indicating a common mode of action. In summary, the mode of action results with artemisinin as a phytotoxin are not definitive.

Artemisinin is highly active as an antimalarial drug,[24] and much mode of action work has been conducted on it as a pharmaceutical. Plasmodium spp. have a plastid, the apicoplast, with much in common with the plastids of plants.[25] In fact, herbicides with a plastid-localized target site, such as glyphosate, are effective against Plasmodium spp.[38] Wang and Wu[46] recently hypothesized that artemisinin's antimalarial activity is due to its reaction with reduced glutathione and Fe(II/III) to give several products, including a free radical adduct. They pointed out that reduced glutathione is very important in the cell cycle and that reduction in glutathione levels could inhibit mitosis. Although the free radical aspect of arteminsin's interaction with heme has been invoked in the mode of action as an antimalarial, the symptoms of plants affected by the compound are not those of oxidative stress. However, the compound might exert its effect through interaction with sulfhydryl-containing compounds.[15]

Sesquiterpenes containing either a methylene-γ-lactone or a cyclopentenone moiety can react with thiol groups to form a covalent linkage. If the thiol group is on a key enzyme, interaction with artemisinin could inactivate the enzyme, disrupting metabolism. Cysteine is a good antidote for artemisinin as a phytotoxin, but there is no evidence that it is due to a direct interaction of the two molecules.[15]

Another sesquiterpene lactone, dehydrozaluzanin C (Fig. 10.1) (DHZ), is a weaker inhibitor of root growth than artemisinin.[19] Like artemisinin, it did not inhibit respiration, but it did not stimulate respiration as much. Histidine and glycine partially (ca. 40%) reverse the growth inhibition effect. However, reduced glutathione will almost completely reverse the effects of DHZ on growth. DHZ has a strong stimulatory effect on cellular leakage in green tissues (Fig. 10.4), apparently due to disruption of plasma membrane function. Electron microscopy revealed that DHZ causes separation of the plasma membrane from the cell wall of green tissue that is leaking electrolytes (Fig. 10.5). No other ultrastructural effects were observed. The effect on cellular leakage is also eliminated by reduced glutathione.

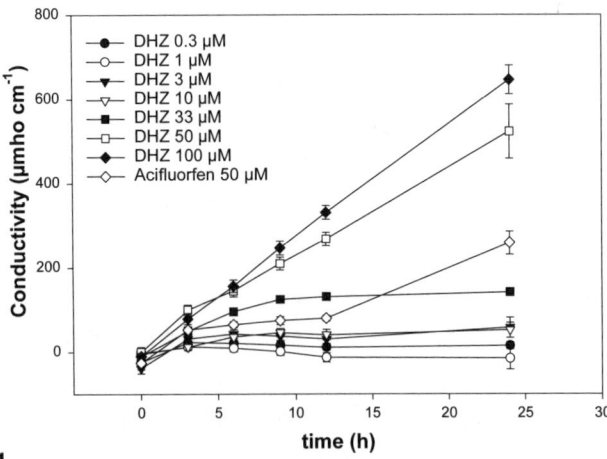

Figure 10.4

Cellular leakage as determined by changes in conductivity of treatments minus control conductivity changes of cucumber cotyledons as affected by exposure to different DHZ concentrations. Error bars are ± 1 SE of the mean of six plates; 50 µM acifluorfen used as positive control. Tissues were incubated in solutions in darkness for 18 h and then exposed to light. (From Galindo, J. C. G. et al. **1999**, *Phytochemistry* **52**, 805-813. With permission.)

DHZ reacts directly with glutathione to form both a monoadduct and a diadduct. Comparison of DHZ effects with those of *iso*zaluzanin C, an analogue without the cyclopentenone group, indicates that the growth-inhibiting effect of DHZ is due to its methylene-γ-lactone moiety, whereas its membrane disruption effect is due to the cyclopentenone group. The molecular target site for neither activity is known.

DITERPENES

There are relatively few reported diterpenoid phytotoxins produced by plants, although there are several potent diterpenes from fungi. The duvatriene-diol diterpenoids from tobacco leaf surfaces are phytotoxic to the serious weed *Echinocloa crus-galli*.[26] Macías et al.[32] summarized the effects of various diterpenes on germination and growth of plants. Some of them are quite active, but no mode of action work has been conducted. More is known of the norditerpenoid podolactones, which inhibit growth and other physiological processes. Some of these compounds have been suggested as herbicide models, due to their strong phytotoxicity when compared to commercial herbicides.[33] Gross effects of these compounds, such as root tip swelling, inhibited chlorophyll synthesis, and inhibited hormone-induced growth, suggest multiple modes of action.[41]

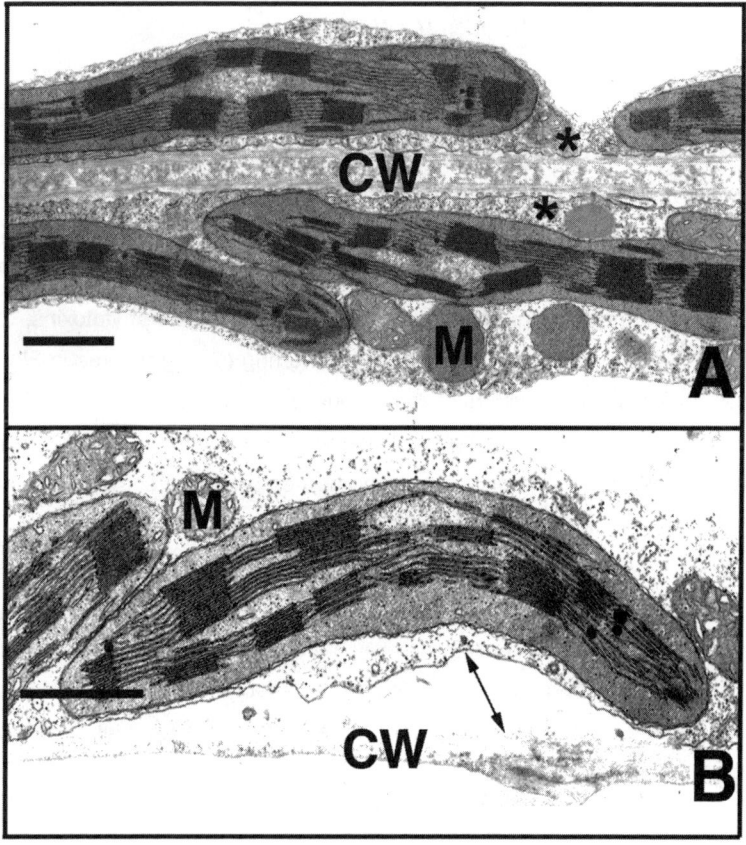

Figure 10.5

Transmission electron micrographs of control (A) and 50 µM DHZ-treated (B) light-grown cucumber cotyledon cells 24 h after treatment. Arrows show plasmalemma separation of the cell wall of a treated cell. Bars = 1 µM. M = mitochondria, CW = cell wall. (From Galindo, J. C. G. et al. **1999**, *Phytochemistry* **52**, 805-813. With permission.)

More specifically, compounds like podolactone A (Fig. 10.1) inhibit proton efflux from plant cells induced by fusicoccin, without affecting ATP levels.[42] The related compound, podolactone E is a strong inhibitor of δ-aminolevulinic acid and chlorophyll synthesis.[34] The authors concluded that this was caused by suppression of synthesis of proteins needed in the porphyrin pathway because podolactones also inhibited gibberellic acid-induced α-amylase synthesis in barley embryos. The molecular target site(s) of this class of terpenoid phytotoxins remains to be determined.

TRITERPENES AND THEIR DERIVATIVES

The quassinoids are degraded triterpenes with a C-20 basic skeleton, but C-18, C-19, and C-25 quassinoids exist. Several of these compounds, such as ailanthone, holacanthone, glaucarubolone, and chaparrinone (Fig. 10.1), are highly phytotoxic,[9,21] as well has having other biological activities (reviewed by Dayan et al.[9]). Some of these compounds have been patented for herbicide use.[e.g.,20] Little is known of the mechanism of action of these compounds as phytotoxins. Growth is reduced by quassinoids with an oxymethylene ring (see structures in Fig. 10.1), whereas those without this functional group were not phytotoxic.[9] There are differences in the activity of these compounds. Holacanthone inhibits prophase, whereas chaparrinone and glaucarubolone do not. All three compounds inhibit all other stages of mitosis. Holacanthone and glaucarubolone reduce chlorophyll accumulation, whereas chaparrinone does not. Molecular modelling revealed that the quassinoids without oxymethylene ring were more planar and had very different electrostatic charge distributions, which might account for the differences in activity. Dayan et al.[9] did not find quassinoids to have an effect on membrane integrity. This was unexpected, because chaparrinone is reported to be an inhibitor of plasma membrane NADH oxidase.[35]

Some saponins from plants are quite phytotoxic,[36] but little is known of their mode(s) of action. Some of these compounds have effects on membrane properties, due to their detergent-like activity. However, all membrane effects are apparently not due to nonspecific effects on membrane lipids. Kauss and Jeblick[23] provided indirect evidence that digitonin effects on plant cell Ca^{2+} uptake are due to effects on membrane protein phosphorylation/dephosphorylation processes. Betulin (Fig. 10.1), a phytotoxic saponin,[22] is known for its pharmaceutical activity. It is an inhibitor of DNA topoisomerase.[45] The natural product-based DNA topoisomerase inhibitor drug podophyllotoxin is also phytotoxic.[37]

METHODOLOGY

Metabolite complementation studies

This method can be used to compensate for inhibition of a biochemical pathway which results in a deficiency of an essential metabolic product. Detailed variations of the method are provided by Dayan et al.[7] and Amagasa et al.[1] The inhibitor concentration should be no higher than that required for strong herbicidal effect. Metabolite concentrations should be below that which is phytotoxic. For example, certain amino acids such at methionine, are growth inhibitors at relatively low concentrations. So, in preliminary work, dose-response studies should be done with amino acids to find the maximum concentrations that do not inhibit growth. Then, seeds of test plants should be imbibed in solutions of the phytotoxin with and without metabolite solutions. Amino acids, tricarboxylic acid cycle intermediates, vitamins, nucleotides, and reducing agents have all been used in complementation studies to elucidate modes of action of a variety of phytotoxins. Examples of each of these is provided by Dayan et al.[7]

One must be careful not to overinterpret the results of complementation studies. Sometimes reversal does not occur, even though the metabolites used are those that are depleted by the phytotoxin. This can be for many reasons, such as deregulation of the pathway by inhibition or accumulation of toxic intermediates. Reversal can also be due to direct interactions of the inhibitor with the reversing compounds.[e.g., 19]

Membrane integrity

Loss of plasma membrane integrity can be a primary or a secondary effect of a phytotoxin. A simple and rapid method of accessing effects on plasma membrane integrity is to measure movement of cellular contents into a solution on which the affected tissue floats or is submerged.[16] We have found electrolyte leakage to be the most useful type of measurement, as measurements can be taken without the need to remove samples from the bathing medium. Conductivity meters that remove one millilitre of bathing liquid for measurement and then return it are commonly available. With this method, one can generate time-course data for the same samples.

Because of differences in starting conductivity between various treatments and the control, results are generally expressed as change in conductivity from the initial reading (e.g., Fig. 10.4). Maximum potential conductivity change is that obtained by boiling the tissues in the bathing solution.

Mitotic index

Many phytotoxins are mitotic inhibitors. However, the effects are seldom due to direct effects on the mitotic apparatus. Onion (*Allium cepa* L.) seeds are germinated in the presence of the natural products under a 14-h photoperiod.[6] Root tips are prepared according to Armbruster et al.,[2] and mitotic analysis is performed with a compound microscrope on 1000 cells per slide (three slides per treatment).[e.g., 37] In addition to the proportions of cells in each stage of mitosis, the number of abnormal mitotic figures should be determined.

ACKNOWLEDGMENTS

We thank Franck Dayan for his constructive comments and for help with the graphics.

REFERENCES

(1) Amagasa, T., Paul, R. N., Heitholt. J. J., and Duke, S. O., **1994**. Physiological effects of cornexistin on *Lemma pauscicostata*. *Pest. Biochem. Phys.* **49**, 37-52

(2) Armbruster, B. L., Molin, W. T. and Bugg, M. W. **1991**. Effects of the herbicide dithiopyr on cell division in wheat root tips. *Pestic. Biochem. Physiol.* **39**, 110-120

(3) Bagchi, G. D., Jain, D. C. and Kumar, S. **1998**. The phytotoxic effects of artemisinin and related compounds of *Artemisia annua*. *J. Med. Aromat. Plant Sci.* **20**, 5-11

(4) Chen, P. K. and Leather, G. R. **1990**. Plant growth regulatory activities of artemisinin and its related compounds. *J. Chem. Ecol.* **16**, 1867-1876

(5) Chen, P. K., Polatnick, M., and Leather, G. R. **1991**. Comparative study on artemisinin, 2,4–D and glyphosate. *J. Agric. Food Chem.* **39**, 991-994

(6) Dayan, F. E., Hernandez, A., Allen, S. N., Moraes, R. M., Vroman, J. A., Avery, M. A., and Duke, S. O. **1999**. Comparative phytotoxicity of artemisinin

and several sesquiterpene analogues. *Phytochemistry* **50**, 607-614

(7) Dayan, F. E., Romagni, J. G., and Duke, S.O., **2000**. Investigating the mode of action of natural phytotoxins. *J. Chem. Ecol.* **26**, 2079-2094

(8) Dayan, F. E., Rimando, A. M., Tellez, M. R., Scheffler, B. E., Roy, T. and Abbas, H. K., and Duke, S. O. **2002**. Bioactivation of the fungal phytotoxin 2,5-anhydro-D-glucitol by glycolytic enzymes is an essential component of its mechanism of action. *Z. Naturforsch.* **57c**, 645-653

(9) Dayan, F. E., Watson, S. B., Galindo, J. C. G., Hernández, A., Dou, J., McChesney, J. D., and Duke, S. O., **1999**. Phytototoxicity of quassinoids: Physiological responses and structural requirements. *Pestic. Biochem. Physiol.* **65**, 15-24

(10) DiTomaso, J. M. and Duke, S. O. **1991**. Is polyamine biosynthesis a possible site of action of cinmethylin and artemisinin?. *Pestic. Biochem. Physiol.* **39**, 158-167

(11) Duke, M. V., Paul, R. N., ElSohly, H. K., Sturtz, G., and Duke, S. O. **1994**. Localization of artemisinin and artemisitene in foliar tissues of glanded and glandless biotypes of *Artemisia annua. Internat. J. Plant Sci.* **155**, 365-373

(12) Duke, S. O. **1991**. Plant terpenoids as pesticides. In: Keeler, R. F. and Tu, A. T. (Eds.), *Handbook of Natural Toxins. Vol. 6. Toxicology of Plant and Fungal Compounds*, Marcel Dekker, NY, 269-296

(13) Duke, S. O., Vaughn, K. C., Croom, Jr., E. M., and Elsohly, H. N. **1987**. Artemisinin, a constituent of annual wormwood (*Artemisia annua*), is a selective phytotoxin. *Weed Sci.* **35**, 499-505

(14) Duke, S. O., Dayan, F. E., and Rimando, A. M. **2000**. Natural products and herbicide discovery. In: Cobb, A. H. and Kirkwood, R. C. (Eds.), *Herbicides and their Mechanisms of Action*, Academic Press, Sheffield, 105-133

(15) Duke, S. O., Paul, R. N., and Lee, S. M., **1988**. Terpenoids from the genus *Artemisia* as potential pesticides. *ACS Symp. Ser.* **380**, 318-334

(16) Duke, S. O. and Kenyon, W. H. **1993**. Peroxidizing activity determined by cellular leakage. In: Böger, P. and Sandmann, G. (Eds.), *Target Assays for Modern Herbicides and Related Phytotoxic Compounds*, Lewis Publishers, Boca Raton, FL, 61-66

(17) Duke, S. O., Dayan, F. E., and Romagni, J. G. **2000**. Natural products as sources for new mechanisms of herbicidal action. *Crop Protect.* **19**, 583-589

(18) Fonné-Pfister, R., Chemla, P., Ward, E., Giradet, M., Kreutz, K. D., Honzatko, R. B., Fromm, H. J., Schär, H.-P., Grüter, M. G., and Cowan-Jacob, W. S. **1996**. The mode of action and the structure of a herbicide

complex with its target: binding of activated hydantocidin to the feedback regulation site of adenylosuccinate synthetase. *Proc. Natl. Acad. Sci. USA.* **93**, 9431-9436

(19) Galindo, J. C. G., Hernández, A., Dayan, F. E., Macías, F. A., and Duke, S. O. **1999**. Dehydrozaluzanin C, a natural sesquiterpenolide, causes rapid plasma membrane leakage. *Phytochemistry* **52**, 805-813

(20) Grieco, P. A., Morré, D. J., Corbett, T. H., and Valeriote, F. A. **1999**. Extraction and preparation of quassionoids with antineoplastic, antiviral and herbicidal activity. U.S. Patent 5,639,712, 29 pp

(21) Heisey, R. M. **1999**. Development of an allelopathic compound from tree-of-heaven (*Ailanthus altissima*) as a natural product herbicide. In: Cutler, H. G and Cutler, S. J. (Eds.), *Biologically Active Natural Products: Agrochemicals*, CRC Press, Boca Raton, FL, 57-68

(22) Hoagland, R. E., Zablotowicz, R. M., and Reddy, K. N. **1996**. Studies of the phytotoxicity of saponins on weed and crop plants. *Adv. Exp. Med. Biol.* **455**, 57-73

(23) Kauss, H. and Jeblick, W. **1991**. Induced calcium uptake and callose synthesis in suspension-cultured cells of Catharanthus roseus are decreased by the protein phosphatase inhibitor okadaic acid. *Physiol. Plant.* **81**, 309-312

(24) Klayman, D. L. **1985**. Qinghaosu (artemisinin): an antimalarlial drug from China. *Science* **228**, 1049-1055

(25) Lang-Unnasch, N., Reith, M. E., Munholland, J., and Barta, J. R. **1998**. Plastids are widespread and ancient in parasites of the phyllum Apicomplexa. *Int. J. Parasitol.* **28**, 1743-1754

(26) Lawson, D. R., Danehower, D. A., Shilling, D. G., Menetrez, M. L. and Spurr, H. W., **1988**. Allelochemical properties of *Nicotiana tabacum* leaf surface compounds. *Amer. Chem. Soc. Symp. Ser.* **380**, 363-377

(27) Lorber, P. and Muller, W. H. **1980**. Volatile growth inhibitors produced by *Salvia leucophylla*: effects on cytological activity in *Allium cepa*. *Comp. Physiol. Ecol.* **5**, 60-67

(28) Lorber, P. and Muller, W. H. **1980**. Volatile growth inhibitors produced by *Salvia leucophylla*: effects on metabolic activity in mitochondrial suspensions. *Comp. Physiol. Ecol.* **5**, 68-75

(29) Lydon, J. and Duke, S.O. **1999**. Inhibitors of glutamine biosynthesis. In: Singh, B. K. (Ed.), *Plant Amino Acids: Biochemistry and Biotechnology*, Marcel Dekker, NY, 445-464

(30) Lydon, J., Teasdale, J. R., and Chen, P. K. **1997**. Allelopathic activity of annual wormwood (*Artemisia annua*) and the role of artemisinin. *Weed Sci.* **45**, 807-811

(31) Macías, F. A., Molinillo, J. M. G., Galindo, J. C. G., Varela, R. M., Simonet, A. M., and Castellano, D. **2001**. The use of allelopathic studies in the search for natural herbicides. *J. Crop Prod.* **4**, 237-255

(32) Macías, F. A., Molinillo, J. M. G., Galindo, J. C. G., Varela, R. M., Torres, A., and Simonet, A. M. **1999**. Terpenoids with potential use as natural herbicide templates. In: Cutler, H.G. and Cutler, S. J. (Eds.), *Biologically Active Natural Products: Agrochemicals*, CRC Press, Boca Raton, FL, 15-31

(33) Macías, F. A., Simonet, A. M., Pacheco, P. C., Berrero, A. F., Cabrera, E., and Jiménez-González, D. **2000**. Natural and synthetic podolactones with potential use as natural herbicide models. *J. Agric. Food Chem.* **48**, 3003-3007

(34) Miller, G. W., Sasse, J. M., Lovelace, C. J. and Rowan, K. S., **1984**. Effects of podolactone-type inhibitors and abscisic acid on chlorophyll biosynthesis in barley leaves. *Plant Cell Physiol.* **25**, 635-642

(35) Morré, D. J., Grieco, P. A., and Morré, D. M., **1998**. Mode of action of the anticancer quassinoids-inhibition of the plasma membrane NADH oxidase. *Life Sci.* **63**, 595-604

(36) Oleszek, W. A., Hoagland, R. E., and Zablotowicz, R. M. **1999**. Ecological significance of plant saponins. In: Inderjit, Dakshini, K. M. M., and Foy, C. L. (Eds.), *Principles and Practices in Plant Ecology: Allelochemical Interactions*, CRC Press, Boca Raton, FL, 451-465

(37) Oliva, A., Moraes, R. M., Watson, S. B., Duke, S. O., and Dayan, F. E. **2002**. Aryltetralin lignans inhibit plant growth by affecting formation of mitotic microtubular organizing centers. *Pestic. Biohem. Physiol.* **72**, 45-54

(38) Roberts, R., Roberts, C. W., Johnson, J. J., Kyle, D. E., Krell, T., Coggins J. R., Coombs, G. H., Milhous, W. K., Tzipori, S., Ferguson, D. J., Chakrabarti, D., and McLoed, R. **1999**. Evidence for the shikimate pathway in apicomplexan parasites. *Nature* **393**, 801-805

(39) Romagni, J. G., Allen, S. N., and Dayan, F. E. **2000**. Allelopathic effects of volatile cineoles on two weedy plant species. *J. Chem. Ecol.* **26**, 303-313

(40) Romagni, J. G., Duke, S. O., and Dayan, F. E. **2000**. Inhibition of plant asparagine synthetase by monoterpene cineoles. *Plant Physiol.* **123**, 725-732

(41) Sasse, J. M., Wardrop, J. J., Rowan, K. S., Aspinall, D., Coombe, B. G.,

Paleg, L. G., and Buta, J. G. **1982**. Some physiological effects of podolactone-type inhibitors. *Physiol. Plant.* **55**, 51-59

(42) Sasse, J. M., Cerana, R., and Colombo, R. **1984**. The effects of podolactone-type inhibitors on fuscicoccin-induced growth and proton efflux. *Physiol. Plant.* **62**, 303-308

(43) Stiles, L. H., Leather, G. R., and Chen, P. K. **1994**. Effects of two sesquiterpene lactones isolated from *Artemisia annua* on physiology of *Lemna minor*. *J. Chem. Ecol.* **20**, 969-978

(44) Vaughn, S. F. and Spencer, G. F. **1993**. Volatile monoterpenes as potential parent structures for new herbicides. *Weed Sci.* **41**, 114-119

(45) Wada, S., Lida, A., and Tanaka, R. **2001**. Screening of triterpenoids isolated from *Phyllanthus flexuosus* for DNA topoisomerase inhibitory activity. *J. Nat. Prod.* **64**, 1545-1547

(46) Wang, D.-Y. and Wu, Y.-L. **2000**. A possible antimalarial action mode of qinghaosu (artemisinin) series compounds. Alkylation of reduced glutathione by C-centered primary radicals produced from antimalarial compound qinghaosu and 12-(2,4-dimethoxyphenyl)-12-deoxoquighaosu. *Chem. Commun.*, 2193-2194

11   Mode of Allelochemical Action of Phenolic Compounds

F. A. Einhellig

CONTENT

Abstract 217
Introduction.. 218
Perspective 221
Action of Cinnamic and Benzoic Acids 222
    Membrane Effects... 222
    Water Relationships 223
    Phytohormone Interactions and Enzyme Effects 224
    Energy Systems... 225
    Flow of Carbon... 227
Action of Other Allelopathic Polyphenols.. 227
    Coumarins 227
    Tannins 229
    Flavonoids 230
Joint Activity and Environmental Interactions 231
Conclusions 232
References 233

ABSTRACT

The most common phenolic compounds in allelopathy are the derivatives of cinnamic and benzoic acids, coumarins, tannins and other polyphenolic complexes, and certain flavonoids. While the level of production and release of these compounds varies significantly among plants, they are among the most widely distributed and least specific in action of the secondary compounds implicated in allelopathy. The different phenolic acids, coumarins, and tannins appear to have quite similar mechanisms of action, inhibiting plant and microbial growth through multiple physiological effects that confer a generalized cytotoxicity. Their initial actions are on cell membranes, resulting in nonspecific permeability changes that alter ion fluxes and hydraulic conductivity of roots. Membrane perturbations are followed by a cascade of physiological effects that include alterations in ion balance, plant-water relationships, stomatal function, and rates of photosynthesis and respiration. These phenolics also interact with several phytohormones and enzymes, causing deviations from typical patterns for biosynthesis and flow of carbon into metabolites. Mechanisms of flavonoid action are less understood than the phenolic acids. Some of the allelopathic flavonoids are potent inhibitors of energy metabolism, blocking mitochondrial and chloroplast functions, but the threshold for these effects for different compounds spans two orders of magnitude. Few studies have evaluated other possible targets of flavonoid action. In addition to having multiple targets of action on plant physiology, effectiveness of the numerous phenolic allelochemicals is mediated, in part, by the fact that in almost all situations a number of different compounds are acting at the same time.

INTRODUCTION

The focus of this chapter — phenolic allelochemicals — has a connotation that is widely used and accepted in discussions of allelopathy. Yet the scope of chemicals referred to in this way does not have concrete limits in the literature on allelopathy. It is an umbrella designation that includes compounds in several of the categories of identified allelopathic compounds. Rice[55] devised a scheme of fourteen, loosely defined chemical categories in his classification of allelopathic agents, and phenolic compounds are found in several of his categories. Specifically, the phenolics active in allelopathy are most commonly accepted to be the cinnamic acids, simple phenols, benzoic acids, and various closely related derivatives such as chlorogenic acid, an ester of caffeic acid. My primary focus will be on the mechanisms of action of these compounds. Allelopathic phenolics also include several coumarins, the highly polyphenolic hyrolyzable and condensed tannins, flavonoids, and a diverse group of closely related structures (Fig. 11.1). Although flavonoids constitute a very large grouping of phenolic substances, only a relatively small number of them have been established as agents in the phenomena of allelopathy.

The cinnamic and benzoic acid families of compounds have been linked to allelopathy in both natural plant communities and agronomic fields. They have routinely been identified in releases from plant residue and present in the soil. These phenolic acids and aldehydes are the most widely distributed compounds that have been identified as inhibitors of plant and microbial growth, with hundreds of studies establishing a linkage to their activity. The evidence comes from such diverse field situations as the role of allelopathy in the successional sequence in 'old fields' in Oklahoma, to being a part of the explanation for the patterning of vegetation in the California chaparral, and more recent reports implicating humus phenolics in the failure of natural regeneration of conifers in the Boreal forests of North America and northern Europe. In the agricultural sector, the scope of allelopathic effects linked to phenolic compounds is equally broad, ranging from yield-reduction problems in sequential cropping with rice (*Oryza sativa*) in Taiwan, to evidence that phenolics are part of the reason for *Sorghum* allelopathy. However, in many of these situations there are additional kinds of allelochemicals that contribute to the complex of compounds causing allelopathy.

Biosynthesis of the cinnamic acids arises through the common phenylpropanoid pathway from phenylalanine. Unlike many secondary compounds, these compounds have internal functions in plants. Their various esters and glycosides are precursors of the complex, heterogeneous lignin polymers of cell walls. The widespread occurrence of the phenylpropanes may in part be explained by their role in lignification, yet the relative proportion of phenylpropane units in lignin varies extensively from different plant sources. It is not surprising that there is a considerable difference in the amounts and kinds of cinnamic acids that are released from plants and, thus, factor into allelopathy. Recognition that dramatic differences occur in concentrations of phenolics at various locations and points in time in a plant community is central to accepting phenolic allelochemicals as active contributors to allelopathy. Equally important is our growing understanding that receiving plants differ in their tolerance and capacity to detoxify phenolic and other allelochemicals.[38,58,59] Certain phenolic acids implicated in allelopathy also have functional importance to internal physiology. Notable of these is the role of salicylic acid as a mediator for the development of systemic acquired (induced) resistance in a plant's defense against disease.[42,53]

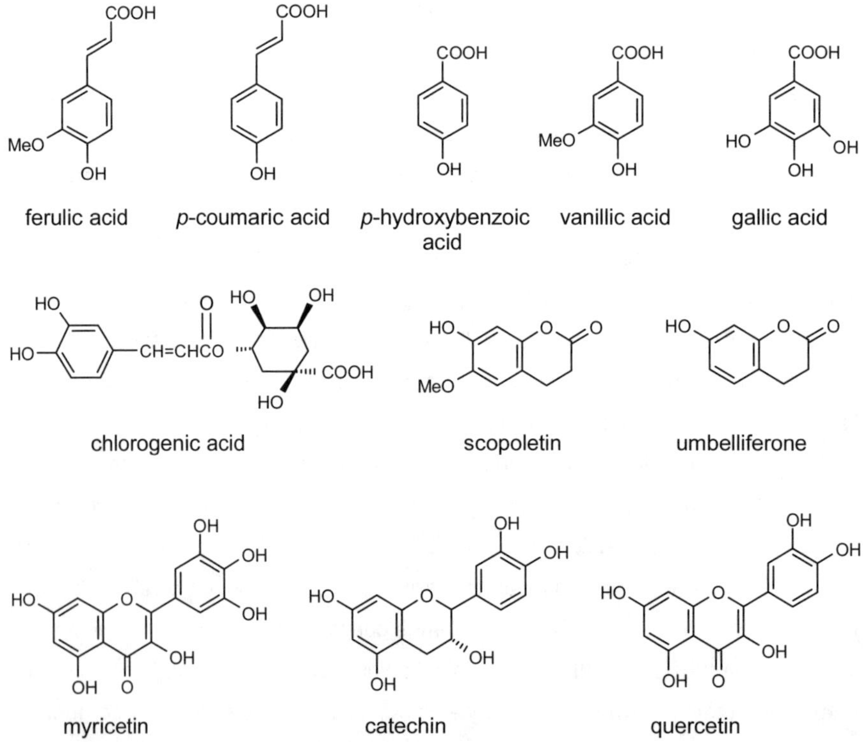

Figure 11.1

Some of the more common cinnamic and benzoic acids, coumarins, and flavonoids implicated as compounds causing allelopathy.

The phenolic acids typify several things that characterize chemical ecology. For one thing, the extent of their inhibitory action is concentration dependent. At the low end of a spectrum, they may even cause stimulatory responses in some bioassays. Second, their action is typically not an herbicidal effect, and the impact on a receiving plant is interdependent with other conditions that affect plant growth. The toxicities of the base compounds, *trans*-cinnamic acid and benzoic acid, are greater than the various methoxy and hydroxylated derivatives.[30] Comparisons of the toxicity of the different cinnamic and benzoic acids show some differences across the group of compounds.[4] But those activity differences are not large, and clear structure-activity patterns have not been established. The concentration of a phenolic acid required to inhibit seed germination is generally higher than it is to inhibit growth in whole seedlings. In most seedling bioassays the inhibition threshold for cinnamic and benzoic acids is in the range from 100 to 1000 µM. This generalization does not fit all situations, and the growth of some species may show greater sensitivity, as shown by a 1.0 µM ferulic acid-inhibition threshold on the cyanobacterium *Oscillatoria perornata*.[57] Finally, the allelopathic action of phenolic acids is characterized by the combined presence and input of multiple different compounds.

PERSPECTIVE

Before launching into exploring the physiological effects of phenolic allelochemicals, several theoretical questions should be posed. It also is worthwhile to reflect on the general evidence showing allelopathy in plant communities is from combinations of secondary products. Further, seldom is allelopathic inhibition an herbicidal effect; most often chemical interference is an outcome that accounts for a relatively small impact on growth of the receiving species. This does not mean that allelochemical effects are insignificant or of no consequence — quite the contrary is true! It is valuable, however, to embrace the perspective that allelochemical action is only one of the stresses acting on plant success.[24] In this context it should not be a surprise that it has proven very difficult to separate primary from secondary physiological effects for the chemicals known to cause plant-to-plant allelopathy. Although there has been some success in this endeavor for a few microbial compounds, determination of a primary target site has been especially challenging for compounds produced by higher plants. The question needs to be asked whether it is logical to expect that there are many situations where action on a primary molecular target is the explanation for how an allelochemical alters the growth, reproduction, or survival of the receiving plant. I think the answer is no, or at a minimum those cases must be rare!

Debate continues on the connection between allelopathy and evolution, but suppose we pose the mechanism question in the context of evolutionary forces and advantage gained. Would it be most likely for phytotoxic compounds to have a primary target or multiple targets that result in a generalized cytotoxicity? The majority of herbicides are reported to have a certain site of action, and one recent classification scheme is by their mode of action.[54] Now we find that in the one-half century of modern agriculture when we have utilized herbicides for weed control, there is a long and growing list of herbicide-resistant weed species. Not surprisingly, evidence of herbicide-resistant weeds is greatest for those compounds that have had long use, a wide spectrum of application, and strong specificity for a molecular target site of action. Using the herbicide scenario for comparison, the relative abundance and frequency of occurrence of specific allelopathic compounds should make a difference. I think it is unlikely that evolutionary pressures would very often favor allelochemicals that are widely distributed to have their action on only one molecular target.

ACTION OF CINNAMIC AND BENZOIC ACIDS

The mechanisms of action of these phenolic allelochemicals have had more scrutiny than any other allelopathic compounds. I will only cite a few of the most relevant studies analyzing the physiological action of the phenolic acids. Literature reports show abundant evidence that phenolic acids interfere with several enzymes and almost all of the major physiological processes — phytohormone activity, mineral uptake, plant water balance and stomatal function, photosynthesis, respiration, organic synthesis of certain compounds and flow of carbon.[19,22] However, they do not seem to alter cell division or directly affect gene translation. Ferulic acid has been the most studied compound. Interestingly, the data available allowing comparisons among the different compounds suggest that the various cinnamic and benzoic acids and aldehydes have commonalties in how they interfere with functions of a receiving plant.

MEMBRANE EFFECTS

Action on the plasma membrane is the first and most fundamental of the bewildering array of deleterious effects of the cinnamic and benzoic acids. They reduce the transmembrane electrochemical potential with the immediacy and extent of that action depending on the concentration and lipid solubility of the compound.[35,37,45,60] Rate of uptake also is concentration and pH-dependent, with transfer into and across the membrane greatest with lower pH conditions and higher external concentrations.[60] Phenolic acid-induced depolarization of membranes causes a nonspecific efflux of both anions and cations accompanying the increased cell membrane permeability, and these membrane effects correlate with an inhibition of ion uptake. The phenolic acids suppress absorption of phosphate, potassium, nitrate, and magnesium ions, and overall changes in tissue content of mineral ions are one of the effects on plants grown with phenolic acids in the medium.[1,3,7,9,11,33,34,43]

Baziramakenga et al.[8] found that benzoic acid and cinnamic acid themselves damaged cell membrane integrity by a decrease in sulfhydryl groups. Both compounds induced lipid peroxidation that resulted from free radical formation in membranes and inhibition of catalase and peroxidase activities. Oxidation or cross-linking of plasma membrane sulfhydryl groups was suggested as their first mode of action. Hence, it is likely that the cinnamic and benzoic acid derivatives cause structural changes in membranes that include alterations in a variety of membrane proteins. I suspect that further work will reveal action on specific channel proteins, proton pumps, or some of the membrane transporters. It also is reasonable to project that these phenolics influence yet-to-be-defined membrane hormone-binding sites and other signal receptors. Any effects on membrane proteins, coupled with the known permeability effects, will interfere with cell regulation mediated through signal pathways and transduction events.

WATER RELATIONSHIPS

The actions of phenolic acids on ion retention and flux across the cell membrane link directly to the effects these compounds have on plant water status. They reduce the hydraulic conductivity of roots as evidenced in changes in plant-water parameters. In our studies with soybean (*Glycine max*) seedlings that ranged from 10 days to 4 weeks, we found that all the phenolic acids and closely related compounds investigated altered normal water balance. This was established by reductions in leaf water potential, turgor pressure, conductance, or a change in tissue carbon-isotope ratio.[4,5,6,25,31,64] The carbon isotope ratio in C-3 plants is an indicator of the extent of stomatal resistance or water stress during the growth period. The compounds studied include ferulic, *p*-coumaric, caffeic, hydrocinnamic, salicylic, *p*-hyroxybenzoic, gallic, and chlorogenic acids, as well as hydroquinone, vanillin (aldehyde), and umbelliferone (coumarin). Consistently in these investigations, we found the threshold concentration causing growth-inhibition also resulted in seedlings that evidenced water stress. Interestingly, Holappa and Blum[39] also found ferulic acid-limited water utilization was accompanied by early induction of abscisic acid biosynthesis and an increase in leaf endogenous abscisic acid levels.

Our investigations with combinations of several phenolics at individual concentrations below their inhibition threshold showed that the mixtures caused water stress similar to individual compounds. For example, after growing soybean for 4 weeks with the nutrient solution containing 500 µM of an equimolar mixture of hydrocinnamic, p-coumaric, caffeic, chlorogenic, ferulic, and gallic acids, the $\delta^{13}C$ of leaf tissue was -25.7, compared to -28.9 for untreated plants. Seedling growth was reduced, and this significant reduction in discrimination against ^{13}C supports the contention that it was sustained, long-term water stress that reduced growth. It should be noted that water stress was one of the physiological effects reported in allelopathic effects caused by *Abutilon theophrasti*, *Kochia scoparia*, and several other weeds.[14,29,31]

PHYTOHORMONE INTERACTIONS AND ENZYME EFFECTS

Indications that phenolic allelochemicals can alter auxin activity in plants were among the first work suggesting their physiological action. The early conclusion was that phenolic acids acted through inhibition or activation of the indoleacetic acid (IAA) oxidase system. Chlorogenic, caffeic, and other polyphenols were reported to synergize IAA-induced growth by counteracting IAA destruction, whereas monophenols such as p-coumaric and p-hydroxybenzoic acid stimulated decarboxylation.[62] Prasad and Devi[50] reported ferulic acid caused an increase in activities of peroxidase, catalase, and IAA oxidase while decreasing polyphenol oxidase in maize (*Zea mays*) seedlings. Although numerous investigations have shown that many natural phenols affect the rate of enzymatic oxidation of IAA, it has not been determined how this action is linked to allelopathic growth inhibition. In my laboratory, we found that ferulate stimulated adventitious root formation, although several other phenolics did not. It is a common observation that misshapen, short, stubby roots occur in test seedlings subjected to phenolic acids. This morphology probably arises, at least in part, from phenolic interactions with IAA-induced growth.

There are scattered reports that phenolic acids inhibit a variety of enzymes, and it is evident that these compounds can block the function of many enzymes if they are sufficiently concentrated at the site of enzymatic functions. Activities of amylase, maltase, invertase, acid phosphatase and protease were suppressed by ferulic acid in tests using maize seeds and seedlings.[17,50] Exogenously applied gibberellic acid reversed the effect of ferulic acid on amylase and acid phosphate.

Salicylic acid at 25 µM strongly inhibited ethylene formation from ACC in cell suspension cultures, and a number of other benzoic acid derivatives suppress ethylene production at higher concentrations.[44] The complex interrelationship of phenolic acids with phytohormones deserves further investigation. The early stage of seedling growth is very sensitive to phenolic acids, and this is a prime time in hormone-mediated growth responses.

ENERGY SYSTEMS

Photosynthesis

At treatment levels that inhibit seedling growth, the cinnamic and benzoic acids reduce net photosynthesis.[22,49] However, suppression of photosynthetic rate in seedling plants appears to be due primarily to reduced stomatal conductance. Chlorophyll reduction also may be part of the explanation. We found 500 µM ferulic and *p*-coumaric acids reduced the amount of chlorophyll a, b and total chorophyll in soybean on a leaf weight basis, albeit no similar chlorophyll loss was measured in grain sorghum.[28] Subsequent reports show benzoic, syringic, protocatechuic, *trans*-cinnamic, and caffeic acids reduced the concentration of chlorophyll in leaves of soybean and cowpea, with the major effect on chlorophyll a.[1,7] By using etiolated rice seedlings, Yang et al.[66] studied the action of phenolic acids on biosynthesis of chlorophyll porphyrin precursors and concluded that suppression of Mg-chelatase caused a slower rate and reduced level of chlorophyll accumulation.

In my laboratory, several different test systems were utilized to try and determine what aspects of the photosynthesis process were affected by allelochemicals. A respirometer was used to evaluate the effects of chlorogenic acid, arbutin, four cinnamic acids (*trans*-cinnamic, ferulic, *p*-coumaric, caffeic), and 13 benzoic acids (benzoic, *p*-hydroxybenzoic, gallic, salicylic, vanillic, gentisic, syringic, protocatechuic, *p*-anisic) and aldehydes (vanillin, isovanillin, benzaldehyde, syringaldehyde) on *Lemna minor* photosynthesis.[48,56] Except for arbutin, all compounds reduced net photosynthesis, and this action was not due to elevated respiration. The different phenolics varied in their growth- and photosynthesis-inhibition thresholds. Most of the compounds reduced photosynthesis in the range from 500 to 1000 µM. Salicylic acid was the strongest inhibitor, being active at 100 µM. The concentration required to inhibit growth was generally slightly below the threshold for significant inhibition of photosynthesis. Chlorophyll reduction closely paralleled growth effects.

If a high enough concentration of a cinnamic or benzoic acid accumulates within cells, these compounds can inhibit the ATP-generating pathway of chloroplasts.[46] But this type of action is probably not central to the physiological mechanisms causing growth inhibition. Yang[65] employed an algal bioassay to follow the action of allelochemicals on photosystem II as indicated by chlorophyll fluorescence. Treatments of 1000 µM ferulic, p-coumaric, p-hydroxybenozic, and vanillic acids reduced efficiency of photosystem II. However, algal growth was reduced by concentrations well below that required to perturb photosystem II. The work with algae, plus the studies with *Lemna minor* and seedling plants, all confirm that the phenolic acid affects net carbon fixation. Yet, the data also provide evidence that interference with photosynthesis is secondary to phenolic acid actions on conductance, chlorophyll content, and other processes.

Respiration

A number of investigations have shown that phenolic acids and mixtures of these compounds alter the rate of respiration in target plants and microorganisms. Mycorrhizal fungi are particularly sensitive.[12,61] Phenolic mixtures of 0.1 µM reduced fungal respiration, in contrast to millimolar levels increasing oxygen consumption. The effect phenolic acids have on respiratory metabolism could arise either from direct action on mitochondrial functions or increased respiratory demand to meet energy requirements for membrane repair and lost efficiency in ATP-mediated events.

Work with isolated mitochondria established that benzoic and cinnamic acids inhibit oxygen uptake with I_{50} values in the 4 to 27 mM range.[16,46,51] This interference appears to be from alteration in the inner membrane, and our studies of salicylic acid, gentisic acid, and p-hydroxybenzaldehye indicated a block in electron transport at the b/c_1 cytochrome complex. By comparison to many inhibitors, the levels of phenolic acids required to depress mitochondrial uptake are quite high. However, even though alterations in mitochondrial functions may not be an important phenolic acid target, they certainly are part of the general cytotoxic effects of these compounds.

FLOW OF CARBON

One result of exposure to cinnamic and benzoic acids is that normal patterns of cellular synthesis are altered. Danks et al.[15] found that cell-suspension cultures treated with 10 μM cinnamic and 100 μM ferulic acid had reduced incorporation of carbon into protein. These phenolics differed in their effects on other cellular constituents. Ferulic acid enhanced production of lipids and reduced organic acids, whereas cinnamic acid did not change the proportion of these fractions. Ferulic and cinnamic acids also inhibited protein synthesis in lettuce seedlings (*Lactuca sativa*).[13] Other work has shown salicylic acid is a strong stimulant of nitrate reduction.[41] The data show that these and other phenolic actions on enzyme functions and regulatory process depend on the specific phenolics present and the concentration of the allelochemicals. Even though they may not be consistent and predictable, changes in patterns of organic synthesis must be counted among the general deleterious cellular effects of the phenolic acids (Fig. 11.2). Modifications in the synthesis and allocation of metabolites are particularly problematic when reserves are being mobilized and converted to compounds necessary for growth during germination and seedling development. Hence, the action of phenolic acids in altering carbon flow helps to explain why early seedling growth is very sensitive to allelochemicals.

ACTION OF OTHER ALLELOPATHIC POLYPHENOLS

COUMARINS

Scopoletin is purported as the most widely distributed coumarin in higher plants, and scopoletin, umbelliferone, and esculetin are the ones most frequently linked to allelopathy. Given their phenylpropane origin, it is not surprising that these simple coumarins have many actions in common with the cinnamic acids. One variance is that coumarin and scopoletin have been reported to decrease mitosis,[2] whereas at least at concentrations that correlate with growth inhibition, the phenolic acids do not appear to affect cell division.

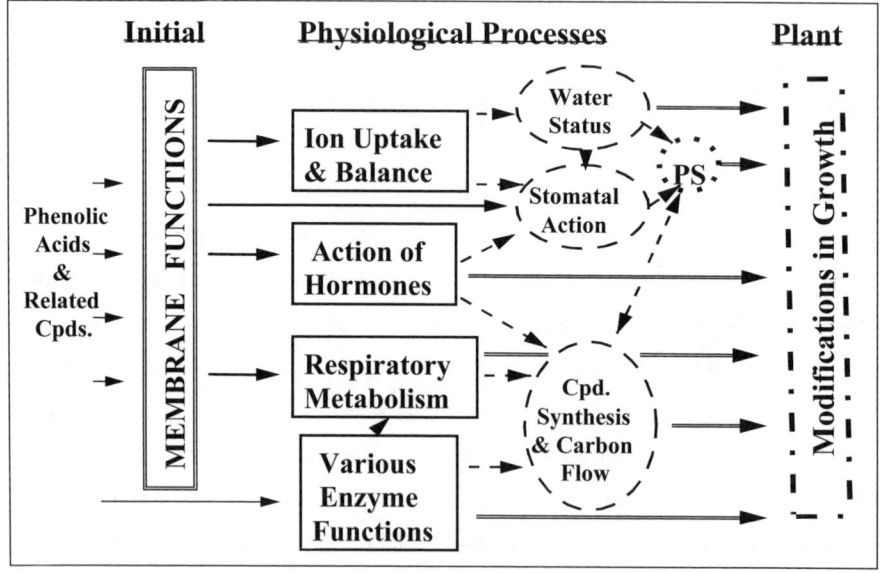

Figure 11.2

Suggested mode of action sequence for cinnamic and benzoic acids and closely related allelochemicals as they affect plant growth and development.

Much like the phenolic acids, early work with scopoletin showed it inhibited oxidation of IAA and thus could affect growth in this manner. Inhibition of several other enzymes by scopoletin and coumarin has been shown. Coumarin was reported to induce ethylene synthesis.[47] Also, it is one of several phenolic compounds that antagonize abscisic acid-induced inhibition of growth and stomatal closure.[52] Undoubtedly, these and possibly other interactions with hormones are part of the physiological action of the coumarins.

I believe that the mechanism of action sequence proposed for phenolic acids shown in Figure 11.2 fits quite well for the allelopathic coumarins. We found that concentrations of umbelliferone that reduce soybean seedling growth also lowered leaf water potential, reduced conductance and the transpiration ratio, and resulted in less discrimination against incorporation of ^{13}C.[25] The $\delta^{13}C$ of soybean leaf tissue after 4 weeks growth with 500 µM umbelliferone was -27.7 compared to -29.2 for untreated soybean — a significant difference indicative of long-term water stress. Hence, impairment of normal functions in root cell membranes that results in water stress is the likely reason for earlier reports of scopoletin causing stomatal closure and reductions in photosynthetic rate.[27,32] Yet interpreting whether there is a direction action of scopoletin on the photosynthesis process is complicated. We found scopoletin and esculetin treatments that suppressed *Lemna minor* growth also reduced the rate of photosynthesis.[48] In contrast, growth reductions in *Chlorella pyrenoidosa* and *Selenastrum capricornutum* caused by coumarin or scopoletin were not accompanied by any change in the efficiency of photosystem II.[65] Moreland and Novitzy[46] did find that umbelliferone inhibits functions in isolated chloroplasts and mitochondria, but only at relatively high concentrations. Even these chloroplast and mitochondrial actions are explainable as perturbations of the organelle membranes, giving further credence to our inference that coumarins act in much the same manner as the phenolic acids (Fig. 11.2).

TANNINS

Hydrolyzable and condensed tannins have multiple roles in chemical ecology, including their defensive value in limiting herbivory, disease resistance, protecting seed decay, and modulation of plant and microorganism growth that alters the dynamics in plant communities. An example of the latter is that tannic acid (gallotannin) has been shown to inhibit both free-living and nodulating nitrogen-fixing bacteria.[10] The biological action of tannins is the result of their polyphenolic molecular structure that causes binding with many different kinds of proteins. The various tannins have a much higher molecular weight than the previously discussed phenolic allelochemicals, and it is reasonable to assume their initial action is on proteins of the plasma membrane. In tests with *tobacco (Nicotiana tabacum)*, we found tannic acid reduced water uptake and caused stomatal closure.[18] Other work shows that tannins antagonize gibberellic acid induced growth.[36,40] Tannic acid is a potent inhibitor of respiratory metabolism in isolated mitochondria, inhibiting oxygen uptake and blocking electron transport with concentrations above 4 µM and an I_{50} of 10 µM.[51] Using *in vitro* assays, Wink and Latz[63] found tannins inhibit DNA polymerase and protein synthesis translation, but provided no evidence that blocking these processes was a mechanism of action in allelopathy. Certainly, since tannic acids interfere with the activity of numerous enzymes, any molecules that enter the cell will cause a generalized toxicity from multiple sites of action.

FLAVONOIDS

The number of flavonoids linked to allelopathic inhibition is not large, and less is known about their physiological action inhibiting growth than the other phenolics that have been discussed. They have been most often reported as inhibitors of energy metabolism. In comparing a number of literature reports of various classes of allelochemicals on mitochondria, we found flavonoids were the second most active class of allelochemicals inhibiting mitochondrial oxygen uptake.[22] Only the quinones were more effective in blocking mitochondrial functions. Moreland and Novitsky [46] reported that quercetin and naringenin applied to isolated mitochondria resulted in I_{50} values for oxidation of malate being 20 and 110 µM, respectively. By comparison, a similar effect from representative phenolic acids and coumarins required concentrations from one to two orders of magnitude higher. They concluded the flavonoids acted primarily as electron transport inhibitors through perturbation in the mitochondrial inner membrane, and there was no indication they acted as uncouplers. Also, the compounds inhibited hydrolysis of ATP catalyzed by mitochondrial Mg^{2+}-ATPase.

The action of flavonoids extends to chloroplasts and inhibition of photophosphorylation. We found myricetin, naringenin, kaempferol and +/- catechin suppressed CO_2-dependent oxygen evolution in isolated pea chloroplasts.[25] The I_{50s} for their actions were 20, 200, 300, and 2000 µM, respectively, thus spanning two orders of magnitude. The primary effect of flavonoids appears to be on the ATP-generating pathway, but higher concentrations can inhibit electron transport.[46] The algal studies cited earlier included work with selected flavonoids tested on photosynthesis.[65] *Selenastrum capriconutum* subjected to 25 µM kaempferol had a dramatic reduction in photosystem II efficiency after 4 days treatment, and there was a good correspondence with reduction in algal cell growth. However, there was less correspondence between effects of quercetin on photosystem II and cell growth.

While the flavonoids suppress oxygen uptake in isolated mitochondria and oxygen evolution from chloroplasts, there has been too little work to establish these organelle effects as the only mechanisms of action. Flavonoids are known to protect membrane lipids against destructive reactions and, based on current evidence, these compounds do not readily fit the model of Figure 11.2. The flavonol rutin did not show an effect on soybean seedling water relations.[64] It is still likely, however, that flavonoids have multiple targets of action that cause a generalized cytotoxicity.

JOINT ACTIVITY AND ENVIRONMENTAL INTERACTIONS

It has become clear in the last two decades that the contribution of phenolic compounds to allelopathy is probably never due to a single compound. Case studies where cinnamic and benzoic acids have been isolated consistently show not one, but a number of different compounds. The presence of a single compound at a given time in the environment is almost always below the growth-inhibition threshold established in bioassays. However, the data demonstrate that phenolic allelochemicals have additive or synergistic inhibitory action on growth, and below-threshold concentrations of individual compounds affect growth through their action in concert.[20,21,24] While most of this evidence has come from investigations with various combinations of phenolic acids, additive inhibitory action was also established with a combination of salicylic acid, umbelliferone, and rutin.[23]

Environmental conditions also impact the extent of activity of allelopathic phenolics, with a number of abiotic and biotic stresses probably accentuating their action.[20,21,23,24] Stress may result from pathogens, herbivory, herbicides, pesticides, various pollutants, ultraviolet irradiance, deficits in moisture or minerals, temperature extremes, or other less than optimum factors of the environment. Perhaps the clearest demonstration of this was our study showing that the ferulic acid concentration for a growth-inhibition threshold in grain sorghum and soybean was reduced by half under higher, more stressful temperatures.[26] No growth reduction was found in grain sorghum grown at an average temperature of 29°C, but the dry weight of seedlings grown at 37°C was reduced by one-third. Similarly, lowering the osmotic potential of the growth medium by -0.15 M Pa in concert with exposure to ferulic acid showed a significant interactive, inhibitory effect.[21] Neither 150 µM ferulic nor -0.15 M Pa from polyethylene glycol in the nutrient solution was inhibitory, but the dry weight of grain sorghum seedlings was reduced more than 40% when both conditions existed. Interactive effects like those described appear to be most evident when both conditions are below the threshold for either one by itself to have a measured impact.

Phenolic acids disruption of normal membrane functions provides a ready explanation as to why several environmental stresses appear to increase the sensitivity to phenolic allelochemicals. Both high temperature stress and moisture deficits place demands on the physiological capacity for a plant to gain and retain water. As we have seen, phenolic acids reduce the efficiency of those same systems. While combinations of allelochemical stress and stress from herbivory, pathogens, and several other environmental conditions may not be acting on the same plant systems, all of these conditions tax the efficiency of carbon gain. These dual stresses divert plant resources or, stated differently, they cause a reduction in available energy to maintain membrane and other cellular functions.

CONCLUSIONS

Since phenolic allelochemicals are broadly found throughout the plant kingdom, their concentration at a point in time and differences in the sensitivity of receiving species are keys to their allelopathic action. At very low concentrations they may be stimulatory, whereas higher concentrations inhibit functions in the receiving species. Whether by chance or selection, derivatives of cinnamic and benozic acid, coumarins, and various polyphenolic compounds like the tannins have multiple target sites whereby they alter the physiology of plants or microorganisms. There is sufficient evidence to conclude that these compounds first act by altering permeability and protein functions of the cell membrane. Subsequently, they impact major physiological processes that normally act to maintain a favorable ion balance, water content, photosynthetic rate, respiratory metabolism, and hormone regulation. One of the consistent effects of phenolic acids on higher plants is water stress that reduces cell expansion and causes more stomatal closure and a reduction in photosynthesis. However, we conclude that it is the general cytotoxicity of the phenolics and the resultant loss of efficiency in several cellular functions that reduce growth. In contrast to the phenolic acids, the mode of action of allelopathic flavonoids is less clear. Flavonoids inhibit electron transport in mitochondria and chloroplasts, yet other physiological effects cannot be ruled out.

A holistic view regarding the mode of action of phenolic allelochemicals recognizes the effects of these compounds as one of several stresses on plants. Complexes of the individual phenolic compounds become the effective unit causing allelopathy. In summary, explanations of the mode of action of phenolic allelochemicals must take into account that these compounds act in concert, the different compounds vary in toxicity but they have similarities in their mechanisms of action, and they all appear to disrupt cellular functions at multiple target sites.

REFERENCES

(1) Alsaadawi, I. S., Al-Hadithy, S. M., and Arif, M. B. **1986**. Effects of three phenolic acids on chlorophyll content and ions uptake in cowpea seedlings. *J. Chem. Ecol.* **12**, 221-227

(2) Avers, C. J. and Goodwin, R. H. **1956**. Studies on roots. IV. Effects of coumarin and scopoletin on the standard root growth pattern of *Phleum pratense*. *Amer. J. Bot.* **43**, 612-620

(3) Balke, N. E. **1985**. Effects of allelochemicals on mineral uptake and associated physiological processes. *ACS Symposium Series* **268**, 161-178

(4) Barkosky, R. R. and Einhellig, F. A. **1993**. Effects of salicylic acid on plant-water relationships. *J. Chem. Ecol.* **19**, 237-247

(5) Barkosky, R. R., Butler, J. L., and Einhellig, F. A. **1999**. Mechanisms of hydroquinone-induced growth regulation in leafy spurge (*Euphorbia esula* L.). *J. Chem. Ecol.* **25**, 1611-1621

(6) Barkosky, R. R., Einhellig, F. A., and Butler, J. L. **2000**. Caffeic acid-induced changes in plant-water relationships and photosynthesis in leafy spurge (*Euphorbia esula*). *J. Chem. Ecol.* **26**, 2095-2109

(7) Baziramakenga, R., Simard, R. R., and Leroux, G. D. **1994**. Effects of benzoic and cinnamic acids on growth, mineral composition, and chlorophyll content of soybean. *J. Chem. Ecol.* **20**, 2821-2833

(8) Baziramakenga, R., Leroux, G. D., and Simard, R. R. **1995**. Effects of benzoic and cinnamic acids on membrane permeability of soybean roots. *J. Chem. Ecol.* **21**, 1271-1285

(9) Bergmark, C. L., Jackson, W. A., Volk, R. J., and Blum, U. **1992**. Differential inhibition by ferulic acid of nitrate and ammonium uptake in *Zea mays* L. *Plant Physiol.* **98**, 639-645

(10) Blum, U. and Rice, E. L. **1969**. Inhibition of symbiotic nitrogen-fixation by gallic and tannic acid and possible roles in old-field succession. *Bull. Torrey Bot. Club* **96**, 531-544

(11) Booker, F. L., Blum, U., and Fiscus, E. L. **1992**. Short-term effects of ferulic acid on ion uptake and water relations in cucumber seedlings. *J. Exp. Bot.* **43**, 649-655

(12) Boufalis, A. and Pelllissier, F. **1994**. Allelopathic effects of phenolic mixtures on respiration of two spruce mycorrhizal fungi. *J. Chem. Ecol.* **20**, 2283-2289

(13) Cameron, H. J. and Julian, G. R., **1980**. Inhibition of protein synthesis in lettuce (*Latuca sativa* L.) by allelopathic compounds. *J. Chem. Ecol.* **6**, 989-995

(14) Colton, C. E. and Einhellig, F. A. **1980**. Allelopathic mechanisms of velvetleaf (*Abutilon theophrasti* Medic., Malvaceae) on soybean. *Amer. J. Bot.* **67**, 1407-1413

(15) Danks, M. L., Fletcher, J. S., and Rice, E. L. **1975**. Effects of phenolic inhibitors on growth and metabolism of glucose-UL-^{14}C in Paul's scarlet rose cell-suspension cultures. *Amer. J. Bot.* **62**, 311-317

(16) Demos, E. K., Woolwine, M., Wilson, R. H., and MacMillan, C. **1975**. The effects of ten phenolic compounds on hypocotyl growth and mitochondrial metabolism of mung bean. *Amer. J. Bot.* **62**, 97-102

(17) Devi, S. R. and Prasad, M. N. V. **1992**. Effect of ferulic acid on the growth and hydrolytic enzymes of germinating maize (*Zea mays* L.) seeds. *J. Chem. Ecol.* **18**, 1981-1990

(18) Einhellig, F. A. **1971**. Effects of tannic acid on growth and stomatal aperture in tobacco. *Proc. South Dakota Acad. Sci.* **50,** 205-209

(19) Einhellig, F.A. **1986**. Mechanisms and modes of action of allelochemicals. In: Putnam, A. and Tang, C. S. (Eds.), *The Science of Allelopathy*. John Wiley and Sons, Inc., New York, NY, 171-188

(20) Einhellig, F. A. **1987**. Interactions among allelochemicals and other stress factors of the plant environment. *ACS Symposium Series* **330**, 343-357

(21) Einhellig, F. A. **1989**. Interactive effects of allelochemicals and environmental stress. In: Chou, C. H. and Waller, G. R. (Eds.), *Phytochemical Ecology: Allelochemicals, Mycotoxins, and Insect Pheromones and Allomones*. Institute of Botany, Academia Sinica Monograph Series No. 9, Taipei, ROC, 101-118

(22) Einhellig, F. A. **1995**. Mechanisms of action of allelochemicals in allelopathy. *ACS Symposium Series* **582**, 96-116

(23) Einhellig, F. A. **1996**. Interactions involving allelopathy in cropping systems. *Agron. J.* **88**, 886-893

(24) Einhellig, F. A. **1999**. An integrated view of allelochemicals amid multiple stresses. In: Inderjit, Dakshini, K. M. M., and Foy, C. F. (Eds.), *Principles and practices in plant ecology: Allelochemical interactions*. CRC Press, Boca Raton, FL, 479-494

(25) Einhellig, F. A. **2001**. The physiology of allelochemical action: Clues and

views. In: Bonjoch, N. P. and Reigosa Roger, M. J. (Eds.), *First European OECD Allelopathy Symposium: Physiological Aspects of Allelopathy.* Vigo, Spain. Printed by Gamesal, S.A., 3-25

(26) Einhellig, F. A. and Eckrich, P. C. **1984**. Interactions of temperature and ferulic acid stress on grain sorghum and soybeans. *J. Chem. Ecol.* **10**, 161-170

(27) Einhellig, F. A. and Kuan, L.-Y. **1971**. Effects of scopoletin and chlorogenic acid on stomatal aperture in tobacco and sunflower. *Bull. Torrey Bot. Club* **98**, 155-162

(28) Einhellig, F. A. and Rasmussen, J. A. **1979**. Effects of three phenolic acids on chlorophyll content and growth of soybean and grain sorghum seedlings. *J. Chem. Ecol.* **5**, 815-824

(29) Einhellig, F. A. and Schon, M. K. **1982**. Noncompetitive effects of *Kochia scoparia* on grain sorghum and soybean. *Can. J. Bot.* **60**, 2923-30

(30) Einhellig, F. A., Schon, M. K., and Rasmussen, J. A. **1982**. Synergistic effects of four cinnamic acid compounds on grain sorghum. *J. Plant Growth Regul.* **1**, 251-258

(31) Einhellig, F. A., Stille Muth, M., and Schon, M. K. **1985**. Effects of allelochemicals on plant-water relationships. *ACS Symposium Series* **268**, 179-195

(32) Einhellig, F. A., Rice, E. L., Risser, P. G., and Wender, S. H. **1970**. Effects of scopoletin on growth, CO_2 exchange rates, and concentration of scopoletin, scopolin, and chlorogenic acids in tobacco, sunflower, and pigweed. *Bull. Torrey Bot. Club* **97**, 22-33

(33) Glass, A. D. M. **1973**. Influence of phenolic acids on ion uptake. I. Inhibition of phosphate uptake. *Plant. Physiol.* **51**, 1037-1041

(34) Glass, A. D. M. **1974**. Influence of phenolic acids on ion uptake. III. Inhibition of potassium absorption. *J. Exp. Bot.* **25**, 1104-1113

(35) Glass, A. D. M. and Dunlap, J. **1974**. Influence of phenolic acids on ion uptake. IV. Depolarization of membrane potentials. *Plant. Physiol.* **54**, 855-858

(36) Green, F. B. and Corcoran, M. R. **1975**. Inhibitory action of five tannins on growth induced by several gibberellins. *Plant Physiol.* **56**, 801-806

(37) Harper, J. R. and Balke, N. E. **1981**. Characterization of the inhibition of K^+ absorption in oat roots by salicylic acid. *Plant. Physiol.* **68**, 1349-1353

(38) Hogan, M. E. and Manners, G. D. **1991**. Differential allelochemical

detoxification mechanisms in tissue cultures of *Antennaria microphylla* and *Euphorbia esula*. *J. Chem. Ecol.* **17**, 167-174

(39) Holappa, L. D. and Blum, U. **1991**. Effects of exogenously applied ferulic acid, a potential allelopathic compound, on leaf growth, water utilization, and endogenous abscisic acid levels of tomato, cucumber, and bean. *J. Chem. Ecol.* **17**, 865-886

(40) Jacobson, A. and Corcoran, M. R. **1977**. Tannins as gibberellin antagonists in the synthesis of α-amylase and acid phosphatase by barley seeds. *Plant Physiol.* **59**, 129-133

(41) Klepper, L. **1991**. NO_x evolution by soybean leaves treated with salicylic acid and selected derivatives. *Pest. Biochem. Physiol.* **39**, 43-48

(42) Klessig, D. F. and Malamy, J. **1994**. The salicylic acid signal in plants. *Plant Mol. Biol.* **26**, 1439-1458

(43) Kobza, J. and Einhellig, F. A. **1987**. The effects of ferulic acid on the mineral nutrition of grain sorghum. *Plant and Soil* **98**, 99-109

(44) Leslie, C. A. and Romani, R. J. **1988**. Inhibition of ethylene biosynthesis by salicylic acid. *Plant Physiol.* **88**, 833-837

(45) Macri, R., Vianello, A., and Pennazio, S. **1986**. Salicylate-collapsed membane potential in pea stem mitochondria. *Physiol. Plant.* **67**, 136-140

(46) Moreland, D. E. and Novitzky, W. P. **1987**. Effects of phenolic acids, coumarins, and flavonoids on isolated chloroplasts and mitochondria. *ACS Symposium Series* **330**, 247-274

(47) Morgan, P. W. and Powell, R. D. **1970**. Involvement of ethylene in responses of etiolated bean hypocotyl hook to coumarin. *Plant Physiol.* **45**, 553-557

(48) Nyberg, P. F. **1986**. Effects of allelopathic chemicals on photosynthetic rate in *Lemna minor*. MS. Thesis. University of South Dakota, Vermillion, SD

(49) Patterson, D. T. **1981**. Effects of allelopathic chemicals on growth and physiological responses of soybean (*Glycine max*). *Weed Sci.* **29**, 53-59

(50) Prasad, M. N. V and Devi, S. R. **2001**. Physiological basis for allelochemical action of ferulic acid. In: Bonjoch, N. P., Reigosa Roger, M. J. (Eds.), *First European OECD Allelopathy Symposium: Physiological Aspects of Allelopathy*. Vigo, Spain. Printed by Gamesal, S.A., 27-45

(51) Quah, S. G. H. **1990**. The effects of five allelopathic chemicals on respiration of soybean mitochondria. MS Thesis. University of South Dakota, Vermillion, SD

(52) Rai, V. K., Sharma, S. S. and Sharma, S., **1986**. Reversal of ABA-induced stomatal closure by phenolic compounds. *J. Exp. Bot.* **37**, 129-134

(53) Raskin, I. **1992**. The role of salicylic acid in plants. *Annu. Rev. Plant Physiol. Plant Mol. Biol.* **43**, 439-463

(54) Retzinger, E. J., Jr. and Mallory-Smith, C. **1997**. Classification of herbicides by site of action for weed resistance management strategies. *Weed Technol.* **11**, 384-393

(55) Rice, E. L. **1984**. *Allelopathy*. 2nd. Edition, Academic Press, Orlando, FL, 422 p

(56) Scholes, K. A. **1987**. Effects of six classes of allelochemicals on growth, photosynthesis, and chlorophyll content in *Lemna minor*. MS Thesis. University of South Dakota, Vermillion, SD

(57) Schrader, K. K., de Regtm N. Q., Tidwell, R. R., Tucker, C. S., and Duke, S. O. **1998**. Selective growth inhibition of the musty-odor producing cyanobacterium *Oscillatoria cf. chalybea* by natural compounds. *Bull. Environ. Contam. Toxicol.* **60**, 651-658

(58) Schultz, M., Schnabl, H., Manthe, B., Schweihofen, B., and Casser, I. **1993**. Uptake and detoxification of salicylic acid by *Vicia faba* and *Fagopyrum esculentum*. *Phytochemistry* **33**, 291-294

(59) Schultz, M. and Wieland, I. **1999**. Variation in metabolism of BOA among species in various field communities — biochemical evidence for co-evolutionary processes in plant communities? *Chemoecology* **9**, 133-141

(60) Shann, J. R. and Blum, U. **1987**. The uptake of ferulic and *p*-hydroxybenzoic acids by *Cucumis sativus*. *Phytochemistry* **26**, 2959-2964

(61) Souto, C., Pellissier, F., and Chiapusio, G. **2000**. Allelopathic effects of humus phenolics on growth and respiration of mycorrhizal fungi. *J. Chem. Ecol.* **26**, 2015-2023

(62) Tomaszewski, M. and Thimann, K. V. **1966**. Interactions of phenolic acids, metallic ions and chelating agents on auxin-induced growth. *Plant Physiol.* **41**, 1443-1454

(63) Wink, M. and Latz-Brunning, B. **1995**. Allelopathic properties of alkaloids and other natural products. *ACS Symposium Series* **582**, 117-126

(64) Wixon, R. **1991**. Effects of several allelochemicals on water status of soybean. MS Thesis. University of South Dakota, Vermillion, SD

(65) Yang, C. F. **1996**. Action of allelochemicals on algal growth and photosystem II efficiency. MS Thesis. Southwest Missouri State University,

Springfield, MO

(66) Yang, C. M., Lee, C. N., and Chou, C. H. **2002**. Effects of three allelopathic phenolics on chlorophyll accumulation of rice (*Oryza sativa*) seedlings: I. Inhibition of supply-orientation. *Bot. Bull. Academia Sinica* **43**, (in press)

12 Mode of Action of the Hydroxamic Acid BOA and Other Related Compounds

A. M. Sánchez-Moreiras, T. Coba de la Peña, A. Martínez,
L. González, F. Pellisier, and M. J. Reigosa

CONTENT

Abstract 239
Introduction 240
Results and Discussion 246
Methodology 248
References 249

ABSTRACT

Certain secondary metabolites are released into the environment and affect the growth and development of different species. The mode/s of action of allelochemicals are diverse, and this knowledge is essential to developmental biology. Previous experiments showed an effect on germination and radicle growth of the hydroxamic acid, 2-benzoxazolinone (BOA), a compound released by plants (mainly grasses) into the environment. Alterations of plant energy metabolism were also reported for BOA. Other effects have been noted, but there are still some doubts about the precise mode of action in the affected plants.

Our research focused on some possible modes of action of hydroxamic acids. Thus, experiments were performed in plants and seedlings of *L. sativa* and membrane permeability correlated with an inhibition in mature plant growth or in cell cycle progression related to a seedling growth inhibition.

INTRODUCTION

Hydroxamic acids are commonly occurring secondary metabolites in cultivated and wild Gramineae.[16,27,43] This family of compounds has not been found in cereal seeds, but they can be detected easily in seedlings and mature plants of cereals such as wheat, corn or rye.[3] The presence of these compounds is species dependent and is also influenced by the age of the plant, temperature, photoperiod and organ assayed.[4,14,19] Their occurrence has been broadly related with the resistance of cereals (corn, wheat, wild rye, giant reed, etc) to insects and disease organisms.[3,23,24,25,43]

Hydroxamic acids are cyclic 4-hydroxy-1,4-benzoxazin-3-ones. The 2,4-dihydroxy-1,4-benzoxazin-3-one (DIBOA) is the major hydroxamic acid found in rye and it readily decarboxylates to form 2(3H)-benzoxazolinone (commonly called BOA). The production of these compounds starts when plant tissue is injured.[20] Other hydroxamic acids and their derivatives were also isolated and reported to play an important role in grass allelopathy.[6,9,11,15,22] The azoperoxide (AZOB), a microbiological conversion product of BOA, the major cyclic hydroxamic acid in corn DIMBOA, its degradation product MBOA, and other products, such as Cl-MBOA, HMBOA, 6-MBOA, and others, were found to have an effect on germination,[9,11,22] radicle elongation,[6,9] root and shoot growth,[21,22] auxin-induced elongation of coleoptiles,[2,29] and other physiological processes.[31,32]

BOA was first discovered by Virtanen and Hietala in 1957[39] as an anti-*Fusarium* compound in 4-days-old rye seedlings. Following this, different research projects were conducted to find the antipest significance of this compound in nature.[18,38,40,41] In recent years the biological activity studies of BOA have also focused on plant metabolism inhibitory activity. A major part of these studies was based on the germination rate and on the root and shoot growth measurements of the species tested.

Several studies have reported the effects of BOA on plant growth at different concentrations. In this way, Aiupova et al. in 1979[1] found that concentrations of 1.5 to 8.2 Kg/ha BOA inhibited 50% radicle elongation of cucumber, oat, radish, and cabbage. An abnormal growth was also observed by Wolf et al.[42] in velvetleaf seedlings when exposed to 5 mM BOA.

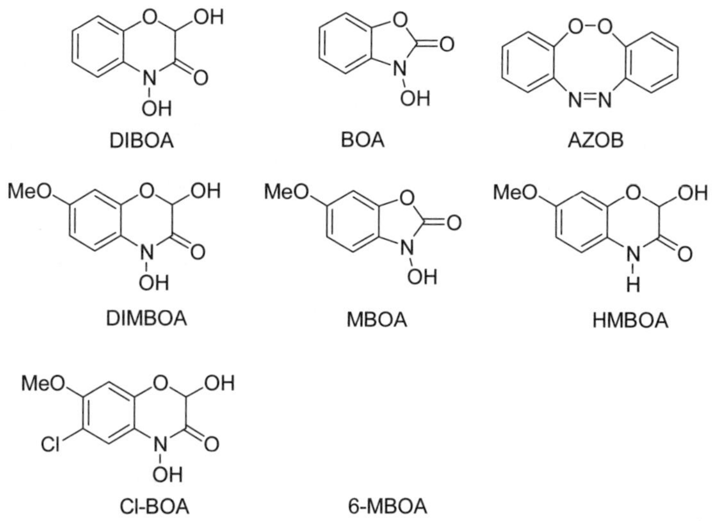

Figure 12.1

Benzoxazinones and related compounds obtained from different natural sources.

Putnam, Barnes and co-workers have broadly studied BOA and DIBOA as the responsible agents for the inhibitory activity of rye residues. In 1986, Barnes et al.[7] showed that concentrations of 25, 50 and 100 Kg BOA/ha tested in soil under field conditions highly inhibited the germination of cress and lettuce, while the seedlings of *Chenopodium album*, which appeared after treatment in the field, were chlorotic and stunted. Comparison of BOA and DIBOA effects at 67 to 250 ppm with PLA (ß-phenylacetic acid) and HBA (ß-hydroxybutyric acid) effects, also present in rye residues,[37] showed hydroxamic acids to be very toxic to the root growth of *Digitaria sanguinalis*, *Echinochloa crus-galli*, *Panicum miliaceum*, *Lycopersicon esculentum* Mill., and *Lactuca sativa*.[8] In general, dicotyledonous species were found to be approximately 30% more sensitive to BOA than monocotyledonous species. Thus, BOA appears to be more active in dicot and small-seeded species and strongly inhibitory to germination and seedling growth of dicotyledonous annual weeds.[6,11,26,33]

Similar results were observed in a more recent work[10] when BOA, DIBOA, and rye extract activities were studied. The results showed that BOA and DIBOA inhibit germination only in the small- to medium-seeded species *Amaranthus palmeri*, *Digitaria sanguinalis*, *Echinochloa indica*, *Lactuca sativa*, and *Lycopersicon esculentum*. Large-seeded crops, including cucurbits and *Zea mays*, were tolerant concluding that, "This bioassay indicated a promising potential for controlling small-seeded weeds in large-seeded crops."

Dose-dependent inhibition on root and coleoptile growth in the presence of BOA was also reported by Pérez and Ormeño-Núñez[30] for *Avena fatua* seedlings, by Chiapusio et al.[12] for *Lactuca sativa* var. Great Lakes seedlings, and by Kato-Noguchi[22] for *Amarantus caudatus*, L., *Lepidium sativum* L., *Lactuca sativa* L., *Digitaria sanguinalis* L., *Phleum pratense* L. and *Lolium multiflorum* Lam. Moreover, with respect to inhibition in radicle growth, Barnes and Putnam[5] could also detect necrosis of the apical root meristem. It closely resembled symptoms evident on lettuce germinating in soil/residue Petri dish bioassays. BOA turned lettuce root meristems black.

Focusing on the mechanisms of action of BOA into the plant cell, Barnes et al.[7] suggested that the chlorotic seedlings observed in the presence of BOA and DIBOA could be the consequence of a benzoxazinone effect on the photophosphorylation and electron transport into the plant metabolism. In this way, Niemeyer et al.[28] studied the effects of BOA on energy-linked reactions in mitochondria and reported an inhibition of the electron transfer between flavin and ubiquinone in Complex I, with complete inhibition of electron transport from NADH to oxygen in SMP. They could also detect an inhibition of BOA on ATP synthesis by acting directly on the ATPase complex at the F1 moiety.

ATPase activity was also studied by Friebe et al. in 1997.[17] They correlated the BOA and DIBOA effects on radicle elongation of *Avena sativa* seedlings with their effects on the activity of plasma membrane H^+-ATPase from roots of *Avena sativa* cv. Jumbo and from *Vicia faba* cv. Alfred. They hypothesized that an alteration in the plasma membrane ATPase activity could be the reason for an abnormal nutrient absorption in plants exposed to hydroxamic acids, because of the role that this enzyme plays in the ion gradient and, therefore, in the ionic transport through plasma membrane. The results of this experiment showed a strong inhibition in the activity of this enzyme in the plasma membrane of chloroplast and mitochondria when it was exposed to BOA and DIMBOA. This alteration implies early interactions with the assayed hydroxamic acids.

Investigations carried out by our group[34] to determine the effect of BOA on membrane permeability of lettuce roots showed an increase in the release of the ions studied (Fig. 12.2). This disruption was time dependent both for anions and for cations, showing a continuous time release and not a bolus effect. Significant increases in NO_3^-, PO_4^{3-}, SO_4^{2-}, Ca^{2+} or NH_4^+ release suggested an alteration in membrane permeability of lettuce root cells.

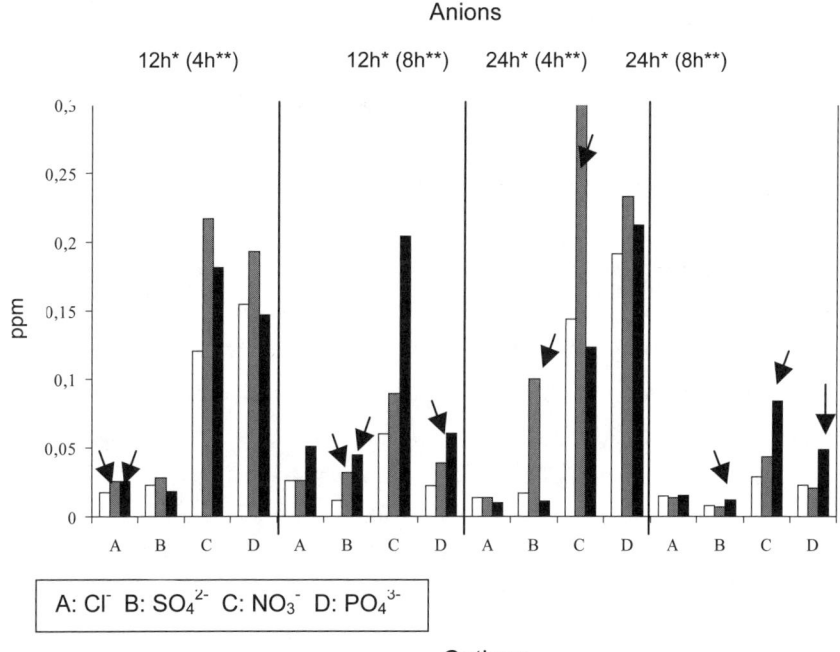

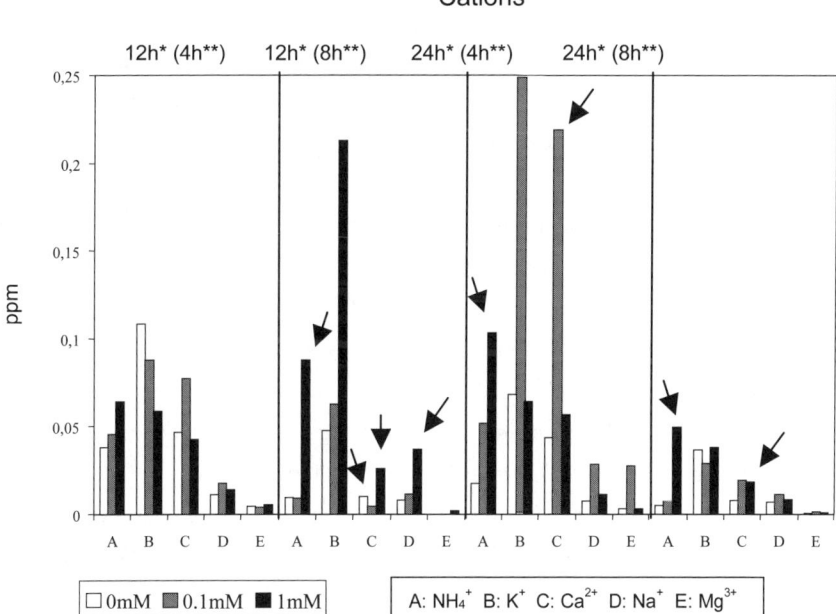

Figure 12.2
Concentration of anions and cations released by roots of mature plants of *L. sativa* exposed to two different time periods (12 h and 24 h) at different concentrations (0; 0.1 and 1 mM) of 2-Benzoxazolinone (BOA). Ions were collected at two washing times (0-4 h and 4-8 h). * Root exposure time (h); ** Root washing time (h); ↓Arrows indicate significant differences at 0,05. (From Reigosa, M. J. et al. **2001**. *Allelopathy J.* **8**, 211-220. With permission).

The objective of this study was also to establish a link between an easily observable allelopathic effect (growth inhibition of roots) and a cellular-based explanation (membrane permeability changes). Several modes of action can explain these results, such as an increase in the lipid peroxidation, a disruption in the intrinsic membrane protein activities, or an alteration in plasma membrane ATPase activity.

Once clear inhibitory effects were detected on germination and radicle growth of lettuce plants exposed to BOA,[12,35] our group developed methodology to show whether there is a detectable BOA effect on the cell cycle progression in lettuce root meristems.[13] This allows a more detailed characterization of the putative direct or indirect effect on the meristems of BOA-treated seedlings.

Kinetics, rate, and cell cycle progression can be adequately studied by flow cytometry. This technique allows recording the length and cell percentage in the G0+G1, S, and G2 phases of the cell cycle. Lettuce root meristems were analyzed after synchronization for 6 h with hydroxyurea (HU), which reversibly inhibits the production of desoxyribonucleotides by inhibiting the ribonucleotide reductase, thereby activity-blocking the cell cycle in G1 and early S phase.[15] The synchronization allows for a higher number of cells in the same phase of the cell cycle, so that weaker or stronger cell cycle specific effects can be detected. In synchronized plant cell cultures, S nuclei can represent more than 50% of the total population.

When the HU synchronization period was finished, and the HU was removed, the seedlings were immediately incubated with 1 mM BOA or water (controls), and the cell cycle was analyzed at different times. Samples (nuclear suspensions) are prepared from root meristems and analyzed by flow cytometry every 2 h during a time period of 12-14 h.[15] Comparison of the synchronized-cell cycle progression of BOA-treated plants with that of the corresponding controls allows for detection of partial or total inhibition of the cell cycle.

Furthermore, cell cycle analysis by flow cytometry can be improved by using other complementary techniques that provide additional information. This is the case of the mitotic index that informs us about the number of cells undergoing mitosis during the experiment. This measurement was also recorded in our investigation, and the results indicated its excellent complementary information.

Mitotic indices were counted in the root tip cells from lettuce root meristems treated with 1 mM BOA (treated-meristems) or with distilled water (control-meristems) after 6 h HU synchronization (Fig. 12.3).[36]

Mode of Action of the Hydroxamic Acid BOA

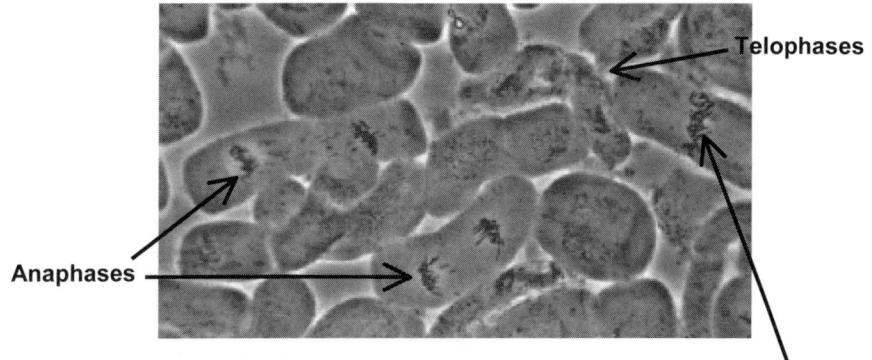

Figure 12.3

Mitotic cells in lettuce root meristems.

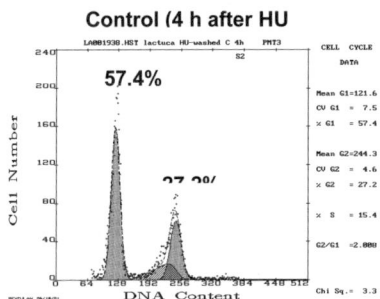
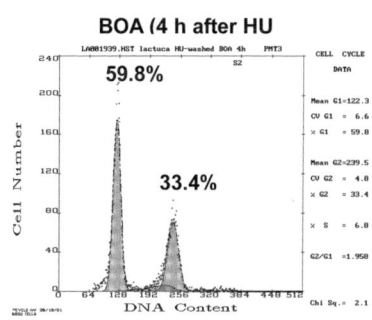
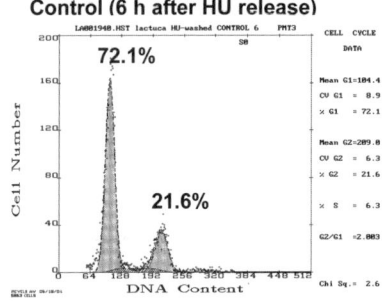
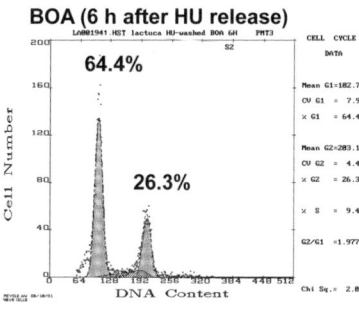

Figure 12.4

Comparative cell cycle analysis of 6 h hydroxyurea-blocked root cells from 1 mM BOA-treated and control lettuce meristems. Seedlings were exposed to treatment after HU release, and then nuclear suspensions were prepared with 40 meristems per treatment and analyzed by flow cytometry every 2 h. (From Coba de la Peña, T. and Sánchez-Moreiras, A. **2001**, *Handbook of Plant Ecophysiology Techniques*, Kluwer Academica Publishers, Dordrecht, the Nederlands, pp. 65-68. With permission).

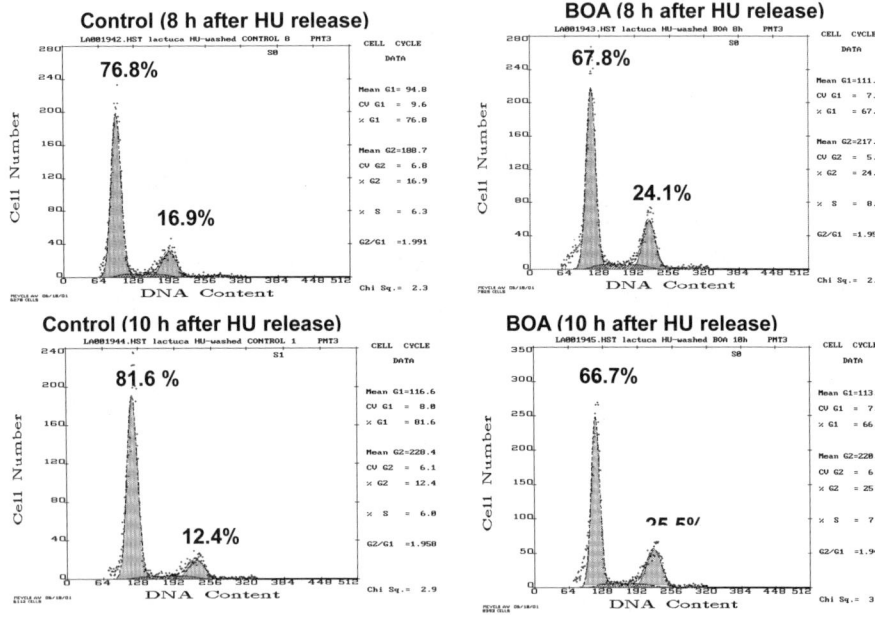

Figure 12.4 (cont.)

Comparative cell cycle analysis of 6 h hydroxyurea-blocked root cells from 1 mM BOA-treated and control lettuce meristems. Seedlings were exposed to treatment after HU release, and then nuclear suspensions were prepared with 40 meristems per treatment and analyzed by flow cytometry every 2 h. (From Coba de la Peña, T. and Sánchez-Moreiras, A. **2001**, *Handbook of Plant Ecophysiology Techniques*, Kluwer Academica Publishers, Dordrecht, the Nederlands, pp. 65-68. With permission).

RESULTS AND DISCUSSION

Figure 12.4 shows the cell cycle progression in treated- (1 mM BOA) and control-meristems (distilled water) in synchronized lettuce meristems (unpublished data). Here we see that after 4 h treatment (the cells progressed in this time from G0+G1, where HU blocked cell cycle, to S and from S to G2) the number of cells in G2 phase in treated meristems is higher than in controls. In the next 6 hours, the control meristems normally continue the cell cycle and more cells went from G2 to G0+G1 phase starting, again, division in an asynchronous way. However, in BOA-treated-meristems during these same 6 hours, a high number of cells in division seem to be blocked at G2 phase. The cell cycle goes so slowly in these meristems that 10 h after treatment the cells in G2 phase are double that of controls.

So, these results showed a double effect of BOA on lettuce meristems: an increasing significant delay in the cell cycle progression and a decrease in the mitotic index. Flow cytometry analysis showed a weak effect of BOA at cell cycle level in the step from G2 to M. BOA effect appeared to retard the cell cycle progression of treated-root meristems, which was very clear after 10 h BOA exposition (Fig. 12.5), and as an inhibition of the number of cells undergoing cell division.

The effects exhibited in the flow cytometry analysis are also clearly detected by the mitotic index technique (unpublished data), which appears to be an excellent complementary technique in the cell cycle studies using flow cytometry. The mitotic index (see Fig. 12.5) revealed that the cell cycle progression goes slower in BOA-treated meristems than in control meristems and also that the maximum number of cells undergoing cell division is fewer and later in treated meristems (18% at 6 h BOA exposure time) than in control meristems (36% at 4 h distilled water exposure time).

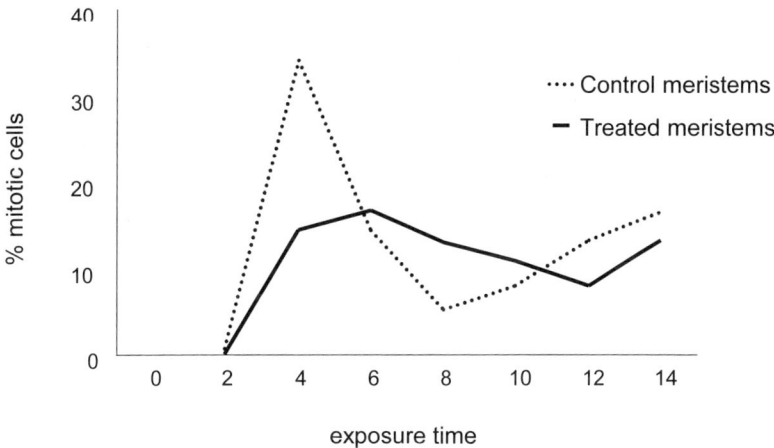

Figure 12.5

Mitotic index in root tip cells from lettuce meristems treated for 2-14 h with 1 mM BOA (treated meristems) or distilled water (control meristems) after 6 h HU synchronization. Mitotic cells were counted for each treatment every 2 h after HU release (From Sánchez-Moreiras, A. M. et al. **2001**. *Handbook of Plant Ecophysiology Techniques*. Kluwer Academic Publishers, Dordrecht, The Netherlands, 81-95. With permission).

So, we can conclude that a clear effect of BOA on lettuce cell cycle can be considered as an important mechanism of action for this compound on the inhibition of seedling growth, in this plant species. This effect could be related, and affected, by the previously reported effects on auxins, which suggested the interference of BOA and other hydroxamic acids with auxins at the cell cycle level.[29]

METHODOLOGY

Cell cycle analysis[13]

Lactuca sativa seeds were germinated at 27°C in the dark for 20 h. Twenty 1-3 mm-root length seedlings were transferred to Petri dishes with 2.5 mM hydroxyurea and incubated for 6 h in the dark. HU was removed by washing with distilled water, and immediately after, seedlings were transferred to other Petri dishes with 1 mM BOA (treatment) or distilled water (control). These seedlings were incubated at 27°C in the dark.

Then, and every 2 h, samples of BOA-treated plants and the corresponding controls were simultaneously processed using flow cytometry analysis. The 1 mm-apical tips of root meristems from forty BOA-treated plants were chopped with a razor blade in Galbraith buffer, supplemented with 100% Tween 20 and 100% beta-mercaptoethanol. Once filtered, the suspension, 5 µL of 1% RNase solution and 30 µL of 10 mg/mL ethidium bromide (EtBr), were added to 500 µl of filtered nuclei suspension and incubated for 30 min at room temperature in the dark.

After incubation started, cell cycle analysis by flow cytometry and cell cycle histograms were recorded for BOA-treated and control plants every 2 h, until 12 or 14 h of incubation with BOA. At least 10,000 nuclei from each sample must be analyzed in the flow cytometer.

Histogram profiles were analyzed using the computer program Multicycle (Flow Systems, San Diego), and G0+G1, S and G2 populations were estimated comparatively in control and BOA-treated plants.

Mitotic index[36]

HU-synchronized root meristems were fixed for 24 h with acetic acid/chloroform/ethanol (6:3:1) and trace iron, then stored before analysis at -20 °C for at least 3 days. After storage, samples were hydrolyzed with hot 1N HCl for 25 min at 60°C to allow dispersion of cells and chromosomes. Samples were stained after hydrolysis with Schiff's reagent. Finally, meristems were embedded in a drop of acetic acid, cut on a slide, and heated in a flame. After covering with the coverslip, the meristem was squashed and heated again.

Once the sample was prepared, meristems were scored with light microscopy using the x40 objective, and the mitotic index was estimated by counting up a total number of 1000 cells in three slides of the same sample.

ACKNOWLEDGMENTS

We are grateful to Dr. Marina Horjales, Dr. Nieves Redondo, and Alfonso Blanco (Faculty of Sciences, University of Vigo) for their help in the optimization of cell cycle analysis and mitotic index technique applied to lettuce. The authors also want to thank Claudia Cárcamo and Oliver Weiss for their continual and valuable help in these experiments. Financial support from Ministry of Science and Technology (Spanish Government) is also acknowledged.

REFERENCES

(1) Aiupova, A. T., Molchanov, L. V., Kadryov, C. S., Aliev, N. A., Giasov, K., Loi, N. P., Tsoi, Z., and Umarov, A. A. **1985**. Benzoxazolinone, 2,4-dihydroxy-1,4-benzoxazin-3-one, and its glucoside from *Acanthus mollis* seeds inhibit velvetleaf germination and growth. *J. Nat. Prod.* **48**, 59-63

(2) Anai, T., Aizawa, H., Ohtake, N., Yamamura, S., and Hasegawa, K. **1996**. A new auxin-inhibiting substance, 4-Cl-6,7-dimethoxy-2-benzoxazolinone, from light grown maize shoots. *Phytochemistry* **42**, 273-275

(3) Argandoña, V. H., Luza, J., Niemeyer, H. M., and Corcuera, L. J. **1980**. Role of hydroxamic acids in the resistance of cereals to aphids. *Phytochemistry* **19**, 1665-1668

(4) Argandoña, V. H., Niemeyer, H. M., and Corcuera, L. J. **1981**. Effect of content and distribution of hydroxamic acids in wheat on infestation by the aphid *Schizaphis graminum*. *Phytochemistry* **20**, 673-676

(5) Barnes, J. P. and Putnam, A. R. **1986**. Evidence for allelopathy by residues and aqueous extracts of rye (*Secale cereale* L.). *Weed Sci.* **34**, 384-390

(6) Barnes, J. P. and Putnam, A. R. **1987**. Role of benzoxazinones in allelopathy by rye (*Secale cereale* L.). *J. Chem. Ecol.* **13**, 889-905

(7) Barnes, J. P., Putnam, A. R., and Burke, B. A. **1986**. Allelopathic activity of rye (*Secale cereale* L.). In: Putnam, A.R., Tang, Ch-Sh. (Eds.), *The Science of Allelopathy*. John Wiley and Sons, New York. pp. 271-286

(8) Barnes, J. P., Putnam, A. R., Burke, B. A. and Aasen, A. J., **1987**. Isolation and characterization of allelochemicals in rye herbage. *Phytochemisry* **26**, 1385-1390

(9) Blum, U., Gerig, T. M., Worsham, A. D., Holappa, L. D., and King, L. D. **1992**. Allelopathic activity in wheat-conventional and wheat-no-till soils: development of soil extract bioassays. *J. Chem. Ecol.* **18**, 2191-2221

(10) Burgos, N. R. and Talbert, R. E. **2000**. Differential activity of allelochemicals from *Secale cereale* in seedling bioassays. *Weed Sci.* **48**, 302-310

(11) Chase, W. R., Nair, M. G. and Putnam, A. R. **1991**. 2,2'–oxo-1,1'-azobenzene: selective toxicity of rye (*Secale cereale* L.) allelochemicals to weed and crop species: II. *J. Chem. Ecol.* **17**, 9-19

(12) Chiapusio, G., Sánchez, A. M., Reigosa, M. J., González, L. and Pellissier, F. **1997**. Do germination indices adequately reflect allelochemical effects on the germination process? *J. Chem. Ecol.* **23**, 2445-2453

(13) Coba de la Peña, T. and Sánchez-Moreiras, A. M. **2001**. Flow cytometry: Cell cycle In: Reigosa, M.J. (Ed.), *Handbook of Plant Ecophysiology Techniques*. Kluwer Academic Publishers, Dordrecht, The Netherlands, 65-80

(14) Copaja, S. V., Nicol, D., and Wratten, S. D. **1999**. Accumulation of hydroxamic acids during wheat germination. *Phytochemistry* **50**, 17-24

(15) Doležel, J., Cíhalíková, J., Weiserová, J., and Lucretti, S. **1999**. Cell cycle synchronization in plant root meristems. *Meth. Cell Sci.* **21**, 95-107

(16) Friebe, A., Schulz, M., Kück, P., and Schnabl, H. **1995**. Phytotoxins from shoot extracts and root exudates of *Agropyron repens* seedlings. *Phytochemistry* **38**, 1157-1159

(17) Friebe, A., Roth, U., Kück, P., Schnabl, H., and Schulz, M. **1997**. Effects of 2,4-dihydroxy-1,4-benzoxazin-3-ones on the activity of plasma membrane H^+-ATPase. *Phytochemistry* **44**, 979-983

(18) Friebe, A., Vilich, V., Henning, L., Kluge, M., and Sicker, D. **1998**. Detoxification of benzoxazolinone allelochemicals from wheat by *Gaeumannonmyces graminis* var. tritici, *G. graminis* var. graminis, *G. graminis* var. avenae and *Fusarium culmorum*. *Appl. Environ. Microbiol.* **64**, 2386-2391

(19) Gianoli, E. and Niemeyer, H. M. **1997**. Environmental effects on the

accumulation of hydroxamic acids in wheat seedlings: the importance of plant growth rate. *J. Chem. Ecol.* **23**, 543-551

(20) Hofman, J. and Hofmanova, O. **1971**. 1,4-Benzoxazine derivative in plants: absence of 2,4-dihydroxy-7-methoxy-2H-1,4-benzoxazin-3(4H)-one from uninjured *Zea mays* plants. *Phytochemistry* **10**, 1441-1444

(21) Kato-Noguchi, H., Kosemura, S., and Yamamura, S. **1998**. Allelopathic potential of 5-chloro-6-methoxy-2-benzoxazolinone. *Phytochemistry* **48**, 433-435

(22) Kato-Noguchi, H. **2000**. Allelopathy in maize II: Allelopathic potential of a new benzoxazolinone. *Plant Prod. Sci.* **3**, 47-50

(23) Klenke, J. R., Russell, W. A., Guthrie, W. D., Martinson, C. A., and Pedersen, W. L. **1987**. Disease resistance in five cycles of BS9 corn synthetic selected for resistance to two generations of European corn borer. *Phytopathology* **77**, 735-739

(24) Leszczynski, B. and Dixon, A. F. G. **1990**. Resistance of cereals to aphids: interaction between hydroxamic acids and the aphid *Sitobion avenae* (Homoptera: Aphididae). *Ann. Appl. Biol.* **117**, 21-30

(25) Long, B. J., Dunn, G. M., Bowman, J. S., and Routley, D. G. **1977**. Relationship of hydroxamic acid content in corn and resistance to the corn leaf aphid. *Crop Sci.* **17**, 55-58

(26) Nair, M. G., Whitenack, C. J., and Putnam, A. R. **1990**. 2,2'–Oxo-1,1'-azobenzene. A microbially transformed allelochemical from 2,3-benzoxazoline: I. *J. Chem. Ecol.* **16**, 353-364

(27) Niemeyer, H. M. **1988**. Hydroxamic acids (4-hydroxy-1,4-benzoxazin-3-ones), defense chemicals in Gramineae. *Phytochemistry* **27**, 3349-3358

(28) Niemeyer, H. M., Calcaterra, N. B., and Roveri, O. A. **1987**. Inhibition of energy metabolism by benzoxazolin-2-one. *Comp. Biochem. Physiol.* **87B**, 35-39

(29) Pérez, F. J. **1990**. Allelopathic effect of hydroxamic acids from cereals on *Avena sativa* and *A. fatua*. *Phytochemistry* **29**, 773-776

(30) Pérez, F. J. and Ormeño-Núñez, J. **1993**. Weed growth interference from temperate cereals: the effect of a hydroxamic-acids-exuding rye (*Secale cereale* L.) cultivar. *Weed Res.* **33**, 115-119

(31) Pethö, M. **1992**. Occurrence and physiological role of benzoxazinones and their derivatives. IV. Isolation of hydroxamic acids from wheat and rye root ssecretions. *Plant Physiol. Agrochem.* **41**, 167-175

(32) Pethö, M. **1993**. Possible role of hydroxamic acids in the iron uptake by

grasses. *Acta Agronom. Hungar.* **42**, 203-214

(33) Putnam, A. R., Nair, M. G. and Barnes, J. P., **1990**. Allelopathy: a viable weed control strategy In: Baker, R.R., Dunn, P.E. (Eds.). *UCLA Symposia on Molecular and Cellular Biology.* Frisco, CL, 317-322

(34) Reigosa, M. J., González, L., Sánchez-Moreiras, A. M., Durán, B., Puime, O., Fernández, A., and Bolaño C., **2001**. Comparison of physiological effects of allelochemicals and commercial herbicides. *Allelopath. J.* **8**, 211-220

(35) Sánchez-Moreiras, A. M., **1996**. Efectos fisiológicos producidos por la acción de aleloquímicos. Minor Thesis. University of Vigo, Vigo, Spain

(36) Sánchez-Moreiras, A. M., Coba de la Peña, T., Martínez Otero, A., and Blanco Fernández, A., **2001**. Mitotic index. In: Reigosa, M.J. (Ed.), *Handbook of Plant Ecophysiology Techniques.* Kluwer Academic Publishers, Dordrecht, The Netherlands, 81-95

(37) Shilling, D. G., Liebl, R. A., and Worsham, A. D. **1985**. Rye (*Secale cereale* L.) and wheat (*Triticum aestivum* L.) mulch: the suppression of certain broadleaved weeds and the isolation and identification of phytotoxins. In: Putnam, A. R. and Tang, Ch.-Sh. (Eds.), *The Chemistry of Allelopathy.* John Wiley and Sons, NY, 243-271

(38) Vilich, V., **1998**. Effect of benzoxazolinone allelochemicals from wheat on selected soil-borne pathogens and antagonists. *Toegepaste Biologishe Wetenshappen Universit. Gernt.* **63**, 971-976

(39) Virtanen, A. I., Hietala, P. K., and Wahlross, O. **1957**. Antimicrobial substances in cereals and fodder plants. *Arch. Biochem. Biophys.* **69**, 486-500

(40) Wahlroos, O. and Virtanen, A. I. **1958**. On the antifungal effect of benzoxazolinone and 6-methoxy-benzoxazolinone, respectively, on *Fusarium nivale. Acta Chem. Scand.* **12**, 124-128

(41) Wilkes, M. A., Marshall, D. R., and Copeland, L. **1999**. Hydroxamic acids in cereal roots inhibit the growth of take-all. *Soil Biol. Biochem.* **31**, 1831-1836

(42) Wolf, R. B., Spencer, G. F. and Plattner, R. D., **1985**. Benzoxazolinone, 2,4-dihydroxy-1,4-benzoxazin-3-one, and its glucoside from *Acanthus mollis* seeds inhibit velvetleaf germination and growth. *J. Nat. Prod.* **48**, 59-63

(43) Zúñiga, G. E., Argandoña, V. H., Niemeyer, H. M., and Corcuera, L. J. **1983**. Hydroxamic acid content in wild and cultivated Gramineae. *Phytochemistry* **22**, 2665-2668

13 Mode of Action of Phytotoxic Fungal Metabolites

H. G. Cutler, S. J. Cutler, and D. Matesic

CONTENT

Abstract 253
Introduction 254
Results and Discussion 254
    Physical Aspects of Allelopathic Induction 254
    Chemical Aspects of Allelopathy 256
References 267

ABSTRACT

A brief outline of the physical aspects of phytopathogenesis is followed by examination of the mode of action of a select series of phytotoxic fungal metabolites. Included is moniliformin from *Fusarium moniliforme*, which appears to act in part by inhibiting spindle microtubule formation in mitosis. A discussion of the cytochalasins selectively refers to cytochalasin B, which influences cytoplasmic streaming. The latter, in turn, is dependent upon the action of compounds on subcortical actin bundles. Chaetoglobosin K, an indolylcytochalasin, inhibits etiolated wheat coleoptiles in an odd manner and is also effective in gap junction physiology. Cytochalasin H is discussed with respect to its overall effects on plant growth, especially flowering in tobacco, *Nicotiana tabacum*. The mode of action of the trichothecenes as either initiation inhibitors or elongation and/or termination inhibitors relative to protein synthesis by acting on peptidyl transferase at the site of the 60S ribosome is also considered.

INTRODUCTION

Any discussion concerning the mode of action of phytotoxic microbial metabolites is fraught with problems for at least two very obvious reasons. First, myriad biologically active natural products (allelochemicals) have been isolated, with their activity reported in considerable detail, and second, their mode of action has been largely ignored. One may well ask the reason for this huge gap in the knowledge, and the answer is quite simple. Most, if not all, of these compounds have not been commercially developed for practical use as agrochemicals. True, some of them have been patented, but by and large the topic of their mode of action has not been germane in the field of plant science. However, odd events happen in research, and microbial metabolites that have medicinal properties have found their way to the marketplace. But, in order to arrive there, their mode of action has been critically determined solely to comply with the Food and Drug Administration requirements necessary to the development of a commercial product.

Sometimes, by sheer luck, a natural pharmaceutical product is accidentally reisolated by a plant scientist whose interest is the discovery of new agrochemicals. And, sometimes, the development of a pharmaceutical is concurrent with work being carried out in an agrochemical laboratory, albeit unknown to either party.

While the mode of action of microbial metabolites may differ between plants and 'medicinal' systems, it is surprising how often the two are relatively identical. We now examine certain classes of compounds and note their mode of action in both plants and therapeutic systems.

RESULTS AND DISCUSSION

PHYSICAL ASPECTS OF ALLELOPATHIC INDUCTION

Often, great focus is placed on an allelopathic agent, generally a bioactive natural product, without much emphasis being placed on the physical events that transpire between the donor and the recipient. Suffice it to say that in special cases, such as in phytopathogenesis where the donor is a pathogen, the recipient a plant, and the lethal agent a discreet molecule, there is the ever present problem of the establishment of the pathogen on the host. While, at first, this may appear to be somewhat trivial, the mechanisms whereby the relationships are established are highly sophisticated. And while this topic would take a great deal of space to explain, a précis of the situation is given.

The spore of a pathogen may land on a leaf and, because of the morphology of the latter, become lodged in a suitable environment. The first event is that the spore germinates sending out a germ tube or, in some cases, an appressorium and in turn a penetration peg that forces its way through the protective cuticle into the underlying cells.[42] The main hyphae will then generate haustoria and the mycelium branches throughout the leaf cells, disrupting them and causing damage and, eventually, death. However, aligned with the rapid mycelial growth may be the introduction of cutinase, an enzyme that has been isolated and experimentally proved[26] without which the establishment of the phytopathogen, for example *Fusarium solani* f sp. *pisi*, cannot be successful; therefore, this is the rate limiting step in the delivery of the allelochemical. The mechanism is a wonderful example of natural systems in that cutinase breaks the cutin polymers into monomers and the monomers elicit the pathogenic spores to produce more cutinase. An analogous situation occurs with pectinase, so that a pathogen will degrade carbohydrate polymers into oligomers and monomers, thereby stimulating pectinase production.

In the case of the rust fungus, *Uromyces appendiculatus*, the mode of action is highly regulated. The spores, upon germinating, produce hyphae that are directionally oriented to the stomates, through which they enter, gaining access to the subcuticular cells. There is a simple ridge on the leaf of the bean plant, *Phaseolus vulgaris*, ideally 0.5 μmeters high, which acts as a sensor for the pathogen. Remarkably, the stomatal lip, which is adjacent to the guard cell, is approximately the same height. All this has been proven experimentally in some very clever experiments wherein silicon wafers were etched with ridges ranging from 0.25 to 1.0 μmeters, spores of *U. appendiculatus* germinated, and the orientation of the hyphae observed. A greatly expanded treatise has been written on this subject and should be read by any serious researcher in allelopathy.[18]

In attempting to be as brief as is concise with clarity, it is obvious that the subject matter has been considerably condensed to a series of fairly simple statements. However, it should be abundantly clear that the vectoring of allelochemicals into plant systems might include specific enzymes, transformation by microorganisms, and, apparently, mechanisms such as vector nematodes.

CHEMICAL ASPECTS OF ALLELOPATHY

Moniliformin

Moniliformin, 3-hydroxycyclobut-3-ene-1,2-dione (Fig. 14.1), was first isolated from cultures of *Fusarium moniliforme* Sheldon, isolated from infected corn, *Zea mays*. Initially cultured on cracked corn in the laboratory, the compound was finally purified by crystallization from aqueous methanol and immediately exhibited odd physical characteristics. First, it decomposed without giving a melting point at temperatures up to 350°C, and second, the ^{1}H NMR indicated a single proton, which, at the time, posed something of a problem because it was impossible to integrate a single proton. Eventually, the absolute structure was unequivocally established by single crystal X-ray crystallography and synthesis.[34] Early examination of the molecule predicted, incorrectly, that the structure would most probably be unstable because of the C3-C4 double bond. Another vexing problem was that the organism whimsically biosynthesized either the sodium or the potassium salt, which meant that the product had to be evaluated from fermentation to fermentation. With a mass difference between the two salts of 16, based on the overall mass, the effect on specific activity had to be taken into account. This problem was resolved by synthesis where the free acid had a constant MW 98.0081.

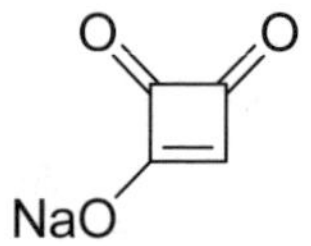

Figure 13.1

Chemical structure of moniliformin, toxin isolated from *Fusarium moniliforme* Sheldon.

It should be understood that the isolation and identification of moniliformin was predicated on its biological activity and that because of the composition of the research team, both animal and plant bioassays were concurrent. Tests with day-old cockerels demonstrated that the compound had an LD50 of 4.0 mg/Kg when administered via intubation in water. Higher doses at 12.5 and 6.25 mg/cockerel induced small hemorrhages in the large and small intestines, gizzard, proventriculus, and skin.[6] Mice required larger doses to induce an effect, and the results were gender dependent so that the LD50 was 29.1 mg/Kg for males and 20.9 mg/Kg for females. Later experiments showed that 5 µM and 4 µM of moniliformin specifically inhibited pyruvate and α-ketoglutarate oxidation,[3,37] critical steps in metabolic pathways in both animals and plants. It is precisely these two vulnerable metabolic events that may modulate the phytotoxic effects in plants.

Results in plant assays indicated that that there existed a potential for developing a practical agrochemical. Primary experiments, using the etiolated wheat coleoptile, *Triticum aestivum* cv. Wakeland, assay, showed that they were significantly inhibited ($p< 0.01$) 57 and 24%, respectively, at 1.6 and 0.16 mM. Most importantly, there was not a 100% inhibition, and the reason for this will become apparent. In turn, this led to greenhouse experiments on intact higher plants using week-old bean, *Phaseolus vulgaris* cv Black Valentine, corn, *Zea mays* cv Norfolk Market White, and six-week-old tobacco seedlings, *Nicotiana tabacum* cv Hick's. Following treatment at 266 and 26.6 g/ha, there was significant growth inhibition in corn, while bean showed some phytotoxic effects. The most dramatic response was obtained in tobacco and the reduction in plant height lasted at least three weeks under ideal growing conditions. The higher concentration of moniliformin created rosette type plants.[6] Because we were searching for a biodegradable substitute for maleic hydrazide (MH 30), 1,2-dihydro-3, 6-pyridazinedione, to control axillary growth in tobacco, application of the metabolites was made to mature field grown tobacco plants from which the floral apices had been removed. The latter is a normal cultural practice, which allows the energy, normally used to produce flowers and seed, to be utilized for larger leaf production. The leaves are the marketable commodities. But a consequence of the action is to destroy apical dominance, causing the axillary leaves to grow, which, in turn, also takes photosynthate from the leaves. Moniliformin performed better than MH 30 at 266 and 26.6 g/ha in controlling axillary shoot growth, but, unfortunately, the effects lasted only three weeks, whereas MH 30 lasted until harvest time. Yields of axillary shoots expressed at ratios, by weight, were MH 30: moniliformin: controls, 1: 2; 7 at three weeks. By way of contrast, MH 30 was used at 3.39 Kg/ha.[12]

The point arose as to the actual site of action of moniliformin in plants. Earlier, it had been stated that the geometrical criteria of hydrogen bond formation suggested that MH could be regarded as either a purine or pyrimidine analogue and, therefore, could substitute as one of the base pairs in nucleic acid.[11] And the fact that it had shown plant growth regulatory activity had led an industrial company to become interested in the product as a potential herbicide.

A series of experiments using corn root tips, *Zea mays* cv Norfolk Market White, was established using the free acid from 10^{-3} to 10^{-4} M. Results definitively showed that mitosis was inhibited at the C-metaphase plate at both concentrations because of inhibition of the spindle microtubules.[35] In this regard, the compound produced the same effect as colchicine. However, moniliformin is hydrophilic, whereas colchicine is lipophilic. This mode of action explains, in part, why etiolated wheat coleoptiles are moderately inhibited, probably because microtubule disruption is very limited in already formed cells, while tobacco axillary shoots, which are undergoing rapid mitosis, are fully inhibited until such time as the compound is metabolized.

Several derivatives of moniliformin were synthesized and patented. Swiss patent #609,836 was awarded to Fischer and Bellus (Ciba Geigy) in 1979. *"The cyclobutene-3, 4-diones, R-alkyl, substituted Ph, CH2CH2SEt, CH2CH:CH2, etc. are herbicides and PGR's. Thus postemergence application of 4Kg 1-octyloxycyclobutene-3, 4-dione/ha controlled* Setaria italica, Lolium perenne, Sinapis alba *and* Stellaria media *more than did the standard moniliformin Na salt"* (*Chem. Abstracts.* Vol 90:1979, Indent 90:1988 82p).

Cytochalasins

Another group of fascinating, biologically active natural products are the cytochalasins and their homologues, the chaetoglobosins. The cytochalasins, first discovered in 1964, were first the domain of zoologists who noted their effects on cytoplasmic cleavage and nuclear extrusion.[32] While several cytochalasins and chaetoglobosins have been discovered, their activity against plants has been limited to only a few compounds. These include cytochalasins A,[10] B,[39] E,[10] H;[40] epoxychalasin H;[8] deacetylepoxycytochalasin H;[8] deacetylcytochalasin H;[7] 7- and 2,7- and 2,7,18-triacetoxycytochalasin H,[10] and chaetoglobosin K.[13] All have shown significant inhibition of etiolated wheat coleoptiles. As an aside, the chaetoglobosins are of interest to plant physiologists because the differing feature between them and the cytochalasins is that the former have an indolyl substitution at the C10 that replaces a phenyl ring at C10 in the cytochalasins. This suggests that chaetoglobosin K may have plant growth regulating properties because of its relationship to indole-3-acetic acid, an ubiquitous plant hormone and, in addition, zoological properties because the indole structure is common to many bioactive natural products, for example, the ergot alkaloids ergotamine and LSD.

There are, perhaps, several reasons as to why far more detailed research has been conducted with both the cytochalasins and chaetoglobosins in zoological models than plant (agricultural) systems. First, cellular activity against animal systems implies that a potential pharmaceutical market may exist, especially if the material has antineoplastic activity, as do some of the cytochalasins and chaetoglobosins against HeLa cells.[38] The cash generated on a gram per unit basis is greater for a pharmaceutical *versus* an agrochemical. Also, the commercial availability of a compound plays a role in research. Hence, of all the cytochalasins, only cytochalasin B has been extensively examined for its mode of action in plants.

Cytochalasin B (CB)

Herth et al. studied the effects of CB against the green alga *Acetabularia mediterranea* prior to cap formation at the filamental stage of development and another marine alga *Caulerpa prolifera* in which the rhizoid tips were treated.[17] The experiments were designed to determine if there were a relationship between tip growth and cytoplasmic streaming. Also included were intact lily pollen tubes, *Lilium longiflorum*, radish root hairs, *Raphanus sativus*, and cress, *Lepidium sativum*. In all cases treatments ranged from 0.07 to 30 µg/mL (1 µg/mL = 2.1 x 10^{-6} M).

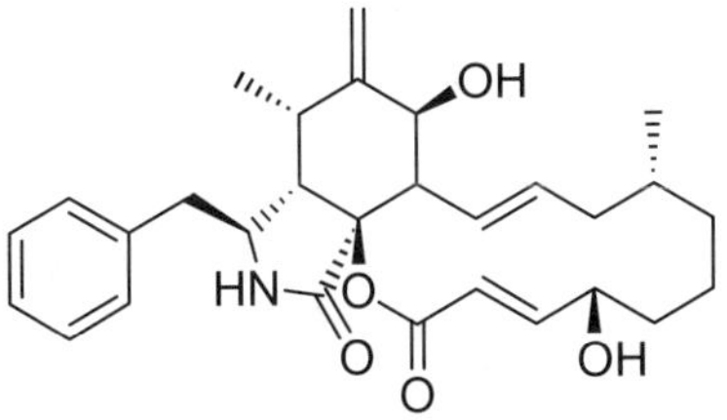

Figure 13.2
  Chemical structure of cytochalasin B.

In lily, 0.1 µg/mL CB stopped tip growth in 5 min, but inhibition of cytoplasmic streaming was immediate: 0.7 µg/mL induced the same responses, but over a greater time span. If the concentration of CB was reduced by addition of water, then normal growth and cytoplasmic flow resumed. In the other species, the same effects were observed, although the concentration of CB and the time for initiation and recovery were species dependent. However, the important message from this set of experiments clearly delineated that not only was tip growth a function of cytoplasmic streaming, but that these phenomena were directly dependent upon microfilaments which may direct the vectorial addition of materials in cell walls.

Another interesting approach in attempting to determine the mode of action of CB was to use colchicine, a classical natural product that fixes multiplying cells at the C-metaphase plate during mitosis, and compare the results obtained by CB in an identical system. In 1977, Sawhney and Srivastava[36] set out to do precisely that. They pointed out that two types of cellular growth occur, one at the surface in epicotyls, coleoptiles, hypocotyls and roots, and the other at tips, for example, in pollen tubes and root hairs. Note the choice of the latter two organs relative to those chosen by Hert et al. (*vide supra*). To put it simply, the two questions being asked were whether CB impedes cell division and whether it stops cytoplasmic streaming. It transpired that under normal growing conditions and, also, in the presence of gibberellic acid (GA3), lettuce hypocotyl growth in seedlings was inhibited with 10^{-3} and 10^{-4} M solutions of colchicines, while CB at 5 and 10 µg/mL had no apparent effect. However, tip growth in lettuce root hairs was not inhibited by colchicine but was inhibited by CB. The answers to the questions were unequivocally given and, essentially, confirmed the earlier results obtained by Hert et al.

While other plant studies were conducted relative to other morphological observations and interactions with exogenously applied plant growth regulators such as indole-3-acetic acid and abscisic acid, the major message was that CB influenced cytoplasmic streaming, but not cell division *per se*, in plants.

The role of the position of microfilaments in cells relative to cytoplasmic streaming were generally conducted in early 1962 as problems to be solved in plant physiology[24,25,30,31,33,41] and in retrospect it seems strange that later work evaluating the potential effects on mitosis were not reported earlier, albeit that the effects were nil!

Streaming and microfilaments in cells have an intimate relationship in that the microfilaments most likely supply the moving force that governs streaming. Importantly, the microfilaments are in bundles and are proximal to the cytoplasmic stream and account for the relationship, and here exogenously applied CB has produced some interesting results.

In 1972, Williamson[43] used two organisms, *Chara corralina* and *Nitella translucens*, to examine the effects of CB on cytoplasmic streaming. Whole cells from rhizoidal and small leaf internode tissue were treated with the metabolite at concentrations of 1-50 µg/mL in dimethylsulfoxide (DMSO). There was a concentration, time dependent result so that 50 µg/mL inhibited streaming in a few minutes, whereas it took 6 h to induce the same effect with 1µg/mL. Washing the cells with a DMSO solution, used at the same concentration in which the CB had been administered, could easily reverse the effects. Another observed correlation was that the metabolite induced an increase in the number of endoplasmic minivacuoles in *Nitella*, and this was concentration dependent. Even low level concentrations of 1 µg/mL resulted in vacuolar increases after 40 min and, again, the vacuolization could be increased or decreased and was dependent on increasing or flushing out the CB. Furthermore, the rate of vacuole formation or deformation reflected the rate of cytoplasmic streaming. Also, the endoplasmic reticulum gives rise to the production of minivacuoles, and these fuse with the tonoplast. Subsequently, the materials contained therein are transferred to the main vacuole,[9] and important ions are consequently transported into the cell. Another important observation in *Nitella* was the effect that interfacial fibrils had on streaming; in spite of the observation that endoplasmic organelles flowed rapidly in proximity to the fibrils, they did not exhibit any oscillatory movement.

The second set of experiments used exuded cytoplasm from *Chara*. Here, streaming was totally inhibited by the addition of CB, and although the fibrils were still present, their motility was significantly reduced relative to controls. Notably, Characean microfilaments are uncharacteristic in that they are operationally different.[41] They fail to shorten to change cell morphology but give rise to a motile force that is parallel to their surface, thereby creating movement.[43]

Further confirmation of these findings were corroborated[5] using rhizoidal tissue from *Nitella* and showed that 3-6 µg/mL of CB inhibited cytoplasmic streaming in 10 to 25 min, the upshot being that the forces involved in the cytoplasmic streaming have their genesis at the interface between the endo- and ectoplasm. Others have observed both looped and linear fibrils[21-25, 27, 43] in isolated and disturbed protoplasm.

The critical controlling of cytoplasmic streaming by CB appears to be due to the action of material on subcortical actin bundles whereby it changes their physical properties, according to experiments conducted on perfused *Chara* cells.[44] As one would expect, concentration plays a vital role. Cessation of streaming in barley, *Hordeum vulgare*, and tomato, *Lycopersicon esculentum*, root hairs was not accomplished with CB at 1 µg/mL. At 5-10 µg/mL not only was streaming stopped, but also cytoplasmic vacuoles formed and were accompanied by aggregation and swelling, all of which could be reversed by washing out the CB. Therefore, it was determined that in root hairs streaming was controlled by contractile protein filaments.[28]

As stated earlier, most of the work with the cytochalasins has been carried out on animal cells, and, consequently, it is cogent to ask if there is some connection between effects noted in plants versus zoological systems. Such a study was completed exposing onion root tips, *Allium sativum*, undergoing mitosis, to CB at 10, 15, and 20 µg/mL for periods of 3 h to 4 days.[15] All the mitotic figures were examined, and no observable effects were seen in the resting stage, prophase, metaphase, anaphase, or telophase. But human diploid fibroplast cells, L809, were morphologically changed. At 0.5 µg/mL mitotic figures showed scattered binucleate cells; 1µg/mL gave mitotic suppression with binucleate cells; 2 µg/mL strongly suppressed mitosis, and there were multiple binucleate cells; at 5 µg/mL there was irreversible mitotic disruption and general toxicity. These events took place in less than 24 h.

Chaetoglobosin K (ChK)

One of the more interesting natural product cytochalasins is ChK, which, as was pointed out earlier, is a 3-indolyl structure. It demonstrated potent activity, even down to 10^{-7}M, in the etiolated wheat coleoptile bioassay, exhibiting significant inhibition of the sections 14% ($p<0.01$) at that level.[13] The sections resembled bananas in shape, with tapered ends, and were unlike normal coleoptile sections that are generally straight and cylindrical. While not enough material was initially isolated to treat higher intact plants, later fermentations gave larger amounts that led to some work with gap junctions. These, in animals, parallel interconnecting plasmodesmata between plant cells.

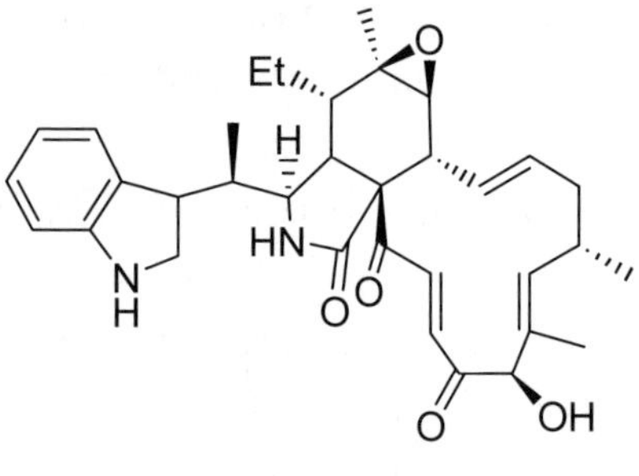

Figure 13.3
Chemical structure of Chaetoglobosin K (ChK).

In cultured rat glial cells, ChK prevented the inhibition of gap junction-mediated intercellular communication by two organochlorine pesticides, dieldrin and endosulfan.[29] In these experiments, pretreatment of glial cells with 1 or 10 µM ChK prevented, in a dose-dependent manner, dieldrin- or endosulfan-induced inhibition of gap junction-mediated communication, assayed by fluorescent dye transfer. Thus, cells preincubated with ChK followed by treatment with an organochlorine pesticide inhibitor demonstrated significantly more fluorescent dye transfer through gap junctions than cells treated with pesticide alone. Biochemical analysis of the major gap junction protein expressed by the glial cells, connexin 43, showed that phosphorylation changes induced by the organochlorine pesticides were prevented by pretreatment with ChK. This suggests that ChK stabilization of the native phosphorylated state of connexin 43 may contribute to its ability to prevent organochlorine-induced inhibition of gap junction-mediated communication.[29] In addition, ChK stabilized gap junction plaques, or clusters, in the membrane, which are disrupted by dieldrin and other organochlorine compounds.[29] The ability of dieldrin and endosulfan to act as tumor promoters, combined with the putative role of gap junctions as tumor suppressor proteins, implicates ChK as a compound of interest for further studies of its antitumor-promoting activity and therapeutic potential.

Cytochalasin H (CH)

While the mode of action of CH has never been explained, the results of its effects on plants lead to some tantalizing speculations. First isolated from a *Phomopsis* sp., it demonstrated significant growth inhibition in a primary wheat coleoptile bioassay and at 10^{-6}M it inhibited sections 10%. Fortunately, enough material was available to test on greenhouse grown plants that included week-old bean *P. vulgaris* cv Black Valentine, corn, *Z. mays* cv Norfolk Market White, and 6–week-old tobacco seedlings, *N. tabacum* cv Hick's. Plants were sprayed in aerosol at rates of 83, 8.3, and 0.83 g/ha. Oddly, no effects were noted on corn, which, like wheat, is a member of the Gramineae. Bean plants exhibited strongly bowed upward petioles of the first true leaves, and the latter were rolled longitudinally as tight cylinders at 83 g/ha. A week later leaves had assumed a normal position and plants flowered at the same time as controls. Tobacco plants gave a different response. One month following treatment, 83 g/ha plants were in the fully vegetative state; 8.3 g/ha plants were in the compact flower state; 0.83 g/ha flowers were starting to open; and controls were in full flower. Concomitantly, the relative heights of the plants were a function of the concentration of CH applied, but, most important, the numbers of leaves were identical from plant to plant and their morphology was also identical. That is, there were no aberrations.[40] Had there been effects on mitosis, certain signs would have been visible, and had there been an effect on cellular expansion, say in the leaves, then cupping would have occurred because of inhibition of the cells at the leaf margins which undergo rapid expansion. The enigma as to why flowering was inhibited cannot presently be explained because the biochemical mechanism-governing flowering has yet to be explained. Unfortunately, the quest for the discovery, isolation, and identification of 'florigen' is as elusive now as it was fifty years ago. But, as a footnote, CH has been considered a candidate as an antineoplastic agent (personal communications).

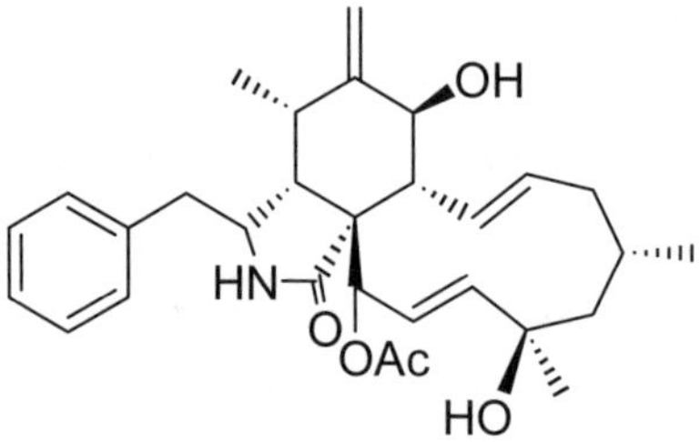

Figure 13.4

Chemical structure of Cytochalasin H.

The trichothecenes

For the greater part, the 12,13-epoxytrichothecenes are notoriously toxic to animals. The family consists of both simple and macrocyclic members, the latter being highly toxic. In 1961, some of the first reports of their phytotoxicity entered the literature[2] when they were applied exogenously. This marked one of the first examples, apart from some phenolics, of different types of inhibitory plant growth regulators and was also some of the last research that Brian and his colleagues did before entering the field of gibberellin research. Essentially, he demonstrated that diacetoxyscirpenol, now available only under license because of its potential use by terrorists, was active at 2.73×10^{-5}M when sprayed on two pea varieties, lettuce, winter tares and other important economic crops. Plants were stunted and there were necrotic lesions on the leaves. But not all plants were affected. Beetroot, carrot, mustard and wheat remained unaffected by the metabolite obtained from *Fusarium equiseti*. To even further confuse the issue, solutions of diacetoxyscirpenol applied at 1.37×10^{-6} to 2.73×10^{-8}M to cress stimulated root growth. In this regard, it had auxin-like properties. Different functional groups on the molecule also produced diverse responses. Scirpentriol has two OH functions that replace the two acetyls, and this was toxic to cress roots at all the concentrations tested.

It is not our intention to discuss all the data concerning the trichothecenes that are active against both plants and animals or to name them. They are metabolites of the fungi *Acremonium, Cylindrocarpon, Dendrostilbella, Fusarium, Myrothecium, Trichoderma* and *Trichothecium*, but only *Cylindrocarpon, Fusarium, Myrothecium*, and *Trichothecium* are considered phytopathogenic. And, as a further note, of 15 evaluated in the etiolated wheat coleoptile bioassay, all were significantly active ($p < 0.01$) to one degree or another, and the most active were the macrocyclic trichothecenes. Of the latter, verrucarin A and J, and trichoverrin B were highly potent.

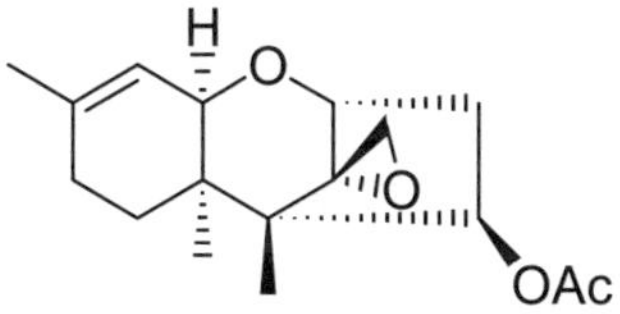

Figure 13.5

Chemical structure of trichodermin.

The least toxic, insofar as mammals are concerned, is the simplest structure of all, trichodermin, isolated from *Trichoderma viride*. It is, nevertheless, inhibitory to etiolated wheat coleoptiles down to 10^{-5}M. The LD50 is 1 g/Kg orally, and 0.5-1 g/Kg subcutaneously in mice, and its low toxicity did, at one time, make it a potential candidate as a pharmaceutical antimicrobial by Leo Pharmaceuticals. It was an effective bacteriocide and, also, exhibited good activity against *Candida albicans*.[16] From the perspective of activity on whole plant growth, it is a remarkable metabolite. Effects on greenhouse grown plants, including 10-day-old bean, *P. vulgaris* cv Black Valentine, corn, *Z. mays* cv Norfolk Market White, and 6-week-old tobacco plants, *N. tabacum* cv Hick's, were dramatic. Aerosols applied to beans at 10^{-2}M induced wilting of the primary leaves within 24 h, but there were no visible effects with 10^{-3} and 10^{-4}M sprays. At 72 h the 10^{-2}M treated leaves were thoroughly desiccated and treatments at 10^{-3}M exhibited lesions resembling virally infected areas, as opposed to necrosis, and these remained visible for one week. Thereafter, plants resumed normal growth and development. Tobacco plants were unlike bean plants in their initial response. At first, there was no response to treatments at 10^{-2} to 10^{-4}M, but at 5 days the 10^{-2}M treated plants showed signs of water stress, wilting, became flaccid, compared to controls, and behaved as though they had been treated with Roundup®. That is, there was a delayed reaction between application and observation of the first symptoms. By one week they were dying. The 10^{-3}M treated plants had water stressed patches on their leaves by this time, with limited necrotic areas, and young shoot growth was inhibited. In contrast, the 10^{-4} treated plants appeared normal. At two weeks, the 10^{-2}M treated plants were dead, and the 10^{-3} and 10^{-4}M treated plants were stunted. By 28 days, the 10^{-3} and 10^{-4}M were inhibited 83% and 72%, relative to controls. It should be emphasized that in these two latter treatments the plants looked morphologically normal.[14]

While an entire chapter could be written about the phytopathogenesis, the selective activity of the trichothecenes relative to specific plant genera, species, and cultivars, it is time to turn our attention to the mode of action of these unique secondary metabolites.

Trichothecenes and their sites of molecular activity

In eukaryotic systems, the trichothecenes act by inhibiting protein synthesis at the site of the 60s ribosome subunit and by inhibiting peptidyl transferase. Protein synthesis may be controlled at any one of three stages: initiation, codon recognition, and termination. In this regard the trichothecenes are divided into two categories and they may be either initiation inhibitors or elongation and/or termination inhibitors. Of the trichothecenes examined, the classes of initiation inhibitors are 15-acetoxyscirpendiol; 4-acetoxynivalenol; diacetoxyscirpenol; HT-2 toxin; nivalenol; T-2 toxin (the one purportedly used in chemical warfare in the 20th century); scirpentriol; and verrucarin A.[4,19] Those that inhibit elongation and/or termination are crotocin, crotocal, trichodermin, trichdermol, trichothecin, trichothecolone, and verrucarol.

In the first stage of protein synthesis, recognition, a single strand of mRNA containing triplets of unique genetic code binds to the 40s ribosome in the presence of co-factors, including Mg^{2+}, eIF^{-3}: Mg^{2+}, GTP, eIF^{-1}, and eIF^{-2}. Then, a single helical strand of t-RNA possessing an (NH)2-methionine at the 3' end and having a complementary genetic sequence to its partner mRNA on the 40s ribosome moves into place, and the bases chemically bond to form base pairs.

During the codon recognition phase another amino acid, for example glycyl-tRNA which has the necessary, correct triplet code moves into the aminoacyl position on the 60s ribosome and bonds with the mRNA and the 40s ribosome when the necessary co-factors, Mg^{2+}, eIF^{-1}, eIF^{-2}, and GTP, are present.

At elongation, methionine is transferred to glycine (in this example) when Mg^{2+}, K^+, and peptidyl transferase are present, followed by release of the spent tRNA (an Mg^{2+} requiring stage) from the 60s ribosome.

Peptidyl transferase is also involved in termination. When the final, or terminal, amino acid has been added to the peptidic sequence, that enzyme releases the neonate peptide from the 60s ribosome. Mg^{2+}, GTP, and other release factors (RF's) are necessary for complete release. Also, at this stage, the new peptide has both at least one terminal and carboxyl function, which may interact with 'foreign' molecules.

Trichothecenes and electrolytic leakage

Some simple trichothecenes, notably 3-acetyldeoxynivalenol, diacetoxyscirpenol, deoxynivalenol, fusarenone, nivalenol, and T-2 toxin, have been evaluated for their effect on electrolyte leakage. Tomato plants, *Lycopersicon esculentum* cv Supermarmande, were the test organism. The most active was T-2 toxin and leakage was both exposure and concentration dependent. Diacetoxyscirpenol was also active, but the values obtained were lower than for T-2 toxin.[20] Surprisingly, the other trichothecenes induced no measurable effects. Of the macrocyclic trichothecenes, roridin E has been tested in muskmelon, and, again, the response was time and concentration dependent[1]. As we noted in the opening paragraphs of this chapter, the effects of trichothecene treatment are species dependent in plants.

REFERENCES

(1) Bean, G. A., Fernando, T., Jarvia, B. B., and Bruton, B. **1984**. The isolation and identification of trichothecene metabolites from plant pathogenic strains of *Myrothecium roridum*. *J. Nat. Prod.* **47**, 727-729

(2) Brian, P. W., Dawkins, A. W., Grove, J. F., Hemming, H. G., Lowe, G., and Norris, G. L. F. **1961**. Phytotoxic compounds produced by *Fusarium equiseti*. *J. Exp. Bot.* **12**, 1-12

(3) Burmeister, H. R., Cieglar, A., and Vesonder, R. F. **1979**. Moniliformin, a metabolite of *Fusarium moniliforme* NRRL 6322: purification and toxicity. *Appl. Environ. Microbiol.* **37**, 11-13

(4) Busby, Jr., W. F., and Wogan, G. N. **1981**. In: Shank, R.E. (Ed.), *Mycotoxins and N-nitroso Compounds: Environmental Risks*. Vol.II., Table 2. CRC

Press, Boca Raton, FL

(5) Chen, J. C. W. **1973**. Observations of protoplasmic behaviour and motile protoplasmic fibrils in cytochalasin B treated *Nitella rhizoid*. *Protoplasma* **77**, 427-435

(6) Cole, R. J., Kirksey, J. W., Cutler, H. G., Doupnik, B. L., and Peckham, J. C. **1973**. Toxin from *Fusarium moniliforme*: effects on plants and animals. *Science* **179**, 1324-1326

(7) Cole, R. J., Wells, J. M., Cox, R. H., and Cutler, H. G. **1981**. Isolation and biological properties of deacetylcytochalasin H from *Phomopsis* sp. *J. Agric. Food Chem.* **29**, 205-206

(8) Cole, R. J., Wilson, D. M., Harper, J. L., Cox, R. H., Cochran, T. W., Cutler, H. G., and Bell, D. K. **1982**. Isolation and identification of two new cytochalasins from *Phomopsis sojae*. *J. Agric. Food Chem.* **30**, 301-304

(9) Costerton, J. W. F. and MacRobbie, E. A. C. **1970**. Ultrastructure of *Nitella translucens* in relation to ion transport. *J. Exp. Bot.* **21**, 535-542

(10) Cox, R. H., Cutler, H. G., Hurd, R. E., and Cole, R. J. **1983**. Proton and carbon-13 nuclear magnetic resonance studies on the conformation of cytochalasin H derivatives and plant growth regulating effects of cytochalasins. *J. Agric. Food Chem.* **31**, 405-408

(11) Cradwick, P. D. **1975**. Is maleic hydrazide a pyrimidine or purine analogue? *Nature* **258**, 774

(12) Cutler, H. G., Cole, R. J., and Wells, J. M. **1976**. New naturally occurring plant growth regulators: potential use in tobacco culture. *Proc. 6th. International Tobacco Scientific Congress*, Tokyo, 124-125

(13) Cutler, H. G., Crumley, F. G., Cox, R. H., Cole, R. J., Dorner, J. W., Springer, J. P., Lattrell, F. M., Thean, J. E., and Rossi, A. E. **1980**. Chaetoglobosin K, a new plant growth inhibitor and toxin from *Diplodia macrospora*. *J. Agric. Food Chem.* **28**, 139-142

(14) Cutler, H. G. and LeFiles, J. H. **1978**. Trichodermin: effects on plants. *Plant Cell Physiol.* **19**, 177-182

(15) Deysson, G., Stetzkowski, E., and Adolphe, M. **1972**. Etude comparée de l'inhibition de la cytodiérèse dans la cellule animale et dans la cellule végétale: action de la cytochalasine B. *C. R. Seances Soc. Biol. Fil.* **166**, 254-256

(16) Godtfredsen, W. O. and Vangedal, S. **1965**. Trichodermin, a new sesquiterpene antibiotic. *Acts Chem. Scand.* **19**, 1088-1102

(17) Herth, W., Franke, W. W., and Vanderwoude, W. J. **1972**. Cytochalasin

stops tip growth in plants. *Naturwissenschaften* **59**, 38-39

(18) Hoch, H. C., Staples, R. C., Whitehead, B., Comeau, J., and Wolf, E. D. **1987**. Signaling for growth orientation and cell differentiation by surface topography in Uromyces. *Science* **235**, 1659-1662

(19) Protection against Trichothecene Mycotoxins. **1983**. National Academy Press, Washington, D.C., 11b., on p 133 (Table 6-5)

(20) Iacobellis, N. S. and Bottalico, A. **1981**. *Phytopath. Medit.* **20**, 129

(21) Jarosh, R. **1956**. Plasmaströmung und chloroplastenrotation bei *Characeen*. *Øyton* **6**, 87-108

(22) Jarosh, R. **1957**. Zur mechanik der protoplasmafibrillenbewegung. *Biochem. Biophys. Acta.* **25**, 204-205

(23) Jarosh, R. **1958**. Die protoplasmafibrillen der Characeen. *Protoplasma* **50**, 93-108

(24) Kamiya, N. **1959**. Protoplasmic streaming. In: Heilbrum, L. V. and Weber, F. (Eds.), *Protoplasmatologia* VIII/3a. Springer Verlag, Vienna

(25) Kamiya, N. **1962**. Protoplasmic streaming. In: Ruhland, W. (Ed.), *Handbuch der Pflanzenphysiologie* XVII/2. Springer Verlag, Berlin. 979-1035

(26) Kolattukudy, P. E., Crawford, M. S., Wolosuk, C. P., Ettinger, W. F., and Soliday, C. L. **1987**. Chapter 10. In: Fuller, G. and Nes, W. D. (Eds.), *Ecology and Metabolism of Plant Lipids*, ACS Symposium Series 325, Washington, D.C.

(27) Kuroda, K. **1964**. The behaviour of naked cytoplasmic drops isolated from plant cells. In: Allen, R. D. and Kamiya, N. (Eds.), *Primitive Motile Systems in Cell Biology*. Academic Press, New York, 31-41

(28) Lazar-Keul, G., Keul, M., and Wagner, G. **1978**. Reversible hemmung der protoplasmaströmung im wurzelhaaren der geste (*Hordeum vulgare* L.) und tomate (*Lycopersicum esculentum* Mill.). *Pflanzenphysiol.* **90**, 461-466

(29) Matesic, D. F., Blomme, I., Sunman, J. A., Cutler, S. J., and Cutler, H. G. **2001**. Prevention of organochlorine-induced inhibition of gap junctional communication by chaetoglobosin K in astricytes. *Cell Biol. Toxicol.* **17**, 395-408

(30) Nagai, R. and Rebhun, L. **1966**. Cytoplasmic microfilaments in streaming *Nitella* cells. *J. Ultrastruct. Res.* **14**, 571-589

(31) O'Brien, T. P. and Thimann, K. V. **1966**. Intracellular fibers in oat coleoptile cells and their possible significance in cytoplasmic streaming. *Proc. Nat. Acad. Sci.* **56**, 888-894

(32) Pendse, G. S. and Mujumdar, A. M. **1986**. Distribution and production

aspects of cytochalasins. In: Pendse, G. S. (Ed.), *Recent Advances in Cytochalasans*. Indian Drugs Research Assoc., India, 1-23

(33) Pickett-Heaps **1967**. The effects of colchicine on the ultrastructure of dividing plant cells, xylem wall differentiation and distribution of cytoplasmic microtubules. *Dev. Biol.* **15**, 206-236

(34) Springer, J. P., Clardy, J., Cole, R. J., Kirksey, J. W., Hill, R. K., Carlson, R. M., and Isidor, J. L. **1974**. Structure and synthesis of moniliformin, a novel cyclobutane microbial toxin. *J. Amer. Chem. Soc.* **96**, 2267-2268

(35) Steyer, C. H. and Cutler, H. G. **1984**. Effects of moniliformin on mitosis. *Plant and Cell Physiol.* **25**, 1077-1082

(36) Sawhey, V. K. and Srivastava, L. M. **1977**. Comparative effects of cytochalasin B and colchicine on lettuce seedlings. *Ann. Bot.* **41**, 271-274

(37) Thiel, P. G. **1978**. A molecular mechanism for the toxic action of moniliformin, a mycotoxin produced by *Fusarium moniliforme*. *Biochem. Pharmacol.* **15**, 483-486

(38) Umeda, M., Ohtsubo, K., Saito, M., Sekita, S., Yoshihira, K., Natori, S., Udagawa, S., Sakabe, F., and Kurata, H. **1975**. Cytotoxicity of new cytochalasans from *Chaetomium globosum*. *Experientia* **31**, 435-438

(39) Wells, J. M., Cole, R. J., Cutler, H. G., and Spalding, D. H. **1981**. *Curvularia lunata*, a new source of cytochalasin B. *Appl. Environ. Microbiol.* **41**, 967-971

(40) Wells, J. M., Cutler, H. G., and Cole, R. J. **1976**. Toxicity and plant growth regulator effects of cytochalasin H isolated from *Phomopsis* sp. *Can. J. Microbial.* **22**, 1137-1143

(41) Wessels, N. K., Spooner, B. S., Ash, J. F., Bradley, M. O., Luduena, M. A., Taylor, E. L., Wrenn, J. T., and Yamada, K. M. **1971**. Microfilaments in cellular and developmental processes. *Science* **171**, 135-143

(42) Wheeler, H. **1975**. In: Thomas, G. W., Sabey, B. R., and Vaadia, Y. (Eds.), *Plant Pathogenesis*. Springer Verlag, New York, 5-16

(43) Williamson, R. E. **1972**. A light-microscope study of the action of cytochalasin B on cells and isolated cytoplasm of the Characeae. *J. Cell Sci.* **10**, 811-819

(44) Williamson, R. E. **1978**. Cytochalasin B stabilises the sub-cortical actin bundles of *Chara* against a solution of low ionic strength. *Cytobiologie* **18**, 107-113

14 Proteomic Techniques for the Study of Allelopathic Stress Produced by Some Mexican Plants on Protein Patterns of Bean and Tomato Roots

R. Cruz-Ortega, T. Romero-Romero, G. Ayala-Cordero, and A. L. Anaya

CONTENT

Abstract 271
Introduction 272
Results and Discussion 273
    Effects of *A. sedillense* Aqueous Extracts (1%) 273
    Effects of *L. camara* Aqueous Extracts (1%) 275
    Effects of *C. acuminata* Aqueous Extracts (1%) 277
Methodology 281
References 284

ABSTRACT

We describe the effects of allelochemical stress caused by aqueous extracts of dry leaves of three tropical allelopathic plants, *Acacia sedillense* (Fabaceae), *Lantana camara* (Verbenaceae), and *Callicarpa acuminata* (Verbenaceae), on the radicle growth and cytoplasmic protein patterns of bean and tomato. The analysis of root cytoplasmic proteins was performed using two-dimensional electrophoresis (2-DE) and gel scan densitometry. *A. sedillense* inhibited bean radicle growth 25% and tomato radicle growth 60%, modifying the expression of 16 and 14 proteins, respectively. *L. camara* aqueous extract induced the greatest inhibition in bean and tomato radicle growth, 41% and 81%, respectively, and modified 15 proteins in bean roots and 11 in tomato roots. Aqueous extract of *C. acuminata* had no effect on bean radicle growth and only modified the expression of 5 proteins. One of the increased proteins of bean revealed 94% identity with the inhibitor of α-amylase subunit from another bean. *C. acuminata* inhibited the radicle growth of tomato 47% and decreased 10 proteins in its roots. In this case, one of the increased proteins of tomato showed 69-95% similarity to the glutathione S-transferase enzymes of other Solanaceae. The allelopathic effect of the three plants evaluated depends on both the quality of the allelochemicals in each aqueous extract and the differential response of bean and tomato to each particular allelochemical stress. Proteomics techniques are valuable tools to evaluate alterations of cell physiological processes caused by allelochemical stress.

INTRODUCTION

Proteomics is an important tool in many research areas, including plant biology. Nowadays, it becomes relevant to give more insights to allelopathic modes of action, particularly the effects of allelopathic stress on protein pattern expression of target plants. This important tool allows reproducible separation and analysis of proteins on a single 2-DE gel through the development of immobilized pH gradients and the optimization of solubilization techniques. Proteins can be identified by Edman sequencing, or by mass spectrometry, if genomic information is available. The former is limited in terms of cost, speed and sensitivity; the latter permits the identification at much higher speed and with very small amounts of protein.[40] The use of 2-DE has given the opportunity to identify proteins that are modified during stress.[9,14,36,39] Modifications of protein pattern, shown by 2-DE, can give indications of alteration of various processes at the physiological level.

Modes of action of allelochemicals are diverse and have been described for isolated compounds, as well as for mixtures. They can affect various physiological processes, such as disruption of membrane permeability,[22] ion uptake,[28] inhibition of electron transport in photosynthesis and respiratory chain,[1,10,33] alterations of some enzymatic activities,[11,12,34,38] and inhibition of cell division,[4,13] among others.

In the Allelopathy Laboratory of the Instituto de Ecología, UNAM, we are performing a project to search for Mexican tropical plants with allelopathic potential from the Ecological Reserve 'El Edén', Quintana Roo. The bioactivity of plants is being evaluated on other plants of economic importance, phytopathogenic fungi, *Artemia salina*, and some pest insects.[3] As a part of this project, we screened the changes induced by the aqueous extracts of *Acacia sedillense* Rico (Fabaceae), *Lantana camara* L. (Verbenaceae), and *Callicarpa acuminata* Kunth (Verbenaceae) in root cytoplasmic protein synthesis expression of bean and tomato, with the aim of contributing to the knowledge of the modes of action of allelopathic aqueous extracts of these plants.

RESULTS AND DISCUSSION

EFFECTS OF *A. SEDILLENSE* AQUEOUS EXTRACTS (1%)

A. sedillense aqueous extract only inhibited the radicle growth of bean 25% but tomato radicle growth 60%. In bean root, *A. sedillense* aqueous extract increased the expression of 16 proteins (Table 14.1, Figs. 14.1A and 14.1B). In treated-tomato roots, 14 proteins were modified. Twelve proteins increased (1-8, 10-11, 13-14), one was detected only in this treatment (9), and another one was repressed (12) (Table 14.1 and Figs. 14.1C and 14.1D).

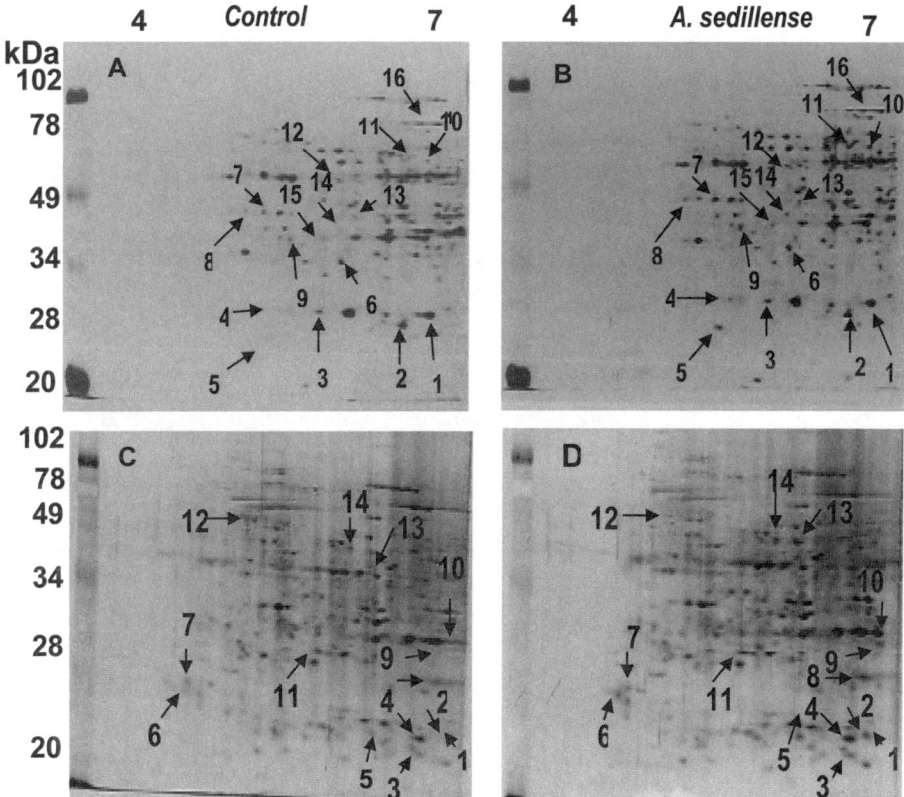

Figure 14.1

2D-PAGE of cytoplasmic proteins from bean and tomato roots: (A) bean control; (B) bean treated with *A. sedillense* aqueous extract; (C) tomato control; (D) tomato treated with *A. sedillense* aqueous extract (From Romero-Romero, M. T. et al. **2002**, *J. Chem. Ecol.* 28, 601-613. With permission).

Table 14.1

Maximum absorbance and relative molecular weights (M.W., kDa) of proteins modified in bean and tomato by aqueous extract of *Acacia sedillense*. The number of each protein corresponds to that on the 2D-gels (Fig. 14.1).

Crop Plant	Protein	M.W. (kDa)	Absorbance Control	Absorbance A. sedillense	% of Change with Respect to Control
Bean					
	1	28.1	0.15	0.59 (I)	393.3
	2	26.0	0.90	1.37 (I)	152.2
	3	27.3	0.56	0.78 (I)	139.3
	4	28.1	0.10	0.52 (I)	520.0
	5	24.0	0.15	0.58 (I)	386.6
	6	36.0	0.48	0.79 (I)	164.5
	7	47.4	0.51	1.07 (I)	209.8
	8	47.4	0.16	0.70 (I)	437.5
	9	41.3	0.25	0.51 (I)	204.0
	10	61.5	0.33	0.92 (I)	278.8
	11	62.3	0.15	0.97 (I)	646.6
	12	51.4	0.14	0.84 (I)	600.0
	13	46.1	0.15	0.53 (I)	353.3
	14	42.4	0.13	0.51 (I)	392.3
	15	41.3	0.18	0.30 (I)	166.6
	16	71.5	0.20	0.85 (I)	425.0
Tomato					
	1	21.7	0.30	0.57 (I)	190.0
	2	24.7	0.36	0.63 (I)	175.0
	3	25.0	0.43	0.55 (I)	127.9
	4	24.7	0.75	1.06 (I)	141.3
	5	27.0	0.43	0.63 (I)	146.5
	6	30.4	0.23	0.50 (I)	217.4
	7	31.2	0.32	0.46 (I)	143.8
	8	34.0	0.48	0.72 (I)	150.0
	9	38.5	0.0	0.44 (N)	100.0
	10	35.8	0.16	1.30 (I)	812.5
	11	37.5	0.56	0.78 (I)	139.3
	12	70.0	0.47	0.0 (R)	0.0
	13	62.2	0.58	0.80 (I)	137.9
	14	62.2	0.69	0.98 (I)	142.0

(I): Increased; (N): Non detected; (R): Repressed

EFFECTS OF *L. CAMARA* AQUEOUS EXTRACTS (1%)

Aqueous extract of *L. camara* inhibited 41% and 81% bean and tomato radicle growth, respectively. However, this treatment modified a similar number of proteins in both plants. In treated-bean roots, the expression of 15 proteins was modified: 6 proteins were increased (2, 8-12), and 8 were decreased (3-7; 13-15) (Table 14.2, Figs. 14.2A and 14.2B). In tomato treated-roots, 11 proteins were modified: six proteins were increased (1-6), and 5 were decreased (7-11) (Table 14.2, Figs. 14.2B and 14.2D).

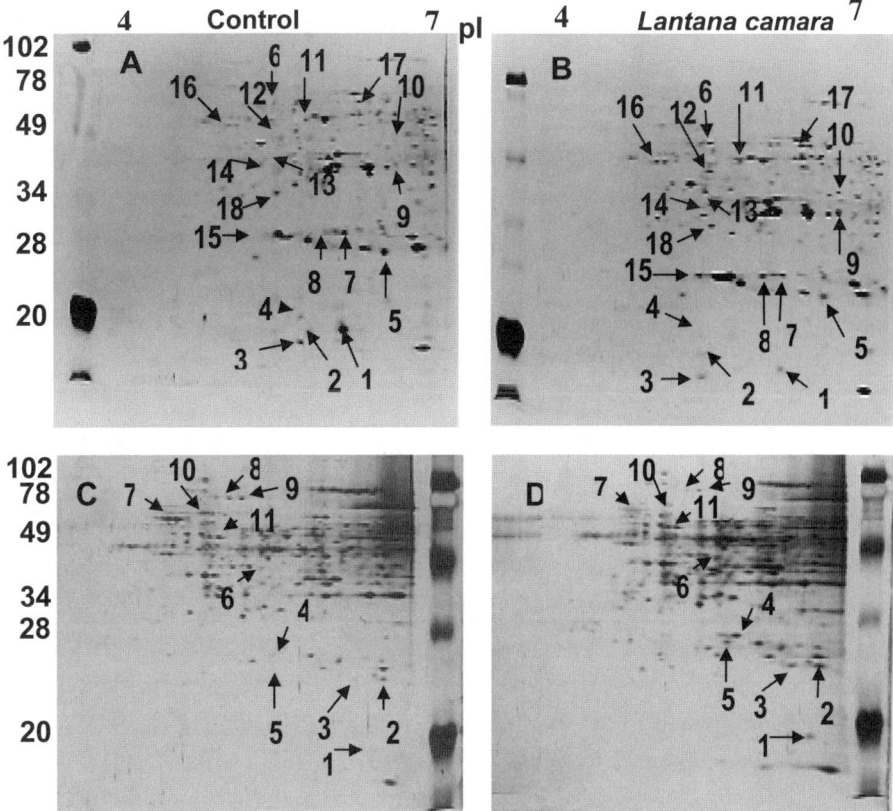

Figure 14.2

2D-PAGE of cytoplasmic root proteins of bean and tomato. (A): bean control; (B): bean treated with aqueous extract of *L. camara*; (C): tomato control; (D): tomato treated with aqueous extract of *L. camara* (From Romero-Romero, M. T. et al. **2002**, *J. Chem. Ecol.* 28, 601-613. With permission).

Table 14.2

Maximum absorbance and relative molecular weights (M.W., kDa) of proteins modified in bean and tomato by aqueous extract of *Lantana camara*. The number of each protein corresponds to that on the 2D-gels (Fig. 14.2). Values in bold correspond to increased proteins.

Crop Plant	Protein	M.W. (kDa)	Absorbance Control	Absorbance L. camara	% of Change with Respect to Control
Bean					
	1	25.0	0.00	0.42 (N)	**420.0**
	2	26.0	0.19	0.39 (I)	**205.2**
	3	27.5	0.34	0.09 (D)	26.5
	4	25.5	0.92	0.71 (D)	77.2
	5	28.3	0.27	0.18 (D)	66.6
	6	35.2	0.49	0.37 (D)	75.5
	7	34.8	0.72	0.36 (D)	50.0
	8	40.7	0.10	0.25 (I)	**250.0**
	9	46.0	0.22	0.44 (I)	**200.0**
	10	56.2	0.25	0.60 (I)	**240.0**
	11	40.2	0.28	0.54 (I)	**192.8**
	12	47.9	0.44	0.13 (I)	29.5
	13	54.1	0.82	0.42 (D)	51.2
	14	17.0	0.57	0.15 (D)	26.3
	15	32.8	0.22	0.09 (D)	40.9
Tomato					
	1	17.2	0.30	0.70 (I)	**233.3**
	2	25.0	0.63	1.15 (I)	**182.5**
	3	26.7	0.24	0.58 (I)	**241.6**
	4	30.3	0.24	1.30 (I)	**541.7**
	5	29.0	0.28	0.86 (I)	**307.1**
	6	51.6	0.25	1.54 (I)	**616.0**
	7	70.0	1.64	0.20 (D)	12.2
	8	77.6	1.08	0.49 (D)	45.4
	9	77.6	1.86	0.26 (D)	14.0
	10	73.0	0.98	0.23 (D)	23.5
	11	66.3	1.76	0.70 (D)	39.8

(I): Increased; (D): Decreased

EFFECTS OF *C. ACUMINATA* AQUEOUS EXTRACTS (1%)

Aqueous extract of *C. acuminata* did not affect bean radicle growth but significantly inhibited tomato radicle growth (47%). In treated bean, only five proteins were modified by this treatment: three increased (1-3) and two decreased (4 and 5). This result suggests that the aqueous leachate of *C. acuminata* does not have a significant effect on the expression of cytoplasmic protein synthesis in root bean. However, the microsequence of protein 2 (11.3 kDa) (4.5-fold increase) revealed 94% identity with the inhibitor of the α-amylase subunit from another *P. vulgaris* (kidney bean) (pir IJC4855[26]), 89% with the α-amylase inhibitor-like protein (dbj BAAA86927[25]); 89% with the α-amylase inhibitor 3 precursor of *P. vulgaris* (kidney bean) (pir S51830[30]), and 84% with the probable lectin precursor of *P. vulgaris* (kidney bean) (ITI2036; Lee, et al. submitted JAN-1997 to the EMBL Data Library) (Fig. 14.3C).

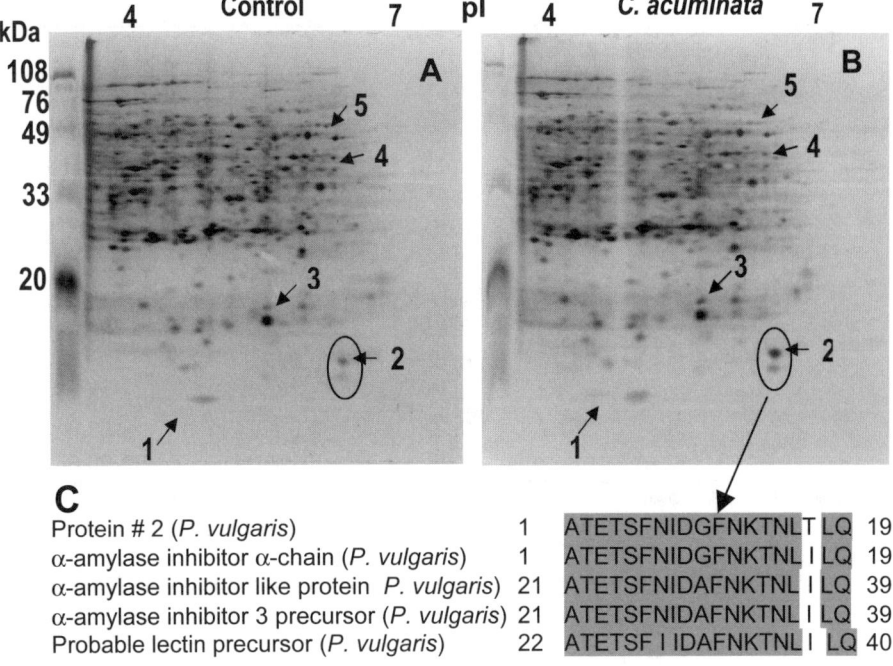

Figure 14.3

2D-PAGE of cytoplasmic root proteins of bean. (A): control; (B): bean treated with *C. acuminata* aqueous extract; (C): comparative microsequences of amino acids of protein 2 (From Cruz-Ortega, R. et al. **2002**, *Physiol. Plantarum*, **116**, 20-27. With permission.)

Table 14.3

Maximum absorbance and relative molecular weights (kDa) of proteins modified in bean and tomato by aqueous extract of *Callicarpa acuminata*. The number of each protein corresponds to that on the 2D-gels (Figs. 14.3 and 14.4). Values in bold correspond to increased proteins.

Crop plant	Protein	M.W. (kDa)	Absorbance Control	Absorbance *C. acuminata*	% of Change with Respect to control
Bean					
	1	8.5	0.06	0.12 (I)	**172.0**
	2	11.3	0.06	0.27 (I)*	**450.0**
	3	13.5	0.13	0.24 (I)	**185.0**
	4	43.8	0.33	0.09 (D)	27.2
	5	52.0	0.27	0.09 (D)	33.0
Tomato					
	1	12.16	0.14	0.04 (D)	35.0
	2	13.56	0.10	0.03 (D)	33.0
	3	17.02	0.26	0.06 (D)	43.0
	4	20.89	0.47	0.22 (D)*	21.0
	5	27.54	0.00	0.21 (I)	**100.0**
	6	29.24	0.26	0.12 (D)	22.0
	7	27.54	0.14	0.07 (D)	19.0
	8	30.20	0.30	0.12 (D)	25.0
	9	25.70	0.12	0.02 (D)	60.0
	10	40.74	0.06	0.27 (I)	**45.0**
	11	28.97	0.16	0.03 (D)	53.0
	12	62.23	0.26	0.10 (D)	26.0

(*): sequenced protein; (I): Increased; (D): Decreased

In treated tomato roots 12 proteins were modified: 10 were decreased (1-4; 6-9; 11 and 12) and 2 were increased (5 and 10) (Fig. 14.4, Table 14.3). Protein 5 (37.5 kDa) was N-terminal microsequenced, and 23 amino acids were obtained. Protein search showed 95% similarity to the glutathione S-transferases (EC 2.5.1.18) (GST), class-phi from *Solanum commersonii*, pir T07906 (Seppanen, 1997; direct submission); 90% similarity with GST from *Hyoscyamus muticus*, pir PQ0744;[6] and 69% with GST from *Nicotiania tabacum*, P46440[21] (Fig. 14.4).

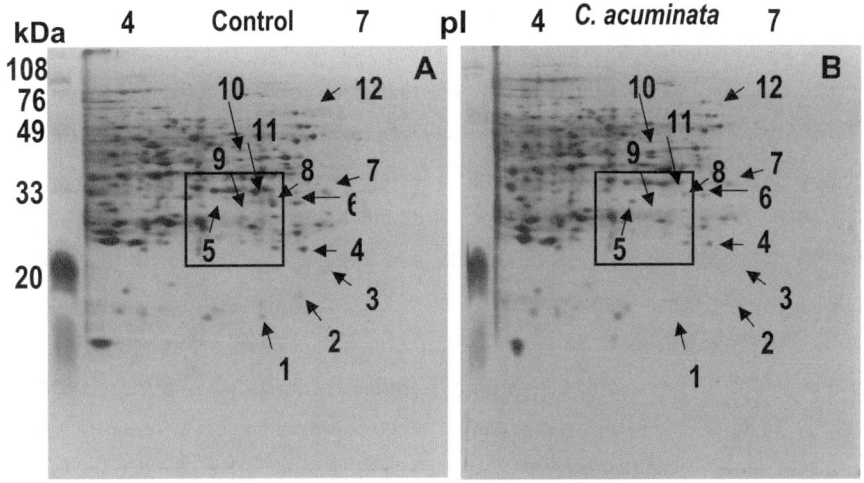

protein # 5   1 AIKVHGPMMSPAVMRVV- TLKEKD 23
GST *Solanum commersonii*   2 AIKVHGPMMSPAVMRVVATLKEKD 25
GST *Hyoscyamus muticus*   3 MKLHGPAMSPAVMRV I ATLKEKD 25
GST *Nicotiana tabacum*   2 AIKVHGSPMSTATMRV AACL IEKD 25

Figure 14.4

    2D-PAGE of cytoplasmic root proteins of tomato. (A): control; (B): treated with *C. acuminata* aqueous extract; (C): protein 5 was N-terminal microsequenced, comparison is shown in panel (From Cruz-Ortega, R. et al. **2002**, *Physiol. Plantarum*, **116**, 20-27. With permission.)

    The allelopathic effect of the three plants evaluated in this study depends on both the kind of chemical compounds in each aqueous leachate and the differential response of bean and tomato to each particular allelochemical stress.[37]

    Aqueous extract of *A. sedillense* inhibited the radicle growth of bean only 25% and tomato radicle growth 60%. In spite of the difference of effects on root growth, both plants showed a similar number of modified proteins (16 and 14, respectively). In both target plants, modified proteins were increased, but this effect is more obvious in bean root.

    Aqueous extract of *L. camara* showed higher phytotoxic effect on bean and tomato (41% and 81%, respectively) compared with the other two plants. However, the number of proteins whose expression was modified by *L. camara* and *A. sedillense* were similar. In addition, bean and tomato modified proteins showed a similar protein pattern change, increasing and decreasing in similar proportions (Tables 14.1 and 14.2).

Finally, the aqueous extract of C. acuminata also affected differentially the radicle growth of the two crop test plants. This treatment did not affect bean radicle growth, but significantly inhibited tomato radicle growth (47%). In treated bean, only five proteins were modified by the aqueous extract, three of them increased significantly. The microsequencing of protein #2 of bean (11.3 kDa, 4.5-fold increase) revealed 94% identity with the inhibitor of the α-amylase subunit from another type of bean. These α-amylase inhibitors are defense storage proteins induced during insect attack.[25] The fact that the α-amylase inhibitor increased during the treatment with C. acuminata aqueous extract suggests that the induction of this kind of protein can be a general defense response and not necessarily a specific mechanism against insect attack. Moreover, some environmental stress can induce expression of proteins not especially related to a particular stress but rather to reactions against cell damage. These include some classes of heat shock proteins,[23] thiol proteases,[41] proteinase inhibitors,[35] osmotin, and other PR proteins.[14,15,27]

C. acuminata aqueous extract modified 12 proteins in tomato roots, decreasing the majority of them. The 23 amino acids of the N-terminal of protein #5 of tomato (37.5 kDa, 100% increase) showed 95% similarity to the glutathione S-transferases of other Solanaceae. In plants, GSTs play roles in normal cellular metabolism, as well as in the detoxification of a wide variety of toxic compounds. GSTs have been involved in numerous stress responses, including pathogen attack, oxidative stress, and others.[18,20,29]

The metabolic stress response is a ubiquitous defense mechanism which is activated when cells are confronted with unfavorable environmental conditions. Induction leads to the expression of proteins known as stress proteins, which are thought to play an important role in protecting various cell processes. Stress and signalling inside the cell lead to protein expression changes, to the activation of new biochemical pathways, and to repression of others that are characteristic of the unstressed state.[7] Dayan et al.[17] asserted that determination of the mode of action of allelochemicals is a challenging endeavor due to the multitude of potential molecular targets. However, the use of proteomic techniques represents a useful tool to discover new sites of action of allelochemicals. Changes in protein expression patterns could be indicative of many physiological alterations within plant cells.

METHODOLOGY

Allelopathic plants

Acacia sedillense, *Lantana camara*, and *Callicarpa acuminata* leaves were collected at El Eden Ecological Reserve at Quintana Roo, Mexico.

Target crop plants

Bean seeds (*Phaseolus vulgaris* L. var. Flor de Mayo) were obtained from El Tinajero, a local supplier at Xochimilco, Mexico, D.F. Tomato seeds (*Lycopersicon esculentum* var. Pomodoro) were obtained from Sun-Seeds, Parma, ID, USA.

Radicle growth bioassays

Aqueous extracts were prepared by soaking dried leaves (2 g/100 ml) in distilled water for 3 h, filtered through Whatman # 4 paper and Millipore membrane (0.45 µm). Osmotic pressure of extracts was measured with a freezing-point osmometer (Osmette A, Precision System, Inc.). Extracts were mixed with agar (2%) to obtain a final concentration of 1% extract. Controls contained only 1% agar. Bioassays were prepared under sterile conditions in a laminar hood. Ten or twelve seeds of bean or tomato were placed on each petri dish and kept in the dark at 27°C. Twenty-five petri dishes were used for each treatment for both crop plants. The terminal 0.5 cm of the primary root was excised after 48 h from bean seedlings, and after 72 h from tomato seedlings. Roots were frozen under liquid nitrogen and kept at −70°C until used for protein isolation. For root growth response, a complete randomized experiment design was performed with four replicates. Bean primary root lengths were measured after 48 h, tomato root length after 72 h. Data were analyzed by ANOVA.

Root tip homogenization, cytoplasmic protein extraction, gel electrophoresis, and densitometry analysis

Cytoplasmic proteins were extracted and purified from seedling roots of controls and treatments. The terminal 0.5 cm of the primary root from about 300 seedlings per treatment (control and the three aqueous extracts of allelopathic plants) were homogenized under liquid nitrogen with a mortar and pestle and subsequently suspended in cold homogenization buffer (50 mM K_2HPO_4 and 50 mM of KH_2PO_4, pH 6.8, 1mM PMSF – phenylmethylsulphonyl fluoride) in a 1:4 w/v ratio. The homogenate was centrifuged at 300 g for 10 min to a pellet nuclear fraction, and then the supernatant was centrifuged at 12000 g for 10 min at 4°C. Proteins from the supernatant were extracted into phenol, precipitated with methanol, and re-dissolved in an isoelectric focusing (IEF) medium (ampholyte solution at pH 4-7) as described by Hurkman and Tanaka.[24] Protein content was determined by the Bradford method.[8]

Two-dimensional gel electrophoresis (2D-PAGE) was performed according to O'Farrell.[32] For the first dimension, samples containing 10 μg of protein were loaded at the base end of the gels (capillary tubes, BIO-RAD). Isoelectric focusing was conducted for 30 min at 300 V, and then for 4 h at 750 V. After extrusion, gels were either frozen at −70°C or loaded onto a second dimension 12% polyacrylamide resolving gel. They were run in a mini-protein gel (BIO-RAD), and then fixed and silver stained according to Morrisey.[31] Each gel was analyzed by scanning with a GelScan XL (release 2.1) to compare control and treated gels. We determined the absorbance of each of the most conspicuous spots on the gels. By subtraction of the background, we determined whether the absorbance of a protein-spot increased, decreased, or was repressed, comparing them with those obtained in the control gels (Fig. 14.5).

Protein microsequencing

Amino acid sequence analysis was performed according to Barent and Elthon[5] and Dunbar et al.[19] Briefly, 400 μg of cytoplasmic protein were used for the first dimension (IEF) and then separated for the second dimension in a 12% acrylamide gel and blotted onto a PVDF membrane using a TE-semidry system (Hoefer Scientific Instruments, San Francisco, CA, USA). Proteins were sent for N-terminal microsequencing to the Protein Chemistry Laboratory of the University of Texas, Medical Branch, at Galveston, TX, USA. The two bean-induced proteins had to be trypsine-digested because of N-terminal blocking. Amino acid sequences obtained were compared with sequences in the nonredundant peptide sequences database of the National Center for Biotechnology Information, using the BLAST program.[2]

1. Protein extraction from control and allelochemical-treated roots of target plants.

⇩

2. Purification and solubilization of root cytoplasmic proteins; separation of protein according to their pI (IEF gels)

⇩

3. Loading of IEF gels onto a second dimension gel

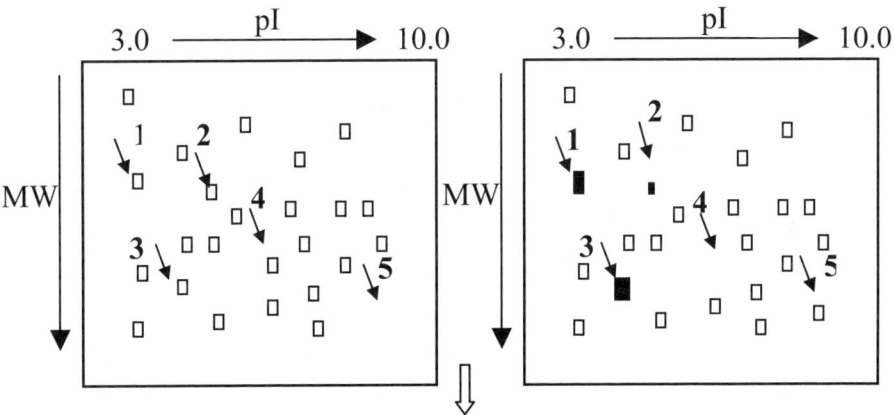

⇩

4. Densitometry analysis to compare between control and treated gels

⇩

5. Blot gel to PVDF membranes and cut spot of interest and sequence either N-terminal amino acid or a trypsine digest fragment.

⇩

6. Amino acid sequencing and searching in protein data bases Bioinformatics (BLAST program) (http://www.ncbi.nlm.nih.gov)

⇩

7. Sequence comparison with those reported in the protein bank; comparison of experimentally determined and predicted MW and pI; amino acid sequence to determine function; oligonucleotides design for gene search.

Figure 14.5

Methodology used for identifying modified proteins of allelochemical stressed plants. It can be used with single compounds or mixtures as found in aqueous extracts.

ACKNOWLEDGMENTS

We acknowledge the technical support of Blanca Estela Hernández-Bautista, and the financial support of the projects CONACyT 25316N and DGAPA-UNAM (PAPIIT) IN217796.

REFERENCES

(1) Abrahim, D., Braguini, W. L., Kelmer-Bracht, A. M., and Ishii-Iwamoto, E. L. **2000**. Effects of four monoterpenes on germination, primary root growth, and mitochondrial respiration of maize. *J. Chem. Ecol.* **26**, 611-624

(2) Altschul, S. F., Gish, W., Miller, W., Myers, E. W., and Lipman, D. J. **1990**. Basic local alignment search tool. *J. Mol. Biol.* **215**, 403-410

(3) Anaya, A. L. and del Amo, S. **1999**. Searching for New Biocides in the Tropical Forests in the El Eden Ecological Reserve, Quintana Roo, Mexico. Final Report. USDA MX-AES6, Grant: FG-Mx- 107. 135 p

(4) Anaya, A. L. and Pelayo-Benavides, H. R. **1997**. Allelopathic potential of *Mirabilis jalapa* L. (Nyctaginaceae): Effects on germination, growth and cell division of some plants. *Allelopathy J.*, 57-68

(5) Barent, L. R. and Elthon, E. T. **1992**. Two-dimensional gels: an easy method for large quantities of proteins. *Plant Mol. Biol.* **10**, 338-344

(6) Bilang, J., Macdonald, H., King, P. J., and Sturm, A. **1993**. A soluble auxin-binding protein from *Hyoscyamus muticus* is a glutathione S-transferase. *Plant Physiol* **102**, 29-34

(7) Bohnert, H. J. and Sheveleva, E. **1998**. Plant stress adaptations–making metabolism to move. *Curr. Opin. Plant Biol.* **1**, 267-274

(8) Bradford, M. R. **1976**. A rapid and sensitive method for the quantitation of microgram quantities of protein utilizing the principle of protein-dye binding. *Anal. Biochem.* **72**, 248-254

(9) Bray, E. **1993**. Molecular responses to water deficit. *Plant Physiol.* **103**, 1035-1040

(10) Calera, M. R, Mata, R., Anaya, A. L, and Lotina-Hennsen, B. **1995**. 5-O-β-D-galactopyranosyl-7-methoxy-3´4´-dihydroxy-4-phenylcoumarin, an inhibitor of photo-phosphorylation in spinach chloroplasts. *Photosynth. Res.* **45**, 105-110

(11) Calera, M. R., Anaya, A. L., and Gavilanes-Ruiz, M. **1995**. Effect of

phytotoxic resin glucoside on the activity of H$^+$-ATPase from plasma membrane. *J. Chem. Ecol.* **21**, 289-297

(12) Cruz-Ortega, R., Anaya, A. L., Gavilanes-Ruíz, M., Sánchez-Nieto, S., and Jiménez-Estrada, M. **1990**. Effect of diacetyl piquerol on the H$^+$-ATPase activity of microsomes from *Ipomoea purpurea*. *J. Chem. Ecol.* **16**, 2253-2261

(13) Cruz-Ortega, R., Anaya, A. L., and Ramos, L. **1988**. Effects of allelopathic compounds of corn pollen on respiration and cell division of watermelon. *J. Chem. Ecol.* **14**, 71-86

(14) Cruz-Ortega, R. and Ownby, J. D. **1993**. A protein similar to PR protein is elicited by metal toxicity in wheat roots. *Physiol. Plantarum* **89**, 211-219

(15) Cruz-Ortega, R., Cushman, J. C., and Ownby, J. D. **1997**. cDNA clones encoding 1,3-β-glucanase and fimbrin-like cytoskeletal protein are induced by aluminum toxicity in wheat roots. *Plant Physiol.* **114**, 1453-1460

(16) Cruz-Ortega, R., Ayala-Cordero, G., and Anaya, A. L. **2002**. Allelochemical stress produced by the aqueous leachate of *Callicarpa acuminata*: effects on roots of bean, maize, and tomato. *Physiol. Plantarum* **116**, 20-27

(17) Dayan, F. E., Romagni, J. G., and Duke, S. O. **2000**. Investigating the mode of action of natural phytotoxins. *J. Chem. Ecol.* **26**, 2079-2094

(18) Dixon, D. P., Cummins, I., Cole, D. J., and Edwards, E. **1998**. Glutathione-mediated detoxification systems in plants. *Curr. Opin. Plant Biol.* **1**, 258-266

(19) Dunbar, B., Elthon, T. E., Osterman, J. C., Whitaker, B. A., and Wilson, S. B. **1997**. Identification of plant mitochondrial proteins: a procedure linking two-dimensional gel electrophoresis to protein sequencing from PVDF membranes using a fastblot cycle. *Plant Mol. Biol. Rep.* **15**, 46-61

(20) Edwards, R., Dixon, D. P., and Walbot, V. **2000**. Plant glutathione S-transferases: enzymes with multiple functions in sickness and in health. *Trends Plant Sci.* **5**, 193-198

(21) Ezaki, B., Yamamoto, Y., and Matsumoto, H. **1995**. Cloning and sequencing of the cDNAs induced by aluminum treatment and Pi starvation in cultured tobacco cells. *Physiol. Plantarum* **93**, 11-18

(22) Galindo, J. C. G., Hernández, A., Dayan, F. E., Téllez, M. R., Macias, F. A., Paul, R. N., and Duke, S. O. **1999**. Dehydrozaluzanin C, a natural sesquiterpenolide, causes rapid plasma membrane leakage. *Phytochemistry* **52**, 805-813

(23) Heikkila, J. J., Papp, J. E. T., Schultz, G. A., and Bewley, J. D. **1984**. Induction of heat shock protein messenger RNA in maize mesocotyls by water stress, abscisic acid, and wounding. *Plant Physiol.* **76**, 270-274

(24) Hurkman, W. and Tanaka, C. **1988**. Polypeptide changes induced by salt stress, water deficit, and osmotic stress in barley roots: A comparison using two dimensional gel electrophoresis. *Electrophoresis* **9**, 781-787

(25) Ishimoto, M., Yamada, T., and Kaga, A. **1999**. Insecticidal activity of an alpha-amylase inhibitor-like protein resembling a putative precursor of alpha amylase inhibitor in the common bean, *Phaseolus vulgaris* L. *Biochim. Biophys. Acta* **1432**, 104-112

(26) Kasahara, K., Hayashi, K., Arakawa, T., Philo, J. S., Wen, J., Hara, S., and Yamaguchi, H. **1996**. Complete sequence, subunit structure, and complexes with pancreatic α-amylase of an α-amylase inhibitor from *Phaseolus vulgaris* white kidney bean. *J. Biochem* **120**, 177-183

(27) Kononowicz, A. K., Raghothama, K. G., Casa, A. M., Reuveni, M., Watad, A. A., Liu, D., Bressan, R. A., and Hasegawa, P. M. **1993**. Osmotin: regulation of gene expression and function. In: Close, T. J, Bray and E. A. (Eds.). *Plant Responses to Cellular Dehydration During Environmental Stress. Current Topics in Plant Physiology.* Vol 10. American Society of Plant Physiologist, Rockville, MD, 144-157

(28) Lehman, M. E., and Blum, U. **1999**. Evaluation of ferulic acid uptake as a measurement of allelochemical dose: effective concentration. *J. Chem. Ecol.* **25**, 2585-2600

(29) Marrs, K. A. **1996**. The functions and regulation of glutathione S-transferases in plants. *Annu. Rev. Plant Phys.* **47**, 127-158

(30) Mirkov, T. E, Wahlstrom, J. M., Hagiwara, K., Finardi-Filho, F., Kjemtrup, S., and Chrispeels, M. J. **1994**. Evolutionary relationships among proteins in the phytohemaglutinin-arcelin-alpha-amylase inhibitor family of the common bean and its relatives. *Plant Mol. Biol.* **26**, 1103-1113

(31) Morrisey, J. **1981**. Silver stain for protein in polyacrylamide gels: A modified procedure with enhanced uniform sensitivity. *Anal. Biochem.* **117**, 307-310

(32) O'Farrell, P. H. **1975**. High-resolution two-dimensional electrophoresis of proteins. *J. Biol. Chem.* **250**, 4007-4021

(33) Peñuelas, J., Ribas-Carbo, and M., Giles, L. **1996**. Effects of allelochemicals on plant respiration and oxygen isotope fractionation by

the alternative oxidase. *J. Chem. Ecol.* **22**, 801-805

(34) Politycka, B. **1998**. Phenolics and the activities of phenylalanine ammonia-lyase, phenol-β-glucosyltransferase and β-glucosidase in cucumber roots as affected by phenolic allelochemicals. *Acta Physiol. Plant.* **20**, 405-410

(35) Reviron, M., Bartanian, P. N., Sallantin, M., Huet, J. C., Pernollet, J. C., and de Vienne, D. **1992**. Characterization of a novel protein induced by rapid or progressive drought and salinity in *Brassica napus* leaves. *Plant Physiol.* **100**, 1486-1493

(36) Riccardi, F., Gazeau, P., de Vienne, D., and Zivy, M. **1998**. Protein changes in response to progressive water deficit in maize. *Plant Physiol.* **117**, 1253-1263

(37) Romero-Romero, M. T., Anaya, A. L., and Cruz-Ortega, R. **2002**. Screening for the effects of phytochemical variability on cytoplasmic protein synthesis pattern of crop plants. *J. Chem. Ecol.* **28**, 601-613

(38) Silva, M. G., Costa, R. A., Ferrarese, M. L. L., and Ferrarese-Filho, O. **1996**. Effects of phenolics compounds on soybean urease activity. *Arq. Biol. Tecnol.* **39**, 677-683

(39) Van Loon, L. C., Pierpoint, W. S., Boller, T., and Conejero, V. **1994**. Recommendations for naming plant pathogenesis-related proteins. *Plant Mol. Biol. Rep.* **12**, 245-264

(40) Van Wiljk, K. J. **2000**. Proteomics of the chloroplast: experimentation and prediction. *Trends in Plant Science* **5**, 420-425

(41) Williams, J., Bulman, M., Huttly, A., Phillips, A., and Neill, S. **1994**. Characterization of a cDNA from *Arabidopsis thaliana* encoding a potential thiol protease whose expression is induced independently by wilting and abscisic acid. *Plant Mol. Biol.* **25**, 259-270

15 Application of Microscopic Techniques to the Study of Seeds and Microalgae under Olive Oil Wastewater Stress

G. Aliotta, R. Ligrone, C. Ciniglia, A. Pollio,
M. Stanzione, and G. Pinto

CONTENT

Abstract 289
Introduction 290
Results and Discussion 292
    Histological and Cytological Features of Radish Radicle in Presence
    of Reverse Osmosis Fraction 295
    Morphological and Cytological Features of *Ankistrodesmus
    braunii* in Presence of Reverse Osmosis Fraction 302
Conclusions 306
Methodology 307
References 311

ABSTRACT

Polluting olive oil mill wastewaters and their filtered fractions were tested for their phytotoxicity on seed germination and seedling growth of radish and the microalga *Ankistrodesmus braunii*. The most potent inhibition was observed with the reverse osmosis fraction. From this, 17 polyphenols with molecular weight less than 300 Dalton were isolated and identified. The inhibitory activity of each polyphenol was much lower than that observed with the initial fraction. Microscopic observations showed different morphological and cytological responses of radish radicle and algal cells in the presence of the reverse osmosis fraction. Light and electron microscopy of radish radicle revealed that cell expansion was reduced, and the apex was wider and coarser than in the control; the mitochondria were the only cellular organelles showing obvious structural alterations relative to the control at ultrastructural level. For algal cells, the treatment caused cellular anomalous shapes, reduction of cellular volume, altered cytoplasmic organization and inhibition of endospore production. TEM observations showed that cytokinesis was differentially affected by the reverse osmosis fraction and catechol. The former prevented the formation of dividing septa whereas catechol altered the geometry of deposition of the septa.

INTRODUCTION

In 1937 the attention of the plant physiologist Hans Molisch was caught by a horticultural problem: the induction of ripening by early-ripening apples and pears on fruits from late-ripening varieties when stored together. Molisch demonstrated that the substance responsible for ripening induction was ethylene. He also demonstrated that root growth of vetch (*Vicia sativa* L.) and pea (*Pisum sativum* L.) seedlings is inhibited when seeds were germinated under a jar together with some apples. In his book, *Der Einfluss einer Pflanze auf die andere Allelopathie*, Molisch reports: "*The described phenomenon that one plant can influence another, plays an important role in* physiology, *so it deserves an appropriate term. For this I coin the word allelopathy from the Greek words "allelon", meaning mutual and pathos: meaning harm or "affection". The shorter word allopathie is appropriate too but it is already present in literature as opposite of homeopathy* ".[25]

Successively, Molisch's definition was adopted in a broader sense by the botanist Rice in his famous book *Allelopathy*, referred the term to "*any process involving secondary metabolites of plants that influence growth and development of other plants and microbes.*" [27] Rice was encouraged to this by some studies that demonstrated the role of allelopathy in the field. Nowadays, effective investigation in allelopathy involves the expert abilities of different specialists in relatively diverse fields: botany, chemistry, plant physiology, and ecology. In only a few cases have these assorted professional talents become incorporated into experimental programs. In this respect, few reports have given detailed proof of allelopathic effects in the field, demonstrating the precise role of allelochemicals and their mechanism of action at the cellular level. Quite often, it has been plant extracts of unknown constituents or an inhibitory growth media that have been used as the basis for physiological-function studies, and few investigators have attempted to quantify the morphological and cytological effects of the allelochemicals on target organisms.[15] Consequently, conclusive proofs of allelopathy in the field remain few, and use of field relevant bioassays has been called for regularly.[22,28] Recent general public dissatisfaction in herbicide and pesticide use in agriculture and forestry has opened up an exciting opportunity for allelopathy scientists in developing pest control methods using allelopathy principles.

In the course of our allelopathic studies to isolate potential bioherbicides from plants, we focused our attention on medicinal plants and vegetable wastes, which represent a primary and neglected source of allelochemicals, assaying their allelopathic phytotoxicity and identifying the site(s) of action of allelochemicals on seeds.[1,3,4,6] The annual production of one hundred million liters and deposition of olive oil mill wastewater is a major environmental problem for agriculture in the Mediterranean basin, where the olive (*Olea europaea* L.) is the most economically important fruit tree, because it has provided valuable storable oil as well as edible fruit since ancient times.[35]

Olive oil is obtained by pressing the ripe fruit. The extraction also generates an aqueous phase formed by the water content of the fruit combined with the water used to wash and process the olives; the combination is the so-called "olive mill wastewater" (OMW). The polluting organic load of this wastewater is considerable because of the sugars, tannins, polyphenols, polyalcohols, pectin and lipid content.[12] There is much evidence that polyphenols, fatty acids and organic acids are phytotoxic and are involved in allelopathy.[11,23,27] A way to reduce OMW pollution in soil is to fractionate wastewater by molecular weight filtration of the components and assay their phytotoxicity. Using this technique, we recently discovered that among the filtered fractions of OMW, the reverse osmosis fraction is responsible for a strong inhibitory activity on germination of radish and algal growth, giving reliable controls and allowing fast bioassay.[5,14]

This chapter deals with the morphological and cytological responses of different plant organisms such as radish germination and seedling growth (*Raphanus sativus* L. cv. Saxa) and the green alga; *Ankistrodesmus braunii*. Target species were treated with OMW fractions and with single polyphenolic compounds isolated from OMW. In accordance with the aim of this book, we have included in the results and discussion some data previously obtained in our labs.[5,14] We have also made every effort to outline as simply as possible some basic microscopic techniques, bearing in mind the many pitfalls one encounters as a beginner.

RESULTS AND DISCUSSION

Olive oil mill wastewater (OMW) sludge free (SF), and all fractions obtained by ultrafiltration (UF1), nanofiltration (NF) and reverse osmosis (RO) were evaluated for their ability to inhibit radish seed germination (Table 15.1) and the growth of the alga *Ankistrodesmus braunii* (Fig. 15.1). No germination was observed with the 1:1 solution of SF, UF1, NF and RO, but only the last two fractions showed strong phytotoxicity at 1:2 dilution. RO was the most active OMW fraction, 6% germination occurred at 1:6 dilution (Table 15.1). On the other hand, the OMW fractions gave different toxicity responses on *A. braunii* cultures inoculated with 1.0×10^5 cells/ml. Fractions UF1 and UF2 did not inhibit algal growth, or had a slight stimulating effect; SF and NF fractions were moderately toxic (less than 20% of inhibition), while the RO fraction almost completely inhibited the growth of *A. braunii* after 4 days exposure. The RO fraction was, therefore, chosen for further study. This fraction, containing low molecular weight compounds (<300 D), was fractionated by chromatography to give twelve polyphenolic compounds (**1-12**), which were identified on the basis of their spectroscopic analysis and by comparison with authentic samples, as previously reported.[5] HPLC analysis of OMW showed five additional peaks besides the signals of the above mentioned compounds. These peaks were attributed to the compounds **13-17** by comparison with commercial samples. The molecular skeletons of the identified compounds were C_6-C_0 (**5**), C_6-C_1 (**8, 12, 15** and **17**) C_6-C_2 (**1, 2, 3, 4, 13, 14**) or C_6-C_3 (**9, 10, 11, 16**), (see Methodology).

The phytotoxicity of the polyphenolic compounds was tested on radish germination, seedling growth (Table 15.2) and algal growth (Table 15.3). Compounds **1, 4, 6** and **7** could not be tested due to insufficient amount for bioassay and were not commercially available.

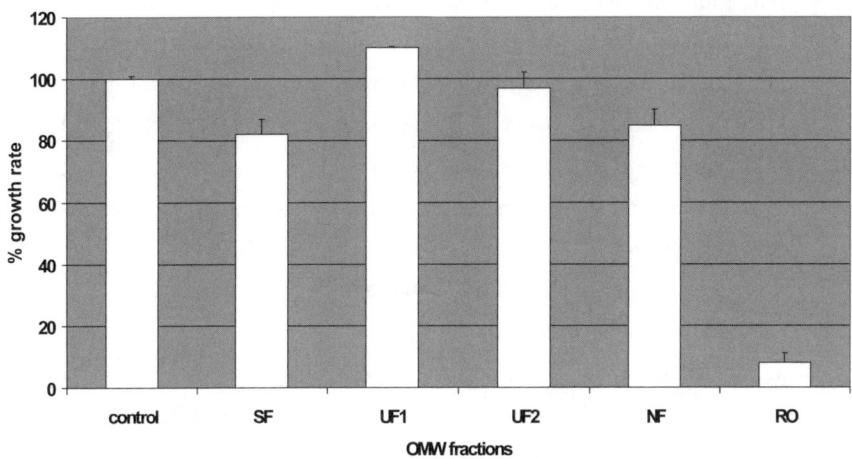

Figure 15.1

Effect of OMW fractions on the growth of *Ankistrodesmus braunii*.

Table 15.1

Effect of OMW fractions and their dilutions on radish germination 48 h after application.

Dilution	Germination (% of control)											
	SF[a]		MF		UF1		UF2		NF		RO	
Control	100	a[b]	100	a	100	a	100	a	100	a	100	a
1:1	0	c	27	c	0	c	14	c	0	c	0	b
1:2	27	b	80	b	74	b	79	b	0	c	2	b
1:6	83	a	95	a	95	a	96	a	72	b	6	b

[a] SF: Sludge Free; MF: Micro Filtered (>120,000 D); UF: Ultra Filtered (UF1 120,000-20,000 D, UF2 20,000-1000 D); NF: Nano Filtered (NF 1,000-300 D); RO: Reverse Osmosis (RO<300 D).

[b] Numbers within columns followed by different letters are significantly different at $P \leq 0.01$ (Tukey's multiple range test). Control = buffered distilled water. (From Aliotta, G. et al. **2002**, *Allelopathy J.* **9**, 9-17. With permission.)

The most phytotoxic polyphenols on radish were **5**, **8**, **9** and **13** with values of GI_{50} ranging from 5.7 to 10.9 mM. Compounds **2** and **10** had the lowest effect on seed germination, but root and hypocotyl growth were affected more as indicated by their low IC_{50} values (8.7 and 17.3 mM for **2**; 13.3 and 14.5 mM for **10**; Table 15.2). Radicle growth and, especially, germination were more sensitive to single polyphenol than hypocotyl growth, as confirmed by I_{50} values.[27,23] This difference was found in all assays.

In the experiments carried out on *A. braunii*, compounds **2**, **3**, **12**, **13**, **15**, and **17** did not inhibit *A. braunii* at the concentrations tested. Compound **10** had slight activity at 10^{-3}M and only compounds **5**, **9**, and **16** were active at the same concentration. Catechol (**5**) was the most active and caused 97% inhibition, while ferulic acid (**9**) and caffeic acid (**16**) induced 50% inhibition.

Table 15.2

Phytotoxic effect of polyphenols on germination, radicle and hypocotyl growth of radish.[a]

Compound	GI_{50} mM	RI_{50} mM	HI_{50} mM
2	925.8	8.7	17.3
3	45.8	3.1	1.3
5	5.7	3.4	6.2
8	8.9	4.1	7.8
9	8.3	3.7	5.2
10	325.8	13.3	14.5
11	10.9	2.4	6.0
12	9.7	5.3	6.7
13	7.9	3.2	4.4
14	13.8	8.3	11.1
15	15.3	5.4	8.9
16	30.1	5.1	7.9
17	13.8	4.8	6.1

[a] Each value is the mean of three experiments and corresponds to the concentration that inhibits 50% of seed germination (GI_{50}), radicle (RI_{50}) and hypocotyl (HI_{50}) growth during seedling stage. (From Aliotta, G. et al. **2002**, *Allelopathy J.* **9**, 9-17. With permission.)

Table 15.3

Effect of seven active phenolic compounds on *Ankistrodesmus braunii*, expressed as percent of inhibition with respect to the control.

	Phenolic compounds						
	5	8	9	11	14	15	16
10^{-4} M	8b[a]	0a	6b	0a	0a	0a	5b
5×10^{-4} M	39c	0a	15c	7b	5b	0a	13c
10^{-3} M	97d	14b	50d	22c	18c	15b	52d
control	0a	0a	0a	0a	0a	0a	0a

[a] Numbers within columns followed by different letters are significantly different at P ≤ 0.01 (Tukey's multiple range test). Control = Bold basal Medium. (From Della Greca, M. et al., **2001**, *Bull. Environ. Contam. Toxicol.*, **67**, 352-359. With permission.)

The concentration of catechol (**5**), the most abundant and active component, is about 10^{-4} M in the RO fraction, while the concentrations of the remaining components are considerably lower. Consequently, none of the phenolic constituents of the RO fraction may account for the toxicity of the whole fraction.[14]

HISTOLOGICAL AND CYTOLOGICAL FEATURES OF RADISH RADICLE IN PRESENCE OF REVERSE OSMOSIS FRACTION

Light microscopy

As shown by measurement of root and hypocotyl elongation, the treatment with the RO fraction did not prevent the initial absorption of water by the seed, which is essentially a passive process. However, it effectively inhibited the ensuing expansion of these organs, which results from a combination of cell expansion and cell division. As late as 46 h after seed imbibition, no sign of cell division was visible in roots from seeds treated with 1:8- and 1:10-diluted RO fraction, and only occasional divisions, mainly in the procambial area, were observed after treatment at 1:14 dilution. By contrast, control roots resumed active cell division within 16 h after imbibition. As a conspicuous consequence of the inhibition of cell expansion/division activity, the apex of roots from treated seeds appeared distinctly wider and coarser than in the controls (Fig. 15.2a,b). Longitudinal sections of the roots, at about 600 μm from the tip, showed that cell expansion was increasingly reduced in specimens treated with 1:14, 1:10 and 1.8-diluted RO fraction relative to the controls (Fig. 15.2c,d and Table 15.4).

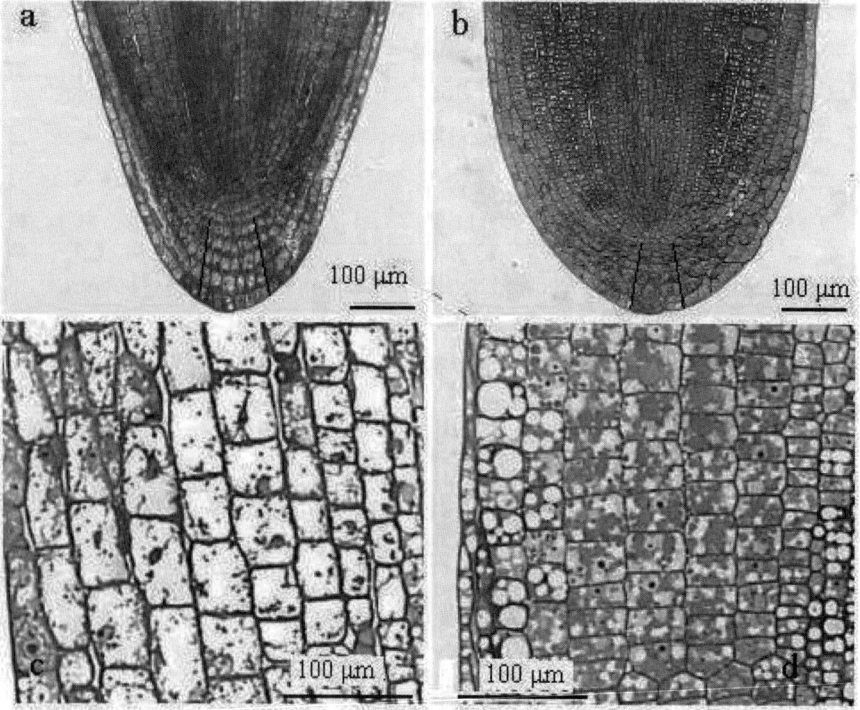

Figure 15.2

Light microscopy of radish radicles from a control seed after 16 h imbibition (a, c) and from a seed treated with 1:14-diluted reverse osmosis fraction, 46 h after imbibition. The root from the treated seed (b) is much shorter and coarser than the control (a). Cell expansion is strongly inhibited in the treated root (d) relative to the control (c).

The root apex is covered with a calyptra extending proximally up to about 600 μm from the tip. The apical meristem is of the closed type, with each of the three root tissue systems, i.e. the vascular cylinder, the cortex, and the calyptra/epidermis, arising from its own independent groups of initials.[16] Cellular differentiation had already started at a distance of about 600 μm from the tip in the 16 h-control while very little differentiation could be seen in roots from treated seeds.

Table 15.4

Length of epidermal and cortex radish radicle cells in presence and absence of reverse osmosis fraction (RO) at different dilutions.

	Length (µm ± SE)	
	Epidermis	Cortex
Control 16 h	54.59 ± 5.73	34.65 ± 8.78
Control 46 h	60.64 ± 10.74	50.55 ± 15.03
RO 1:8	21.41 ± 5.38	21.39 ± 4.92
RO 1:10	20.79 ± 5.12	19.49 ± 5.11
RO 1:14	38.44 ± 10.99	23.18 ± 6.69

Electron microscopy

When not indicated otherwise, our observations refer to cells in the sub apical area between 300 and 600 µm from the root tip. In the actively growing root, this area is the site of active cell division along with the first stages of cell differentiation, depending on the tissue. Root cells from 2 h-imbibed seeds contained numerous protein bodies[19,24] of spheroidal shape, about 1.5-3 µm in diameter and nearly completely filled with highly omiophilic protein material; they also contained abundant lipid reserves in the form of minute droplets, mainly concentrated at the cell periphery. The nucleus had spheroid or ellipsoidal shape and showed a distinct nucleolus. The cytoplasm contained numerous mitochondria with a dense matrix as well as relatively small and scarcely differentiated plastids with no or very little starch (Fig. 15.3a,b).

The appearance of root cells had changed dramatically 16 h after the start of imbibition in control seeds (Fig. 15.3c-e). The protein bodies had lost most of their electron-opaque protein content and had conspicuously increased in size, thus looking as normal vacuoles. The nucleus was highly pleiomorphic, the lipid droplets were scattered throughout the cytoplasm, and the plastids in epidermal cells contained starch (Fig. 15.3d). Numerous mitochondria were visible in the cytoplasm along with elongated microbodies with dense contents (Fig. 15.3e). Cell divisions were present (Fig. 15.4a) and the cell walls were associated with extensive arrays of cortical microtubules, whose presence is considered to be an indication of active cell wall growth[7,17] (Fig. 15.4b). In the 46 h-imbibed root the cells showed almost completely electron-transparent vacuoles and had dense cytoplasm rich in organelles including mitochondria, microbodies, dictyosomes, and rough endoplasmic reticulum, with relatively few lipid deposits. Starch was present in small amounts in the epidermis and in differentiating xylem parenchyma cells (Fig. 15.4c,d).

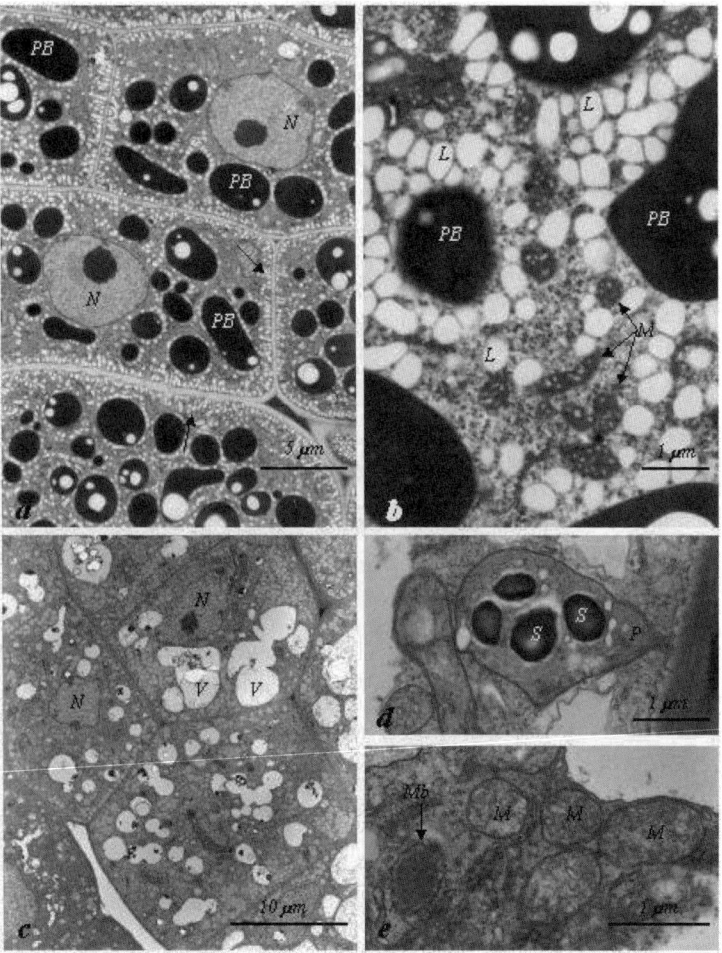

Figure 15.3

Electron microscopy of radish radicle. a) and b) Details of cortical cells from radicles of 2 h-imbibed control seeds. (a) Note the abundance of protein reserves in protein bodies (PB) and lipid reserves in the peripheral cytoplasm (arrows). (b) Numerous mitochondria (M) are visible among lipid bodies (L). (c-e) Details of cells from radicles of 16 h-imbibed control seeds. (c) Cortical parenchyma cells. The protein bodies have converted into normal vacuoles (V) and lipid reserves have been partially depleted. (d) Detail of epidermal cell showing a plastid (P) containing starch deposits (S). (e) Detail of cortical cell, showing several mitochondria (M) and a microbody (Mb).

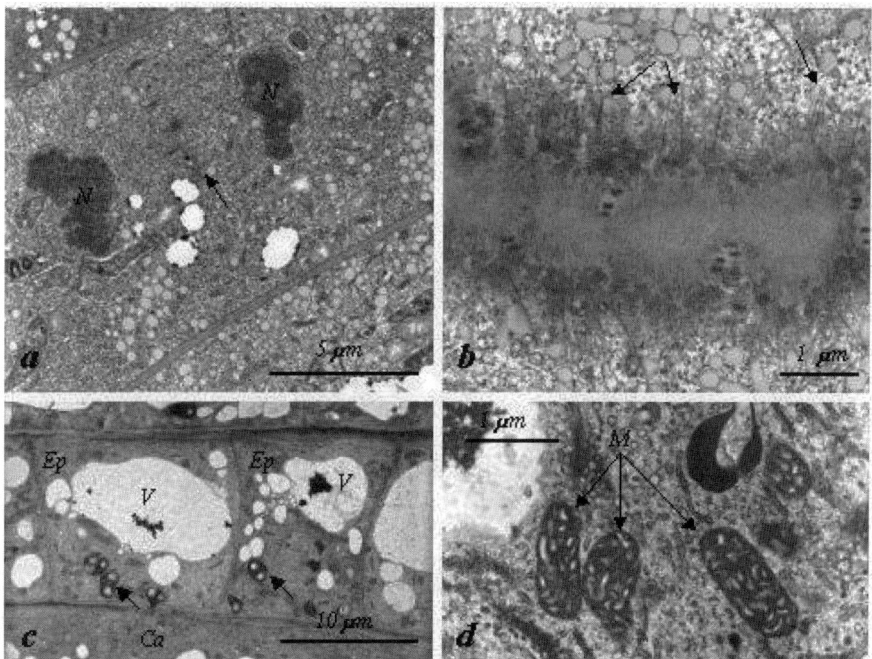

Figure 15.4

Electron microscopy of radish radicle. Details of roots from 16 h- (a, b) and 46 h-imbibed (c, d) control seeds. (a) Dividing procambial cell showing a fragmoplast between the two daughter nuclei (N). (b) Cortical microtubules (arrows) associated with an expanding cell wall. (c) Epidermal cells (Ep) covered with the calyptra (Ca) and containing large vacuoles (V) and starch-filled plastids (arrows). (d) Mitochondria (M) in a calyptra cell.

The treatment with RO fraction slowed down post-imbibition cell development and was directly proportional to the concentration applied. In 1:14-treated roots the protein and lipid reserves were almost completely depleted (Fig. 15.5a). By contrast, in 1:10- and 1:8-roots the vacuolar protein deposits were still abundant, notably in the latter, and lipid reserves appeared to be nearly as abundant as in the 2 h-control roots.

A remarkable effect of RO treatment was the appearance of swollen mitochondria with electron-transparent matrix and very few or no cristae. At low magnification these structures can easily be mistaken for vacuoles (Fig. 15.5a), but at higher magnification their two-membrane envelope is clearly visible (Fig. 15.5b). These bodies were identified as swollen mitochondria based on the fact that the cells containing them had no normal mitochondria while containing plastids much similar to those in the controls. Moreover, remnants of cristae were often visible, notably in the 1:14-treated roots (Fig. 15.5c). While in the 1:10- and 1:8-treated roots, no cell was found containing mitochondria of normal appearance, these were encountered occasionally in procambial cells in the 1:14-treated roots.

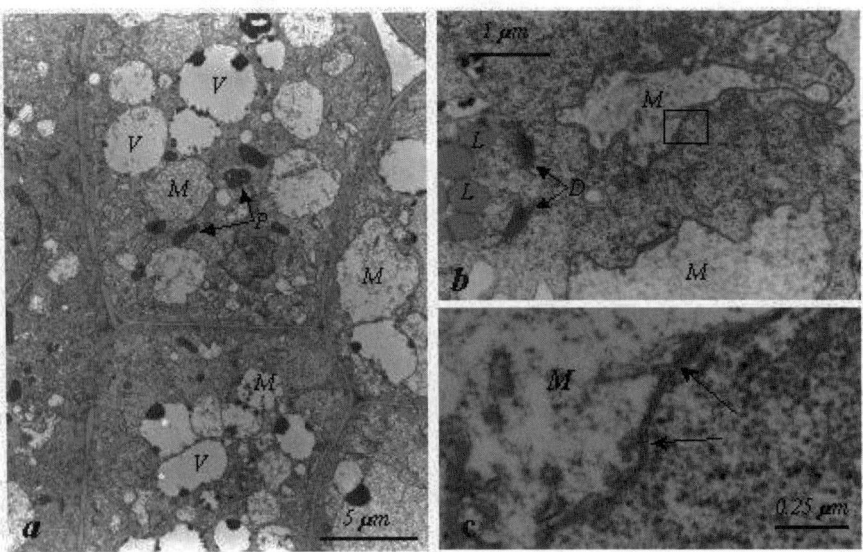

Figure 15.5

Electron microscopy of radish radicle. Details of roots from seeds treated with 1/14-diluted reverse osmosis fraction. (a) Cortical cells showing protein-body-derived vacuoles (V) with remnants of electron-opaque protein material. Extremely swollen mitochondria (M) look like vacuoles with fine granular contents. (b) Detail of epidermal cell, showing swollen mitochondria (M), lipid droplets (L) and two dictyosomes (D). (c) The area enclosed in the rectangle in (b) is enlarged to show the two-membrane envelope and residual cristae (arrows) in a swollen mitochondrion.

Examination of the root apex showed much the same effects as reported above (Fig. 15.6a-d). The calyptra was a particularly sensitive tissue, showing dramatic mitochondrial swelling at all dilutions tested (Fig. 15.6b-d). A conspicuous effect of the treatment with the RO fraction on the root apex was the inhibition of amyloplast development as statoliths in the columella cells of the calyptra (Fig. 15.6b-c). The almost complete lack of starch and of lipid-associated microbodies in the roots from RO-treated seeds is an indication of the absence of gluconeogenetic activity.

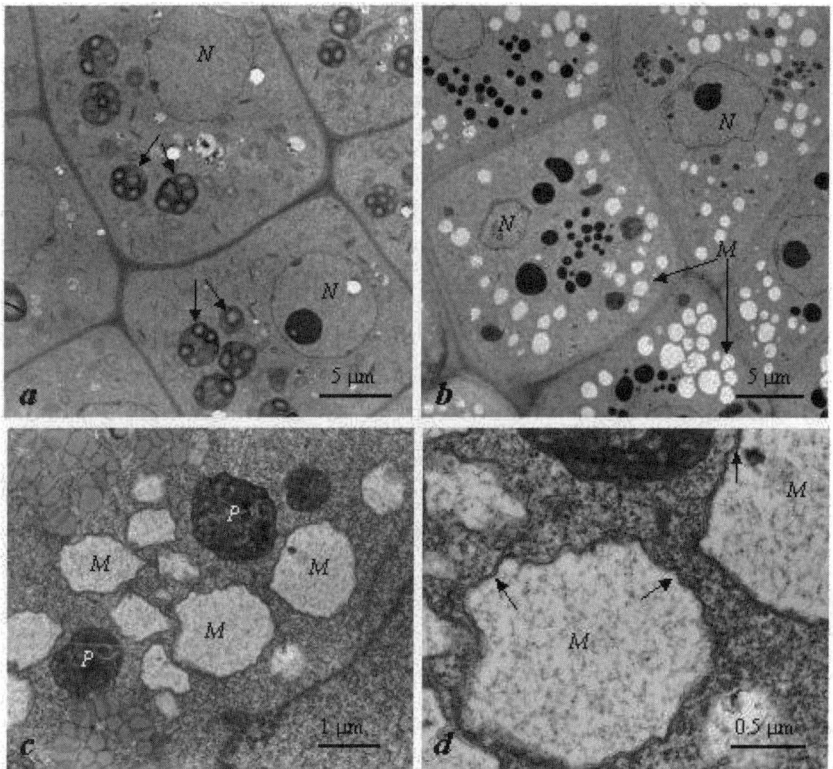

Figure 15.6.

Electron microscopy of radish radicle. Details of columella cells from 16 h-control seed (a) and seed treated with 1/14-diluted reverse osmosis fraction (b-d). (a) Columella cells in the control are distinctly polarized and contain large amyloplasts (arrows). Nucleus (N). (b) Columella cells in treated roots are not polarized and contain no amyloplasts. The numerous electron-transparent vesicles are swollen mitochondria (M). (c) Detail showing swollen mitochondria (M) and starch-less plastids (P). (d) High magnification of swollen mitochondria showing the two-membrane envelope (arrows).

In terms of general appearance, including histological and cytological features, the roots from treated seeds were much more similar to roots from 2 h-imbibed seeds than roots from the corresponding 46 h controls. Therefore, a nonspecific effect of RO fraction appears to be the general inhibition of processes leading to seed imbibition and germination. By contrast, mitochondria appear to be a specific target for toxic component(s) of the RO fraction. The mitochondria are the only cellular organelles showing obvious structural alterations relative to the controls, at least at the ultrastructural level. High-amplitude mitochondrial swelling has been reported in root cells of wheat and other plants as a consequence of anoxia stress.[10,33]

Mitochondrial swelling in animal cells is induced by the opening of the permeability transition pore (PTP).[9] The PTP is controlled by several ligands as well as by the transmembrane electrical potential, whose dissipation induces the opening of the pore and consequent mitochondrial swelling. The physiological function of the PTP is not fully understood, but circumstantial evidence suggests that it is involved in Ca++ homeostasis[8,21] and linked to stress sensing through programmed cell death. As for the well-known phenolic DNP, an effect of phenolic components of the RO fraction tested in the present study might be depolarization of the inner mitochondrial membrane. There is some evidence that a PTP is present in plant mitochondria and that it may be involved in mitochondrial swelling.[13,30] The possibility that RO-induced mitochondrial swelling in radish involves the opening of a PTP-like channel is now open to investigation.

MORPHOLOGICAL AND CYTOLOGICAL FEATURES OF *ANKISTRODESMUS BRAUNII* IN PRESENCE OF REVERSE OSMOSIS FRACTION

Light microscopy

Cells of *A. braunii* are spindle-shaped, straight or nearly straight, about 4 μm wide in the middle, with narrowed or pointed ends. Interphase cells contain a large parietal chloroplast with a prominent central pyrenoid (Fig. 15.7). The cells divide by transverse or oblique septa into two, four or eight autospores which are liberated by rupture of the mother cell wall. Treatment with RO completely stopped endospore production. Sporangia in catechol-treated cultures were observed, though a reduction of their frequency (48±3%) with respect to the control was found during the course of the experiment. Moreover, they occasionally showed a spheroidal rather than an elongated shape. Cells with anomalous shapes and altered cytoplasmic organization became frequent after both RO and catechol treatment (Fig. 15.7b,c). A statistically significant (Student, Newman & Keuls test, P= 0.1) reduction of cellular volume occurred after 96 h exposition to catechol, and similar effects were observed after exposure to RO. In catechol-treated cells, the chloroplast showed a tendency to form large lobes interconnected by thin strands

(Fig 15.7b). In RO-treated cultures, the cells often showed a centrally located chloroplast and colorless cytoplasm at either end (Fig. 15.7c).

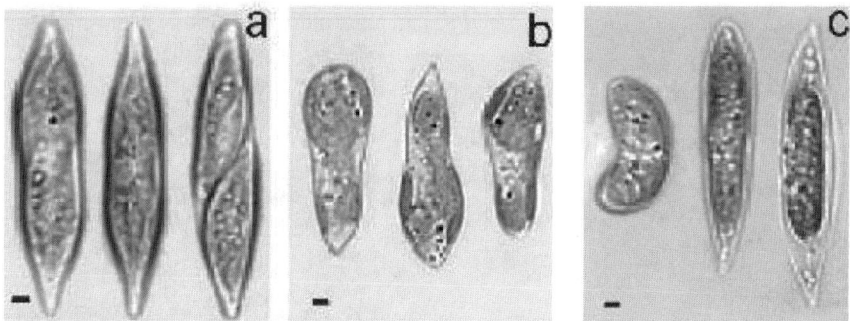

Figure 15.7

Morphological effects of catechol and R.O. on the green alga *A. braunii*. Scale bar =1μm. (a) Control cells. (b) Catechol treated cells. (c) RO treated cells.

Electron microscopy

Interphase cells of *A. braunii* from control cultures have the typical appearance shown in Fig. 15.8a when observed by electron microscopy. Apart from the large parietal chloroplast, the cytoplasm contains numerous small vacuoles, mitochondria and a central nucleus with a single prominent nucleolus. The pattern of mitosis and cytokinesis is similar to that in *Cylindrocapsa* (type III in the Chlorophyta[32]), except that the chloroplast divides after the completion of mitosis. Each mitosis is immediately followed by cytokinesis, which is affected by a cell plate of smooth endoplasmic reticulum associated with a phycoplast, i.e., an array of microtubules lying on the plane of division (Fig. 15.8d). The first two divisions are transverse or oblique relative to the main axis of the mother cell and give rise to a 4-cell chain (Fig. 15.8b). A third round of divisions, if present, is by longitudinal septa and produces an 8-cell biseriate filament (Fig. 15.8c).

The RO fraction and catechol affected the same process in *Ankistrodesmus*, i.e., cytokinesis, but had different consequences. In cells from RO-treated cultures the mitotic activity was not disturbed, but the cell plates developing between daughter nuclei failed to produce dividing septa; consequently, in 72-h-old cultures multinucleate cells became highly frequent (Fig. 15.8a,b). Incomplete cell plates between the nuclei in multinucleate cells were highly irregular and associated with microtubules apparently lacking an ordered orientation (Fig. 15.8c). Moreover, as chloroplast division in *Ankistrodesmus braunii* is associated with cytokinesis, multinucleate cells did not keep a 1-to-1 chloroplast/nucleus ratio.

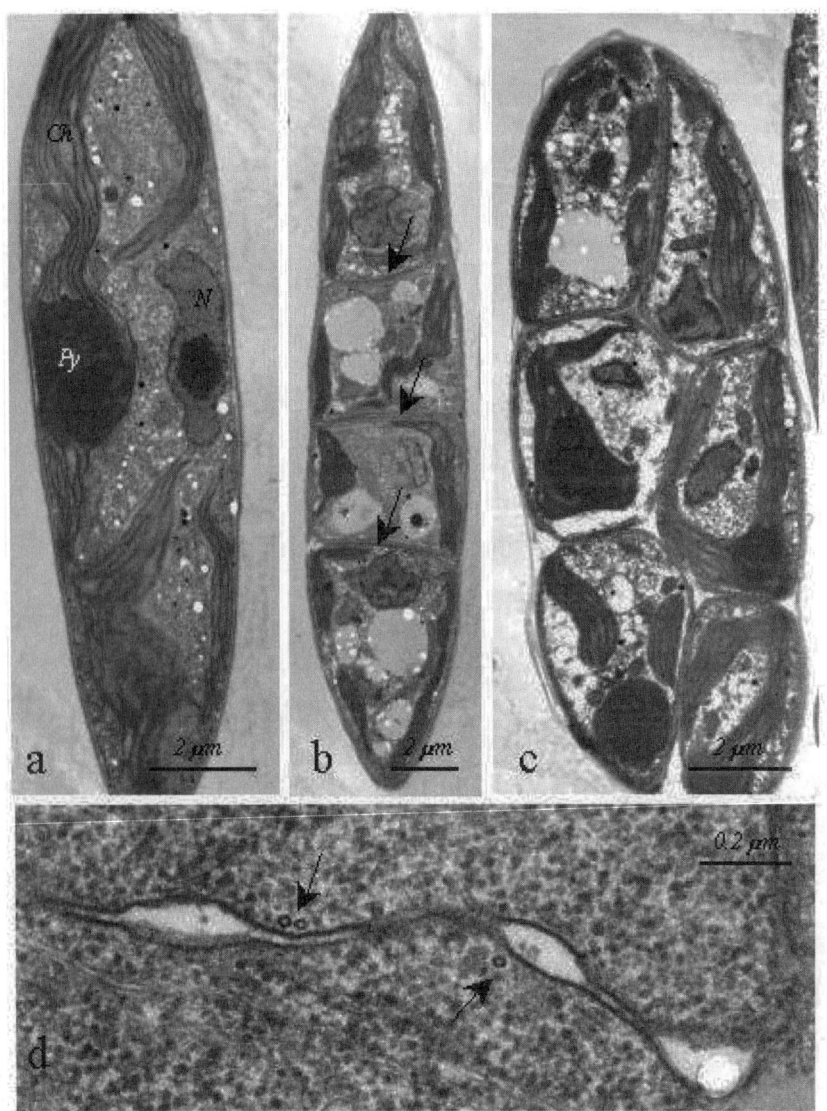

Figure 15.8

Electron microscopy of *Ankistrodesmus braunii*, control culture. (a) Undivided cell showing a single large chloroplast (Ch) with a prominent pyrenoid (Py) and the nucleus (N). (b) Sporangial cell divided into four endospores. The arrows point to transverse septa. (c) Six-celled sporangial cell. (d) Detail of a developing cell plate associated with phycoplast microtubules (arrows) lying in the plane of the developing septum.

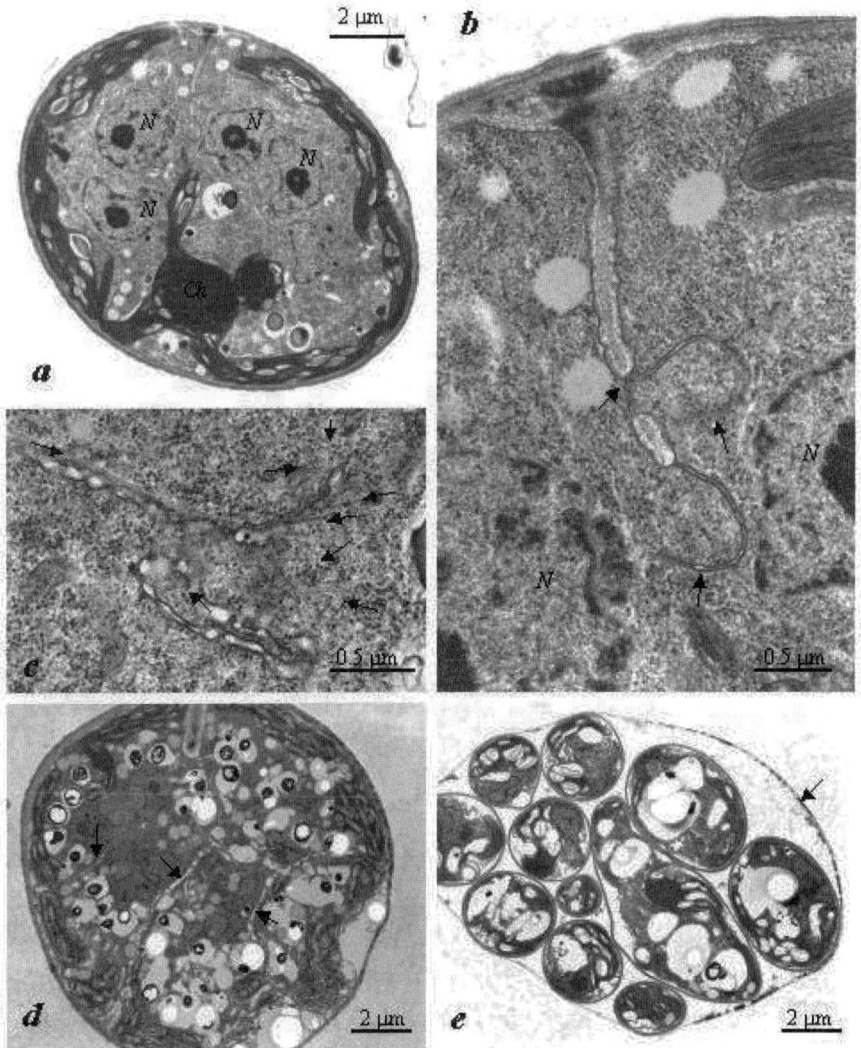

Figure 15.9

Electron microscopy of *Ankistrodesmus braunii*. (a-c) Cells from cultures treated with reverse osmosis fraction; (d, e) cells from cultures treated with catechol. (a) cell containing several nuclei (N) and a single chloroplast (Ch) with abundant starch. (b) Detail from (a) showing an abortive cell plate (arrows) between two of the nuclei (N). (c) The cell plates are associated with irregularly oriented microtubules (arrows). (d) Multinucleate cells with randomly oriented developing septa (arrows) and several chloroplasts with abundant starch. (e) A cluster of fully developed endospores still enclosed in the mother cell wall (arrow).

Unlike RO fraction, catechol did not impair the cell plate ability to form complete dividing septa, but instead caused the geometry of division to shift towards a random pattern. As a result, dividing cells produced clusters of up to 16 cells (Fig. 15.9d,e). The development of septa, however, was much slower that in controls, and therefore cells with two or four nuclei in a cytoplasmic mass not yet completely divided were encountered frequently .

In conclusion, neither RO fraction nor catechol prevented the formation of a cell plate but rather affected late stages of cytokinesis, involving the fusion of the cell plate with the membrane/wall system of the mother cell. Apparently, this process is completely stopped by RO fraction, while catechol only impairs the mechanism responsible for the fusion to take place at certain sites. There are interesting analogies between these effects and the mechanism of action of caffeine, a drug known to inhibit cell division in higher plants by stopping the final maturation of the cell plate.[29,31]

Finally, both the RO fraction and catechol induced a pronounced increase in the amount of starch in the chloroplast (cf. figs. 15.8a, 15.9a and 15.9d). Different from that observed in radish, the treatment with RO fraction in *A. braunii* induced no apparent alteration in mitochondria. The same was observed in catechol-treated cultures.

CONCLUSIONS

It is noteworthy that phenolic compounds contained in the reverse osmosis fraction of olive oil mill wastewater causes different alterations in cells of radish radicle and in the alga *Ankistrodesmus braunii*, for radicle cell mitochondria are the only cellular organelles showing structural damage. In algae, more changes such as shape of the cell, reduction of cellular volume, and prevention of dividing septa can be observed. These results indicate that the comparative study of evolutionary distant organisms can give important clues to the cytological effects of allelochemicals, thus opening new avenues to the understanding of their mechanisms of action.

METHODOLOGY

General experimental procedures

Olive oil mill wastewaters (OMW) were collected from a mill located in Genoa (Italy).To separate the components of the sludge-free OMW by molecular weight, the waters were divided into five fractions by microfiltration (MF>120.000 Daltons), ultrafiltration on different cut-off membranes (UF1 120.000-20.000 Daltons, and UF2 20.000-1000 D), nanofiltration (NF 1.000-300 Daltons) and reverse osmosis (RO<300 Daltons). From the reverse osmosis fraction (250 ml) containing constituents with molecular weight <300 Daltons the following polyphenols were isolated by chromatographic processes and identified and characterized by ^{1}H-NMR and EI-MS analysis as previously reported:[5] 4-hydroxy-3-methoxyphenylethanol **1** (5 mg), 4-hydroxyphenylethanol (tyrosol) **2** (40 mg), 3,4-dihydroxyphenylethanol (hydroxytyrosol) **3** (35 mg), 3,4-dihydroxyphenylglycol **4** (5 mg), catechol **5** (50 mg), the two dimers **6** (12 mg) and **7** (10 mg), 4-hydroxybenzoic acid **8** (3 mg), 4-hydroxy-3-methoxycinnamic acid (ferulic acid) **9** (3 mg), 4-hydroxy-3,5-dimethoxycinnamic acid (sinapic acid) **10** (1 mg), 4-hydroxycinnamic acid (*p*-coumaric acid) **11** (5 mg) and 3,4-dihydroxybenzoic acid (protocatechuic acid) **12** (3 mg). 4-Hydroxyphenylacetic acid **13**, 3,4-dihydroxyphenylacetic acid **14**, 4-hydroxy-3,5-dimethoxybenzoic acid **15**, 3,4-dihydroxycinnamic acid (caffeic acid) **16**, 4-hydroxy-3-methoxybenzoic acid (vanillic acid) **17**.

Light and electron microscopy

Whole and longitudinally cut moistened seeds were observed directly with a stereomicroscope, Wild M3Z. For scanning electron microscopy (SEM), seeds were cut into pieces and fixed in 3% glutaraldehyde in 0.065 M phosphate buffer (pH 7.4) for 2 h at room temperature. The specimens were then placed in 2% OsO_4 in 0.1 M phosphate buffer, pH 6.8, overnight at 4°C before being dehydrated in ethanol and critical-point dried. The specimens were finally coated with carbon and gold in a sputter-coater and observed with a Cambridge 250 Mark 3 scanning electron microscope operating at 20 KV.

Cytological investigation was focused on the effects of the RO fraction on the green microalga *Ankistrodesmus braunii* (at the dilution 1:20) and seeds of *Raphanus sativus* (at dilutions 1:8, 1:10, and 1:14).

The primary fixative was a mixture of 3% glutaraldehyde, 1% formaldehyde freshly prepared from paraformaldehyde, and 0.75% tannic acid in 0.05 M piperazine-N,N'–bis(2-ethanesulfonic acid) (PIPES) buffer, pH 7.0. As tannic acid may lower the pH of the fixative, it is recommended that this is checked before use and adjusted to 7.0 with 1N NaOH if necessary. The addition of low-molecular weight tannic acid to the primary fixative consistently improves the preservation of membranes and microtubules.[2]

The microalgal cultures were mixed with RO fraction to the final dilution 1:20. Algal samples were collected by mild centrifugation 72 h after addition of the RO fraction, resuspended in 5 mL of primary fixative and fixed for 2 h at room temperature. For easier manipulation, after primary fixation the algal samples were collected by centrifugation, and the pellet was resuspended in an equal volume of 4% agar in 0.1 M PIPES, pH 7.0, at about 40°C. When solid, the agar block was cut into pieces about 1 mm^3.

Radish seeds were treated with RO fraction at dilutions 1:8, 1:10 and 1:14 as described above and were processed for electron microscopy 46 h after the start of the treatment. Roots from control seeds, moistened with distilled water, were fixed after 2, 16 and 46 h in order to get several developmental stages as references for treated seeds. The root apex, about 1 mm long, was cut under a dissecting microscope from at least 10 seeds and fixed in the primary fixative for 2 h at room temperature under mild vacuum. Since the control seeds had not yet started germination after 2 h imbibition and nearly 100% of the seeds treated with 1:8-diluted had not germinated after 46 h, in these two cases the seed integument was cut open with a razor blade in order to dissect the root.

After primary fixation the algal and root samples were rinsed in 0.1 M PIPES buffer, pH 7.0, then twice in 0.1 M sodium cacodylate buffer, pH 6.8, and post-fixed in 1% osmium tetroxide in 0.1 M cacodylate buffer, pH 6.8, overnight at 4°C. The osmium solution prepared for the root samples was added with 0.7% potassium ferricyanide in order to improve osmium penetration in the root tissues. This was particularly necessary for the 2 h control roots and roots from RO-treated seed, due to their relatively low water content. The replacement of PIPES buffer with cacodylate buffer before osmication was necessary, as PIPES reacts with osmium, producing a dark precipitate.

Following secondary fixation, the samples were rinsed with distilled water and dehydrated in an increments gradient of ethanol with three exchanges, 20 min each, of anhydrous ethanol. The samples were then transferred to propylene oxide and slowly infiltrated by addition of Spurr's resin at 10% increments for 5 days at 4°C. The samples were then transferred into BEEM capsules with freshly mixed resin and polymerized for 18 h at 68°C.

Thin sections cut with a diamond knife were stained with 3% uranyl acetate in 50% methanol for 15 min, followed by Reynold's lead citrate for 10 min and observed with a Philips electron microscope. Half-µm-thick sections of the roots were cut with a diamond histoknife, stained with 0.5% toluidine blue in borax, and photographed with a Zeiss Axioskop light microscope equipped with a Sensicam photocamera. Living microalgae were photographed with a Nikon Eclipse E800 microscope equipped with a digital photocamera. Measurements of cellular volume were carried out according to the method of Hillebrand et al.[20]

Seed bioassay

Seed germination and seedling growth of radish in petri dishes were designed to evaluate the phytotoxicity of OMW fractions, thirteen polyphenols isolated from RO fraction and a mixture of these compounds. Radish seed (Raphanus sativus L. cv Saxa), collected during 1999, were purchased from Improta Co., Naples. To test the inhibitory effect of OMW fractions and RO, 20 seeds of radish were placed on two layers of filter paper (Whatman No. 1) in Petri dishes (90 mm diameter). The paper was wetted with 4 mL of buffered distilled water (BDW) with MES (2-N-[morpholino]ethanesulfonic acid) 10 mM, or test solution (undiluted fraction and a series of dilutions 1:2, one part of fraction to two parts of DW, 1:6; 1:8; 1:10 and 1:14). All pH values were adjusted to 6.0 before bioassay with MES. Experiments were made in triplicate.

In separate experiments, polyphenols **2-3**, **5**, **8-9** and **10-17**, were dissolved in a minute quantity of acetone to obtain a stock solution. Test concentrations ranging from $5 \cdot 10^{-5}$ to 10^{-3} M were prepared by diluting portions of the stock solution with bacto-agar medium (9 g/L; BA). For this experiment agar bioassay was suitable for the lower quantity of acetone-soluble compounds (**13**). Additionally, a mixture of the same polyphenols was prepared at the concentration at which they were recovered in the reverse osmosis fraction and tested following the agar dilution bioassay. The agar was allowed to solidify in a 9 cm petri dish, and 20 radish or 10 wheat seeds were germinated on 20 mL of BA or distilled water. Control seed plates contained the same number of seeds, volume of BA, and acetone as the test solutions. In all seed germination tests, the dishes were sealed with Parafilm® and placed in a growth chamber at 25°C, 60% relative humidity, (RH) and a 12 h photoperiod. Germination percentage was counted in all Petri dishes daily at noon for 7 days (no more germination occurred after this time). The amount of RO, single phenols, and their mixture required to cause 50% inhibition of germination (GI_{50}), root (RI_{50}) and hypocotyl/coleoptile development (HI_{50}) was determined by interpolation from the curve obtained by plotting the means of the replicates against the amount of substances applied.

Radish (radicle and hypocotyl) seedlings were measured to the nearest millimeter after 72 and 96 h, respectively, and compared with those of untreated seedlings. Seeds that did not germinate were considered to have a radicle length of 0 millimeters.

Algae source and bioassays

Ankistrodesmus braunii, strain 202.7a from the Culture Collection of Algae and Protozoa (CCAP), Cumbria, U.K., was cultivated in Bold Basal Medium (BBM)[26] in 1 L Erlenmeyer flasks placed on a shaking apparatus at 20± 1°C, under continuous illumination, with a total irradiance of 100 µE $m^{-2}s^{-1}$, provided by a daylight fluorescent lamp (Philips TLD 30w/55). For the microscopic observations the OMW fractions were diluted by adding sterile BBM to restore the initial concentrations UF1 (1:14), UF2 (1:16), NF(1:13), RO (1:20). These solutions were aseptically transferred in test tubes of (6 mL each) and algal inocula corresponding to a final concentration of 1×10^5 cells/mL were added. The inhibition was monitored daily for 96 h by measuring the absorbance at 550 nm, with a Baush & Lomb (Spectronic 20) colorimeter or by counting the cell numbers with a Burker blood-counting chamber. Algal growth experiments were carried out in quadruplicate and repeated two times. For each experiment, four replicates of a control, without toxic substances, were also prepared. To test statistical significance of results, one way ANOVA was performed at $\alpha=0.05$. Each comparison among means was performed using Tukey's multiple range test. The SPSS statistical package was used.

ACKNOWLEDGMENTS

This work was supported by a grant from the Italian Ministry of University and Technological Research (PRIN). We thank Dr. Joseph Sepe and Dr. Antonio

Aliotta for their helpful suggestions. The ultrastructural observations were carried out at the CISME (University of Naples "Federico II"), whose technical staff is gratefully acknowledged.

REFERENCES

(1) Aliotta, G., Cafiero, G., De Feo. V., Di Blasio, B., Iacovino, R., and Oliva, A. **2000**. Allelochemicals from Rue (*Ruta graveolens* L.) and olive (*Olea europaea* L.) oil mill waste waters as potential natural pesticides. *Curr. Top. Phytochem.* **3**, 167-177

(2) Aliotta, G. and Cafiero, G. **2001**. Seed bioassay and microscope in the study of allelopathy: radish and purslane responses. In: Reigosa Roger, M. J. (Ed.), *Handbook of Plant Ecophysiology Techniques*. Kluwer Academic Publishers, 1-20

(3) Aliotta, G., Cafiero, G., De Feo, V., and Sacchi, R. **1994**. Potential allelochemicals from *Ruta graveolens* L. and their action on radish seeds. *J. Chem. Ecol.* **20**, 2761-2775

(4) Aliotta, G., Cafiero, G., De Feo, V., Palumbo, A. D., and Strumia, S. **1996**. Infusion of rue for control of purslane weed: biological and chemical aspects. *Allelopathy J.* **3**, 207-216

(5) Aliotta, G., Cafiero, G., Fiorentino, A., Oliva, A., and Temussi, F. **2002**. Olive oil mill wastewater: isolation of polyphenols and their action on radish and wheat germination *in vitro*. *Allelopathy J.* **9**, 9-17

(6) Aliotta, G., Della Greca, M., Monaco, P., Pinto, G., Pollio, A., and Previtera, L. **1996**. Potential allelochemicals from aquatic weeds: their action on microalgae. In: Narwal, S. S. and Tauro, P. (Eds.), *Allelopathy: Field Observations and Methodology*. Scientific Publishers, Jodhpur, India, 243-254.

(7) Baskin, T. I. **2001**. On the alignment of cellulose microfibrils by cortical microtubules: a review and a model. *Protoplasma* **215**, 150-171

(8) Bernardi, P. and Petronilli, V. **1996**. The permeability transition pore as a mitochondrial calcium release channel: a critical appraisal. *J. Bioenerg. Biomembr.* **28**, 131-138

(9) Bernardi, P., Broekmeyer, K. M., and Pfeiffer, D.R. **1994**. Recent progress on regulation of the mitochondrial permeability transition pore: a cyclosporin-sensitive pore in the inner mitochondrial membrane. *J. Bioenerg. Biomembr.*

26, 509-517

(10) Blokhina, O. 2000. Anoxia and Oxidative Stress: Lipid Peroxidation, Antioxidant Status and Mitochondrial Functions in Plants. Ph.D. Dissertation, Faculty of Sciences, University of Helsinki

(11) Blum, U., Wentworth, T. R., Klein, K., Worsham A. D., King L. D., Gerig, T. M., and Lyu S. W. 1991. Phenolic acid content in soil from wheat-no till, wheat-conventional till, and fallow-conventional till soybean cropping systems. *J. Chem. Ecol.* 17, 1045-1067

(12) Capasso, R. 1997. The chemistry, biotechnology and ecotoxicology of the polyphenols naturally occurring in vegetable wastes. *Curr. Top. Phytochem.* 1, 145-156

(13) Curtis, M. J. and Wolpert, T. J. 2002. The oat mitochondrial permeability transition and its implication in victorin binding and induced cell death. *Plant J.* 29, 295-312

(14) Della Greca, M., Monaco, P., Pinto, G., Pollio, A., Previtera, L., and Temussi, F. 2001. Phytotoxicity of low molecular weight phenols from olive mill waste-waters. *Bull. Environ. Contam. Toxicol.* 67, 352-359

(15) Einhellig, F. A. 2001. The physiology of allelochemical action: clues and views. *Proceedings of the First European Allelopathy Symposium*, Vigo, Spain, 3-25

(16) Esau, K. 1977. *Plant Anatomy*. Wiley & Sons, Inc., Singapore

(17) Hasezawa, S. and Nozaki, H. 1999. Role of cortical microtubules in the orientation of cellulose microfibril deposition in higher-plant cells. *Protoplasma* 209, 98-104

(18) He, L. and Lemasters, J. J. 2002. Regulated and unregulated mitochondrial permeability transition pores: a new paradigm of pore structure and function? *FEBS Lett.* 512, 1-7

(19) Herman, E.M. and Larkins, B.A., 1999. Protein storage bodies and vacuoles. *Plant Cell* 11, 601-613

(20) Hillebrand, H., Durselen, C., Kirschtel, D., Pollingher, U., and Zohary T., 1999. Biovolume calculation for pelagic and benthic microalgae. *J. Phycol.* 35, 403-424

(21) Ichas, F., Jouavillem, S., and Mazat, J. P. 1997. Mitochondria are excitable

organelles capable of generating and conveying electrical and calcium signals. *Cell* **89**, 1145-1153

(22) Inderjit **1995**. On laboratory bioassay in allelopathy. *Bot. Rev.* **61**, 28-44

(23) Inderjit **1996**. Plant phenolics in allelopathy. *Bot. Rev.* **62**, 186-202

(24) Lott, J. N. A. **1980**. Protein bodies. In: Tolbert, N. E. (Ed.), *The Biochemistry of Plants. Vol. 1. The Plant Cell*. Academic Press, New York, 589-623

(25) Molisch, H. **1937**. Der Einfluss einer Pflanze auf die andere Allelopathie. Verlag G. Fisher, Jena. (English version by S.S. Narwal, 2001, Scientific Publishers, Jodhpur, India)

(26) Nichols, H. W. **1973**. Growth media-freshwater. In: Stein, J. R. (Ed.), *Handbook of Phycological Methods, Culture Methods and Growth Measurements*. Cambridge University Press, U.K.

(27) Rice, E. L. **1984**. *Allelopathy*. 2nd ed., Academic Press, Orlando, FL

(28) Romeo, J. T. and Weidenhamer, J. D. **1998**. Bioassay for allelopathy in terrestrial plants. In: Haynes, K. F. and Millar, J. G. (Eds.), *Methods in Chemical Ecology, Vol.2: Bioassay Methods*, Kluwer Academic Publishers, The Netherlands, 179-211

(29) Samuels, A. L. and Staehelin, L. A. **1996**. Caffeine inhibits cell plate formation by disrupting membrane reorganization just after the vesicle fusion step. *Protoplasma* **195**, 144-155

(30) Saviani, E. E., Orsi, C. H., Oliveira, J. F., Pinto-Maglio, C. A., and Salgano, I. **2002**. Participation of the mitochondrial permeability transition pore in nitric oxide-induced plant cell death. *FEBS Lett.* **510**, 136-40

(31) Valster, A. H. and Hepler, P. K. **1997**. Caffeine inhibition of cytokinesis: effect on the phragmoplast cytoskeleton in living *Tradescantia stamen* hair cells. *Protoplasma* **196**, 155-166

(32) Van den Hoek, C., Mann, D. G., and Jhans, H. M. **1995**. *Algae: an Introduction to Phycology*. Cambridge University Press, U.K.

(33) Vartapedian, B. B. and Zakhmilova, N. A. **1990**. Ultrastructure of wheat seedling mitochondria under anoxia and postanoxia. *Protoplasma* **156**, 39-44

(34) Vianello, A., Macri, F., Braidot, E., and Mokhova, E. N. **1995**. Effect of cyclosporin A on energy coupling in pea stem mitochondria. *FEBS Lett.* **371**,

258-260

(35) Zohary, D. and Spiegel-Roy, P. **1975**. Beginnings of fruit growing in the Old World. *Science* **187**, 319-327

16 Bioassays– Useful Tools for the Study of Allelopathy

R. E. Hoagland and R. D. Williams

CONTENT

Abstract 315
Introduction 316
Role and Utility of Bioassays in Allelopathy 317
    General Perspectives 317
    General Criteria for Bioassay Selection in Allelopathy Studies 319
    General Plant Bioassays 323
    Bioassays Relevant to Microorganisms 328
    Biooassays Using Aquatic Plants 337
    Allelopathic Stimulatory Actions and Some Relevant Bioassays 337
    Other Considerations Related to Bioassays for Allelopathy 338
    Immunological Methods as Potential Bioassay Tools 339
Concluding Remarks 339
References 341

ABSTRACT

    Bioassays, using plants or plant tissues, have been successful in detecting the biological activity of numerous synthetic compounds and natural products (allelopathic/allelochemical activity). Relatively rapid and inexpensive, bioassays can provide qualitative or quantitative data. A multitude of bioassays have been designed to evaluate interactions of plant compounds on plants (phytotoxicity) and on microbes (plant defense and antibiotic activity), microbial compounds and pathogens on plants (phytotoxicity/pathogenicity), and microbes on microbes (antibiotic activity). The scale of test subjects ranges from whole organism (plant or microbial cell) down to molecular constituents. Typically bioassays for such allelopathic studies have included seed germination, coleoptile growth tests, whole seedling/plant tests, membrane effects via detection of electrolyte leakage and ethane production, photosynthetic activity tests (oxygen evolution, chlorophyll production, etc.), and others. Because of their utility and diversity, bioassays will remain major tools for screening compounds for allelopathic activity and for determining the qualitative and quantitative biological activity of allelochemicals.

INTRODUCTION

In nature, many factors are involved in an allelopathic interaction. These factors are linked, either simultaneously or sequentially, and the overall process(es) is dynamic. Due to the difficulty of separating competitive from allelopathic interactions under field conditions,[78] allelopathic studies have been based heavily upon biological assays conducted under laboratory or controlled conditions. Laboratory, greenhouse, and growth chamber bioassays are important to sort out such allelopathic phenomena when suspected in nature, as has been documented.[70,71,119] Such bioassays, conducted under controlled conditions, are critical to the understanding of and in the demonstration of allelopathy. Bioassays are also important in following the phytotoxicity of active fractions of allelopathic compounds during isolation, purification and identification. Bioassays using multiple plant species can provide information on the phytotoxic selectivity or species sensitivity to allelochemicals. Specific molecular assays can be used to screen allelochemicals that have modes of action at a particular molecular site.

In this review we point out the value of bioassays in the study of allelopathic interactions but do not attempt to critically evaluate or prioritize bioassay techniques that have been published. Since it is also not possible to propose a general bioassay for researchers, we present selected examples of bioassays that have been used to understand these plant-plant and plant-microbe chemical interactions. Furthermore, we attempt to provide an overview of some bioassays that are useful or that may be adaptable to allelochemicals. The general focus will be on laboratory bioassays since they are paramount to determine quantitative effects of allelochemicals and to ascertain actual mechanisms of allelopathy in nature.

ROLE AND UTILITY OF BIOASSAYS IN ALLELOPATHY

GENERAL PERSPECTIVES

Molisch,[77] in 1937, coined the term "allelopathy" as chemical interactions among plants including microorganisms. In 1984, Rice[89] defined allelopathy as the effect(s) of one plant (including microorganisms) on another plant via release of chemicals into the environment. Here we also take this broader view of allelopathy, and discuss plant vs. plant, microbe vs. plant, and microbe vs. microbe allelopathic interactions in regard to bioassay systems.

The presence of allelopathic compounds in plant tissues does not necessarily imply allelopathic potential in an ecological setting.[39] Allelopathic action in the field depends on factors including the effects of released plant compounds on soil microorganisms or on surrounding plant species that may decrease positive effects of growth-promoting microorganisms on plants and/or possibly increase pathogenic microbial activity. Microbial action, or the inherent labile nature of allelochemicals, can also result in degradation of released phytotoxins, rendering them nonphytotoxic. Alternatively, microbial enzymes may transform nontoxic allelochemicals into active phytotoxins. There are at least three fundamental reasons to study allelochemicals. Identification of allelochemicals may help explain plant or microbial allelopathic interactions in ecological systems. A second reason is to identify allelochemicals which may have potential as herbicides or plant growth regulators (PGRs) or that could be used as templates for the design of novel synthetic herbicides or PGRs. Third, the use and/or development (breeding or genetic transformation) of crops or cover crops that produce allelochemicals may be used to suppress weed growth as an alternative management practice. In each of these cases, bioassays can play a significant role in determining the activity, species selectivity, mode of action, and persistence of the compound(s) in question.

The first plant bioassay [*Avena sativa* L. (oat) coleoptile test] was employed by F.W. Went in the 1920's to demonstrate the existence of and to quantitatively assess the first growth-modifying substance [indole-3-acetic acid (IAA)] isolated from plants.[122] Plant bioassays have been extremely useful and intimately linked to the discovery and characterization of the major classes of plant hormones. In fact, many of the bioassays used now were developed for PGRs. Bioassays have been used to screen, evaluate phytotoxicity or plant growth promotion, study mode of action, and assess structure-activity relationships of hundreds of thousands of compounds (natural products, allelochemicals, synthetic and naturally occurring herbicides, and PGRs). If designed and used properly, bioassays can be very useful in studying the biological effects of secondary plant and microbial compounds. Multiple interactions of allelochemicals with other compounds can be more easily determined by bioassay methods than by physicochemical analyses. Bioassays are essential to evaluate allelochemical effects on the growth and development of plants, particularly when determining possible effects on organelles or enzymes, or to determine the molecular mode of action.

Bioassays, like other techniques, have inherent limitations. Bioassays exhibit large response curves compared to data from physicochemical methods. Log-linear concentration response curves do not allow assessment of concentration differences of allelochemicals. Interfering substances in nonpurified extracts may have greater effects in bioassays than in physicochemical analyses. Improved techniques/instrumentation [improved isolation, separation and detection techniques (HPLC, GC, mass spectrometry, NMR, immunological methods, etc.)] have provided greater sensitivity/specificity, and such techniques are less variable than bioassays. Ideally, these tools should be used in tandem with bioassays in allelopathy studies, because intact plants, excised plant parts, and cellular and sub-cellular plant constituents are entities that provide direct biological responses to test chemicals. However, the main difficulty with bioassays is their misuse by failure to run standard calibration curves with known compounds, lack of proper experimental design, and neglect of proper statistical analyses.

The limitations of bioassays can be minimized if the proper experimental design, test material (and treatment thereof), test methodology (seed germination, hypocotyl elongation, etc.), replication, and statistical analysis are chosen. These factors are dependent upon the nature of the compound in question, the research objective(s), and the imagination of the investigator. Some of the best examples of successful bioassays of chemicals can be obtained from the methodologies used to detect plant growth regulator activity, herbicidal activity, and effects of pathogens on plants. Below, we examine some of these specific bioassays for consideration by researchers studying allelopathy.

Herbicide bioassays

Bioassays have been successfully used for over 50 years to screen and evaluate compounds for herbicide development. Herbicidal bioassay protocols as research tools for measuring agrochemicals in soil and water have been reviewed and discussed, as in selected citations.[1,55,69,83,95,110] A compilation of bioassays, including algal, higher plant, and cell-free enzyme systems for the evaluation of chemical inhibitors, has been published.[9] All bioassays described therein represent known target sites of herbicides or other chemical inhibitors. The use of bioassay techniques, such as those outlined in these publications, can and should be incorporated in the development of allelochemical bioassays.

Bioassays for plant growth regulators

Since the use of techniques to measure IAA, bioassays have been important to discover PGR activity of many other compounds. Several bioassays for the PGR gibberellic acid (GA_3) have been developed. One bioassay was based on reduction of amaranthin levels in *Amaranthus caudatus* (tassel flower) seedlings.[63] This method was sensitive to GA_3 from 0.01 to 1 mg L^{-1}. GA_3 was also bioassayed, based on anthocyanin reduction in tomato (*Lycopersicon esculentum* L.).[62] Reduction of anthocyanin in tomato seedlings was linear from 10^{-5} to 10 mg L^{-1}, and thus, this latter plant bioassay method was more sensitive. A multitude of bioassays for nonvolatile and volatile plant growth regulator compounds have been developed and their uses and limitations have been discussed.[129] Since many allelochemicals have been shown to have relatively weak phytotoxicity (especially compared to herbicides), some of these bioassays that have been developed for detecting and quantitatively measuring PGR activity may be useful in allelopathy.

GENERAL CRITERIA FOR BIOASSAY SELECTION IN ALLELOPATHY STUDIES

Aside from the primary research objective, several other criteria should be considered carefully in selecting and designing a bioassay. These include sensitivity, specificity, response time, reproducibility, plant material, cost, labor, equipment, and space requirements. The bioassay should be highly sensitive to the allelochemical(s) in question. Sensitivity and selectivity are dependent on the plant species and plant material being used as the indicator and on the efficacy of the allelochemical studied. In practice, several plant species should be used as indicator species, some of which are highly sensitive to the test compound(s) and some of which appear to be more resistant. Reproducibility is another key factor. The bioassay should be easily replicated in time and show consistent responses. If data from the bioassay are not reproducible or the results are highly variable among replications, then the results should be scrutinized. When this happens, the design or expectations of the experiment should be reconsidered. The bioassay should have a relatively rapid response time. The time required for the bioassay depends on the type of assay (e.g., seed germination vs. whole plant growth), but the total time for the bioassay should be minimized to avoid extraneous interfering factors. This eliminates potential degradation of the allelochemical(s), variation in the plant response due to possible changes in conditions, and contamination by microorganisms. The requirements for the bioassay set-up, including the handling of the plant material, should be simple. This will reduce cost and labor requirements and increase the ease of treatment replication and reproducibility. Finally, an effective bioassay should require minimal equipment and space. Here again this depends on the type of assay and the objectives of the study.

As pointed out above, the bioassay design depends on the objective(s) of the study. A bioassay to determine allelopathic interactions in the field or in an ecological setting may have a quite different design than one used to determine PGR activity of a compound or to determine its molecular mode of action. Specific bioassays can be used to follow the isolation/purification of allelochemicals, evaluate their phytotoxic (or growth simulation) effects (i.e., visual effects), determine their host range/selectivity, evaluate allelopathic action of volatile compounds, or examine physiological/biochemical effects, such as photodynamic and membrane effects, effects on photosynthesis, specific enzyme sites, and effects at the ultrastructural level to locate receptor sites or sites of injury. Several examples of useful bioassays will be presented later.

Another weak point of many allelopathic research studies reported in the literature is the lack of proper statistical analysis. Probably the two most critical errors are the lack of proper controls and insufficient replication. Appropriate controls need to be included even when a minimal amount of a solvent [e.g., dimethyl sulfoxide (DMSO), acetone, etc.] is used to solublilize a compound or for extraction of the test plant material. In this situation, it is also helpful to include positive controls (known compounds at similar concentrations) for comparison to the unknown or suspected allelochemicals. As discussed earlier, it is also useful to include various species so that a range of sensitivity to the test material can be observed. For example, a bioassay using seed germination might include lettuce seed, generally a sensitive species, and other species which might vary in response to the allelochemical(s) or extract. Such selection can demonstrate plant selectivity to the active ingredient(s). This is important in the search for bioherbicides and eventually could provide important knowledge about species-selective allelopathic effects in natural ecosystems.

Replication is also extremely important. All of the treatment and controls (including neutral and treatment controls) need to be replicated. Furthermore, the amount of plant material (e.g., the number of seeds, leaf discs, coleoptiles, etc.) needs to be sufficient for each replication. A critical consideration in seed germination tests is the number of seeds used per replication. Generally, fewer large seeds are used to insure sufficient extract or test solution for imbibition and germination, but more seeds are used when evaluating a smaller-seeded species. For example, 30 or 50 small seeds compared to perhaps only 15 seeds of a large-seeded species might be used per Petri dish. The use of lower seed numbers reduces the number of observations for the larger-seed species and places inordinate statistical weight on the nongerminated seeds when the results are expressed as percent germination. One way to overcome this problem is to use the same seed weight to liquid volume ratio in experiments where seed size and weight vary. This is pointed out in a standard bioassay for seed germination of parasitic weeds[74] as discussed later. Expression of germination data as percentage may influence the results of the analysis. In any event, care should be taken with percentage data. Usually percentage data are either transformed (using an arc-sin square root or log transformation) before analysis, or probit analysis may be used. Detailed discussion on the analysis of percentage data and probit analysis are beyond the scope of this review, but are discussed in textbooks on statistics.[29,56,107]

Often, there is interest in the effects of combinations of two or more allelochemicals on test species. Sometimes chemical combinations produce interactions that either increase or decrease the plant response. The terminology describing "interactions" of mixtures has been misleading or confusing when the literature is considered as a whole. The terms synergism, antagonism, and additive effect are often used, but many reported studies lack the proper statistical analysis to draw such conclusions. Synergism is defined as the cooperative action of two (or more) agents when the observed response of the test organism to their combined application is greater than the predicted effect, based on the expected response to each component separately. Antagonism is defined as the combined action of two agents, such that the observed response of their combination is less than predicted. An additive effect is defined as the cooperative action of two agents, resulting in an observed response equal to the sum of that predicted by each independently. Examples of interactions of chemicals and/or pathogens are provided in a review,[45] and excellent reviews concerning definitions, experimental design, and statistical analyses related to these terms are available.[35,36,38] In some cases, allelochemicals applied in combination have been shown to interact on a statistically significant basis in a seed germination bioassay using combinations of two compounds on three plant species (Table 16.1).[123] However, these interactions were not analyzed for antagonistic or synergistic interaction.

Table 16.1

Summation of the statistical interactions of two phenolic compounds on seed germination of three plant species.

Combination[a,b]	Source of variation						
	Day	Species[c]	Compound	Day-Species	Day-Compound	Species-Compound	Day-Species-Compound
Coumarin + p-hydroxybenz-aldehyde	**	**	**	**	**	**	**
Ferulic + p-coumaric acid	**	**	*	**	NS	*	NS
Chlorogenic acid + hydrocinnamic acid	**	NS	**	**	**	NS	NS
Chlorogenic acid + p-hydroxy-benzaldehyde	NS	NS	NS	NS	NS	NS	NS

[a] each compound at 10^{-3} M.
[b] significance levels: **, 0.01 level of probability; *, 0.05 level of probability; NS, non-significant.
[c] *Sorghum bicolor* (L.) Moench; *Sesbania exaltata* (Raf.) Rydb.; *Cassia obtusifolia* L.
Source: Williams, R. D. and Hoagland, R. E. **1982**, *Weed Sci.*, **30**, 206-232. With permission.

Treatment and preparation of test materials and bioassay specimens

The biochemicals identified as allelochemicals are generally metabolites resulting from secondary plant and microbial metabolism. The levels of such compounds can be elevated in plant tissues by stress (temperature, nutrient, water stress) or as a response to pathogenic infections. Because of this, care needs to be taken in deciding how and when samples to be used in allelochemical tests are extracted and tested. It is important to determine if the stress caused by competition among plants is elevating production of the allelochemical that could cause increased amounts of allelochemical to be released into the environment.

Also, precautions during the extraction processes are necessary to ensure that the allelochemicals are not destroyed or that artifacts are not produced. Excessive heating or drying of the samples can destroy heat-liable compounds or inactivate compounds. Often artifacts are also produced by oxidation and enzymatic action on allelochemical substrates. The choice of a suitable solvent (aqueous or various organic compounds) is also important and determines which class of allelochemicals is extracted. Proper storage of extracts and/or ground plant preparations is also important since wet nonrefrigerated samples can be contaminated by microorganisms that may degrade active compounds or have allelochemical activity themselves. These and other associated factors have been discussed elsewhere.[8,126] Bioassay systems should also be void of extraneous microorganisms and chemicals that could cause interfering interactions. One example of such a problem is contamination of leaching stands by microorganisms, including algae, bacteria and fungi.

GENERAL PLANT BIOASSAYS

Germination bioassays

Seed germination bioassays are probably the most common tests reported in the allelopathy literature. Seeds from test plants are placed in Petri dishes with various concentrations of extracted plant material and/or known allelochemicals on filter or germination blotter paper, followed by incubation at controlled temperature. Germination (usually defined as radicle emergence) is determined at various time intervals and expressed as percent seed germination at a specific period or rate of germination. When testing extracted plant material, care should be taken to ensure that seed germination is not delayed by the osmotic potential of the extract solution. Also when examining the effect of plant extracts, bioassays are more meaningful if a known allelochemical (at known concentrations) is included in the assay as a standard comparison. The drawback to germination bioassays is that seed germination may not be as sensitive to the effects of allelochemicals as plant growth or other plant process. Assays can be performed under dark growth conditions that allow more rapid stem elongation and, thereby, may increase bioassay sensitivity. Alternatively, if objectives are to study photodynamic effects, the system can be performed under light conditions.

Photosensitization is also a property of some compounds that are light-activated. Many natural and synthetic products can photosensitize viruses, and bacterial and mammalian cells.[84] Several furocoumarins act as photosensitizers and are widely distributed among some plant species.[75] Some of these compounds may also have allelopathic action on plant tissues. A photobiological bioassay based upon the inhibition of growth in a DNA repair-deficient $E.\ coli$ strain B_{s-1} was developed as a ultra-sensitive method for detection of photosensitization.[2] Psoralen (and several analogues), angelicin, 5-nitroxanthotoxin, oxypeucedanin, and others were detectable at a range of levels from 10 µg to 50 µg. All compounds were detectable at several orders of magnitude higher in $E.\ coli$ strain B_{s-1} than in wildtype $E.\ coli$.

Because germination may not be the primary target of most allelochemicals, many researchers make additional observations of the emerging seedlings.[123] Radical, coleoptile, and hypocotyl length are often recorded. If seed germination is delayed by an allelochemical, seedling length may reflect the delay in seedling development, rather than be due to a direct effect by the active compound(s). This problem can be eliminated or checked by pregerminating the seed, placing the developing seedling in contact with the extract or compound, and growing the seedlings in the dark for several days.[44] In this way, direct observations can be made on cellular growth, cell elongation and tissue development. Any of the tissue bioassays (e.g., oat coleoptile elongation) recommended for growth regulators can be adapted to allelochemicals.[129]

Plant growth bioassays

Bioassays using whole plant growth are not used as frequently as bioassay systems with short-term turnaround. In some cases, plants are grown hydroponically in a porous medium (sand or soil) which is amended with the allelochemicals or extracts. Since the plants are not grown under sterile conditions, the influence of microbes in the growing media cannot be discounted. Microbial metabolism of the compounds or conversion of the compounds to a nonactive state is always a possibility. However, such long-term studies do aid in the understanding of the overall effect allelochemicals have on plant growth. Blum and his co-workers have used this technique to determine the long-term effects of exposure of plants to allelochemicals.[54,120] The use of plant bioassays to evaluate the fungal bioremediation of contaminants in soil from a hazardous waste site has been reported.[4] Higher plants [lettuce (*Lactuca sativa* L.), oat (*Avena sativa* L.) and millet (*Panicum miliaceum* L.)] were used (seed germination in soils and root elongation in aqueous soil eluates) as monitors of contamination reduction before and after treatment. Bioassays showed significant detoxification with time and that (in this case) seed germination in soil was the more sensitive bioassay to measure these toxicants. This was suggested to be due to a lack of efficient extraction of toxic components in soil and/or more availability of contaminants to seeds germinating in the soil. Similar methodologies have been (or should be) applied to field sites when allelopathy is suspected and/or to study degradation or molecular turnover of potent allelochemicals at sites where allelopathy has been proven. As stated earlier, allelopathic interactions (in field experiments especially) are dynamic and may be mediated by the rhizoshpere organisms of both the donor and host plant, an area often ignored in the allelopathy literature.

Bioassays based on pigment analysis

Bioassays can also be useful to determine the relative allelopathic potency of allelochemical analogues. For example, fumonisins are secondary metabolic analogues produced by *Fusarium moniliforme*. A range of phytotoxicity of fumonisins A_1, A_2, B_1, B_2 and B_3, aminoalcohols HB_1 and HB_2 (derivatives of B_1 and B_2, respectively) and AAL-toxin (produced by *Alternaria alternata*[112]) was found using quantification of chlorophyll content in duckweed (*Lemna parisicotata*) as a bioassay plant. A simple bioassay to measure allelochemical activity was used to determine the effects of albizziin and mimosine on chlorophyll development in hemp sesbania and sicklepod.[124] Cotyledons were excised from 4-day old seedlings and imbibed in test solutions in the dark for 2 h; then tissue was exposed to continuous, low intensity (70 $\mu E\ m^{-2}\ sec^{-1}$) light for 48 h. Chlorophyll was then extracted with DMSO and measured spectrophotometrically. Both compounds inhibited cotyledon greening at 10^{-3} and 10^{-4} M. Compounds that cause photo-bleaching in plant tissues were studied by measuring carotene precursor accumulation.[94] Cells of the alga *Scenedesmus acutus* were chosen due to the ease of cellular extraction and noninterference of constituents during absorbance spectral analysis of pigments.

Electrolyte leakage as a bioassay

Conductivity measurement of cellular electrolytes caused by a variety of compounds that disrupt membranes has also been a useful and sensitive bioassay method. Fumonisin analogue phytotoxicity has been determined using duckweed (*Lemna parisicotata*) as a bioassay specimen.[112] This duckweed species was also useful in measuring peroxidizing activity of several herbicides that caused peroxidation in plant tissues.[21] Similar bioassays can be constructed to also measure cellular leakage of specific compounds such as amino acids or phenolic compounds.[96] Tests are generally run on tissue slices or leaf disks, but plant cell suspensions have also been used to screen growth retardants and herbicides.[37]

Isolated systems as bioassay specimens

Isolated cells or organelles as bioassay subjects may have advantages and be more applicable than whole plant systems. The use of isolated cells and protoplasts in the study of phytotoxins and for breeding pathogen resistance in crop plants has been examined and reviewed.[32] Various parameters of isolated cell and protoplast bioassays can be measured to provide reliable, reproducible, and quantitative data. These include measurements of cell vitality, cell-division potential, cell membrane permeability, effects on cellular organelles, and effects on activities of specific isolated enzymes or on enzyme activities in isolated organelles. These authors provide specific protocols and discussion of the bioassay of *Xanthomonas compestris* culture filtrates on peach mesophyll cells, *Phytophthora citrophthora* culture filtrates on *Citrus* sp. protoplasts, cercosporin on tobacco cells, sirodesmin on *Brassica* sp. cells and protoplasts, and on *Helminthosporium maydis* race T-toxin (*Hm* T) using maize protoplasts.

Specific molecular bioassays can be indispensable in screening compounds that act on certain biosynthetic pathways or enzymes. For example, recent patents report a screening bioassay for detection of herbicides that specifically inhibit the plant purine biosynthetic pathway.[104,111] Many enzymes are involved in this pathway and may be targets of test compounds. *Arabidopsis* seeds express the enzymes adenylosuccinate synthase and adenylosuccinate lyase; this is a suitable plant species. Such site-specific assays could be and are currently used by industry, to help to screen synthetic compounds and plant and microbial allelochemicals, and to discover new herbicides that can disrupt sensitive plant pathways that result in potential phytotoxicity. Several books have reviewed these strategies for use in herbicide evaluation, and obviously allelochemical evaluations could use such specific target site bioassay systems.

Other enzymes have been used to test a multitude of compounds for herbicidal activity. The enzyme acetolactate synthase (ALS) [also called acetohydroxy acid synthase (AHAS)] is involved in the biosynthesis of the amino acids leucine, isoleucine, and valine. Inhibition of ALS is a lethal herbicidal target in plants. Sulfonylurea[14] and imidazolinone[98] herbicides inhibit activity in ALS enzyme preparations when treated with these herbicides and their active analogues. Glutamine synthetase (GS) preparations have been useful to detect compounds that inhibit glutamine biosynthesis, another lethal herbicidal target site. Inhibition of GS is the target site of the herbicide phosphinothricin (glufosinate), and GS has been used to screen many compounds for herbicidal activity.[91] Thus, if it is desirable to search for allelochemicals from microbes or plants that inhibit these specific phytotoxic sites, such enzymes associated with these sites can be used in specific screening programs. If other sensitive enzymes are required, then preparation of those enzymes could be used to examine large numbers of allelochemicals for possible *in vitro* inhibition. One fundamental requirement is that these enzymes must possess stability in crude preparations. Such enzyme target bioassays are sensitive, rapid and inexpensive. Many other enzymes are known target sites of commercial herbicide molecular mode of action and their preparation and use as specific bioassay systems have been summarized.[9]

Bioassays to detect inhibition of photosynthesis

Leaf discs have commonly been used for bioassays to determine if herbicides inhibit photosynthesis (Table 16.2). The simplest leaf-disc bioassay uses small discs cut from fully expanded cucumber or pumpkin cotyledons, floated in the light on a phosphate buffered medium containing suspected photosynthesis inhibitors.[115] Qualitatively, if photosynthesis is inhibited, the leaf disc sinks. There are several variations of this method that can provide quantitative data. Evolution of O_2 in the test solution can be measured with an oxygen electrode, CO_2 induced pH changes colorimetrically determined with bromothymol-blue, or electrolyte leakage monitored with a conductivity meter. Leaf strips, algae, isolated chloroplasts, and duckweed (*Lemna minor*) have been used as test subjects. Although the bioassays presented in Table 16.2 are fairly easy to perform, few allelochemicals have been tested as possible inhibitors of photosynthesis. Many allelochemicals can inhibit plant growth, but in most cases, the modes of action of these compounds have not been investigated. Even if an allelochemical does inhibit photosynthesis, such inhibition may be an indirect response due to changes in leaf-water potential, membrane induced changes, stomatal effects, or general effects on plant metabolism. Thus, the choice of bioassay is important so that secondary vs. primary effects on photosynthesis can be differentiated.

Table 16.2

Selected examples of methods to measure the effects of compounds on photosynthesis in several species.

Species	Method
Capsicum annuum	Leaf slices w/ O_2-electrode[67]
Spinacia oleracea	Leaf slices w/ O_2-electrode[57]
Selenastrum capricornutum	O_2-electrode[6]
Glycine max	Leaf discs w/ O_2-electrode[92]
Myriophyllum spicatum	Plant segments w/ O_2-electrode[97]
Cucurbita pepo	Leaf discs; sinking disc technique[115]
Citrullus vulgaris	Leaf discs; sinking disc technique[19]
Chenopodium album Senecio vulgaris Amaranthus hybridus	Leaf discs; buoyancy technique[40]
Cucumis sativus	Leaf discs; buoyancy technique[93]
Cucumis sativus	Leaf discs; alleviation of paraquat-induced electrolyte leakage[128]
Chlorella sp.	O_2 evolution, using luminous bacteria[113]
Chlorella sp. Scenedesmus sp.	CO_2 induced pH changes, detected w/ bromothymol blue[72]
Lemna minor	Visual injury[81]

BIOASSAYS RELEVANT TO MICROORGANISMS

Bioassays to detect pathogenicity

There are numerous publications on the allelopathic interactions of phytopathogens and plants. Most of these deal with the interactions of pathogens that attack crop plants. However, over the past 30 years, considerable information has been published concerning pathogens which have weeds as primary hosts. Plant defense mechanisms and plant-microbe interactions with regard to weed control and allelopathy have been discussed.[52] Plants protect themselves from attack by microorganisms (especially phytopathogens) by a variety of physical and biochemical defense (allelopathic) mechanisms. There are many biochemical mechanisms of plant defense against pathogens or against microbial secondary compounds.[52] One biochemical mechanism is via phytoalexin production and accumulation at or near infection sites. Induction of many of these defense responses is caused by compounds called elicitors that are produced by pathogens. Most fungal elicitors are oligo- or polysaccharides from the mycelia of fungal pathogens.[61,99] Bioassays of elicitor activity in rice cell-suspension cultures for several compounds, i.e., chitin,[87] *N*-acetylchitosaccharides,[127] and mycelial extract preparations from *Phytopthora infestans*[64,65] have been developed. More recently, a new bioassay to measure elicitor activity in rice leaves was reported.[65] Fungal elicitors of phytoalexins are important because of their role in plant defense against pathogens and their phytotoxicity. These compounds may have potential in agriculture as stimulators of plant defense and as herbicides.[82]

Many bacteria and fungi have been evaluated as agents for the biological control of weeds.[49] Such pathogens infect and decimate weeds by producing enzymes that degrade plant tissues and/or by secreting phytotoxins that translocate to sensitive sites and disrupt essential metabolic or physiological processes resulting in mortality. Collectively, pathogens and microbial phytotoxins used for weed control are referred to as bioherbicides or biological weed control agents. There is interest in these bioherbicides for use in weed control, for the discovery of new chemistries for direct herbicide application, and in the design of new synthetic herbicides. This research area has been summarized in books and reviews (see selected citations).[41,43,46,48,49,51,90,114] These interactions of pathogens and phytotoxins on plants are indeed allelopathic interactions, and various bioassays have been used to monitor these effects.

Some plant pathogenic bacteria and their phytotoxins have been screened in bioassays that monitor the effects of their toxins (antibiotic and phytotoxic) on other sensitive bacteria. For example, several fluorescent *Pseudomonas syringae* pvs. produce extracellular phytotoxins.[76,106,116] Tabtoxin is produced by *P. syringae* pv. *tabaci* and pv. *coronafacines*, and this natural product inhibits glutamine synthetase.[34,46,116] Phaseolotoxin, produced by *P. syringae* pv. *phaseolicola* inhibits L-ornithine carbamyltransferase, an enzyme of arginine biosynthesis.

Enzyme inhibition can be reversed by supplementation of arginine or citrulline.[34,106] Pathovars of *P. syringae* were shown to inhibit *Escherichia coli* growth, an effect reversed by L-arginine, but not by L-citrulline or L-glutamine.[121] This suggested that the site of action of the toxin produced is involved in the conversion of citrulline to arginine in the urea cycle.

Various nonphytopathogenic microorganisms from soils produce compounds from new chemical classes that exhibit phytotoxicity toward higher plants and act as antibiotics against other microorganisms. Actinomycetes have been widely studied because they produce a wide range of antibiotic chemicals that possess phytotoxic activity. These organisms are ubiquitous in soils and some of their secondary metabolites have been isolated, purified, and shown to have herbicidal potential.[50] The most successful of these is the unique tripeptide bialaphos that has been isolated from *Streptomyces* spp. Reviews on this natural product chemistry and biological activity are available.[47,53] The intact bialaphos molecule is not herbicidal, but plant peptidase action on this peptide releases the highly herbicidal compound phosphinothricin. Bialaphos is marketed in Japan as a commercial herbicide, and glufosinate is the synthetically produced ammonium salt of phosphinothricin which is marketed globally.

Although various bioassays have evaluated the injurious effects of phytotoxins from pathogens and other microorganisms, comparatively few bioassays have been reported for testing pathogen virulence, host range, effects of nutrition and culture production, and interactions with other chemicals on pathogen-plant interactions. Bioassays to measure bioherbicidal (microbes and/or their phytotoxins) efficacy differ somewhat compared to those used for herbicide screening and evaluation. Some aspects of such bioassays have been considered.[51] Generally, a living plant bioassay system is required in research and developmental studies of pathogens for weed control. A living host plant system can be used to maintain pathogen virulence via periodic inoculation and re-isolation of the pathogen from this host, examine the pathogen host range, and measure efficacy (pathogenicity) in host plants at various growth stages and environmental conditions (temperature, humidity, light quality and quantity, etc.). Briefly, greenhouses or growth chambers are used to grow test plants and also for incubation and growth after pathogen inoculation. Disease symptomology and/or weed injury are monitored over several days or in some cases up to a week or more. At the end of the incubation/exposure period, parameters such as visual rating and plant fresh weight/dry weight are measured and pathogen virulence (weed control efficacy) is calculated.

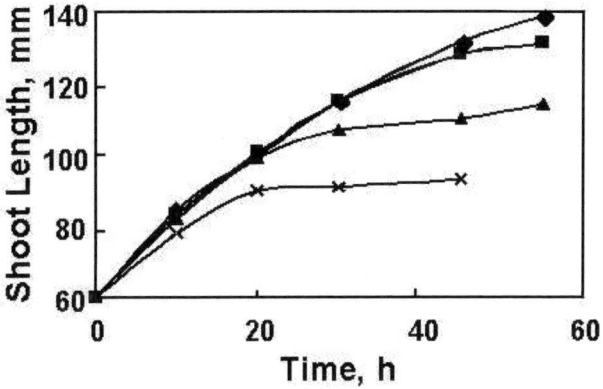

Figure 16.1

Effects of *Colletotrichum truncatum* spore concentration on shoot elongation of dark-grown hemp sesbania seedlings in a pathogen-weed bioassay. Diamond symbols = control (H_2O), squares = 10^2, triangles = 10^3, and cross marks = 5×10^5 spores/ml respectively. (From Hoagland, R. E. **1995**, *Biocontrol Sci. Technol.*, **5**, 251-259. With permission.)

A more rapid and sensitive bioassay method was developed to assess the pathogenicity of weed pathogens; e.g., a hydroponic seedling bioassay was developed to measure bioherbicidal efficacy of two weed fungal pathogens, *Colletrotrichum truncatum* (Schwein.) Andrus & W. D. Moore, and *Alternaria cassiae* Jurair & Khan, on hemp sesbania [*Sesbania exaltata* (Raf.) Rybd.] and sicklepod [*Senna* (formerly *Cassia*) *obtusifolia* L.], respectively.[44] *Colletotrichum truncatum* has high bioherbicidal activity on hemp sesbania,[11] and *Alternaria cassiae* can control sicklepod.[117,118] The bioassay used 4-day-old dark-grown seedlings, grown hydroponically. Uniform seedlings were measured, placed in paper towel cylinders, and sprayed with spore suspensions of their respective pathogens. Shoot lengths were monitored nondestructively and recorded over time under dark growth. Shoot growth inhibition (Fig. 16.1) and stem collapse (mortality) (Fig. 16.2) were directly related to the spore concentrations applied.

The pathogens (10^3 to 10^4 spores mL^{-1}) caused significant shoot growth inhibition within 25 to 30 h and seedling death within 40 to 50 h. Stem collapse time, as a function of various spore concentrations, was also a useful bioassay parameter. Nonlinear regression analysis[86] was used to model stem collapse time as a function of spore concentration (Fig. 16.2, solid lines). The trend used for this model was an exponential decay model of the form:

$$Y = A + Be^{(-k/c \times X)}$$

where A = minimum value of Y (time of stem collapse), approached as the common log of spore concentration (X) nears its maximum; B = total time that Y can decrease as X = increased to maximum concentration; c is the value of X when Y has decreased by 70% of its total decrease (i.e., B); and k is a constant (1.204) required to make the rate parameter c equal to 70%. Nonlinear regression provided estimated values of three parameters, A, B and c. The value of k was chosen; not predicted by the model. Values other than 70% could have been chosen for c, but intermediate values, rather than those near the ends of the actual spore concentration range, are more statistically valid. This equation provided a good fit to the actual data (Fig. 16.2) acquired for both of these pathogen-weed interactions under these conditions.

This model provided a tool for predicting the spore concentration required to achieve a certain degree of efficacy, i.e., 70%. R^2 values, calculated as 1 minus the residual sum of squares divided by the corrected total sum of squares, were 99.06 and 97.41 for *C. truncatum* and *A. cassiae*, respectively. Although the bioassay was not designed to compare the efficacy of different pathogen species on different host species, it was observed that in this system, *C. truncatum* reached 70% efficacy at a lower concentration ($10^{2.786}$) than *A. cassiae* ($10^{3.771}$); i.e., the rate of decrease in slope for *C. truncatum* was faster. But, overlap of the confidence intervals indicated that they were not significantly different. A statistically significant comparison of these two pathogens could undoubtedly have been made by choosing a somewhat lower level for the rate parameter c or by increasing the range of spore concentrations used for *A. cassiae*. In a similar fashion, more meaningful comparisons could be made to assess the efficacy of con-specific pathovars, pathogens produced under various conditions or grown under different nutrient regimes, and genetically engineered pathogens. Stem collapse time data and modeling, coupled with growth inhibition vs. time data, can provide the most accurate overall comparison of differences in pathogen virulence or spore concentration.

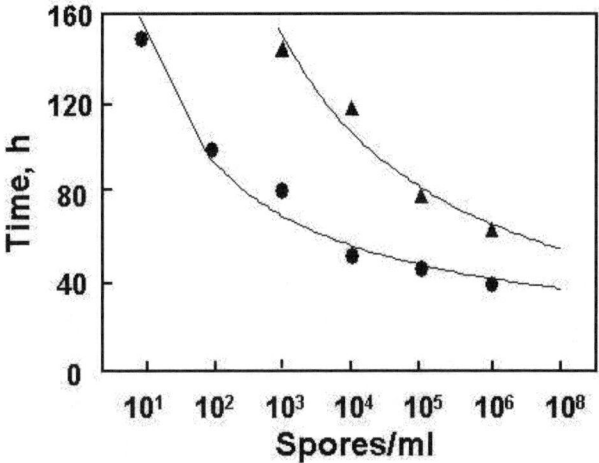

Figure 16.2

Effects of different *C. truncatum* and *A. cassiae* spore concentrations on seedling stem collapse bioassay of hemp sesbania and sicklepod, respectively. Solid lines for each data set = predicted trends for *C. truncatum*/hemp sesbania and for *A. cassiae*/sicklepod interactions, respectively, based on the exponential decay model described in the text. Triangles = recorded values and closed circles = recorded values for *A. cassiae*/sicklepod and *C. truncatum*/hemp sesbania, respectively. (From Hoagland, R. E. **1995**, *Biocontrol Sci. Technol.*, **5**, 251-259. With permission.)

The bioassay described above can be used to study herbicide-pathogen interactions and may be extended to determine the bioherbicidal efficacy of different pathogen isolates, pathovars or spore formulations of these pathogens. The method can be adopted to some other pathogen-plant interactions, and pathogenic effects may also be monitored during light growth conditions to evaluate photomorphogenic effects. This technique is more rapid and sensitive, uses lower inoculum volume, requires less space, and is performed under more controlled conditions than conventional greenhouse bioassay methods. These data are more quantitative than those obtained from bioassays relying on visual rating systems.[44]

Other bioassays have been developed to evaluate weed pathogen effects using leaf disks. Culture extracts of three fungal pathogens (*Fusarium oxysporum*, *Cylindrocarpon destructans* and *Colletotrichum dematium*) were bioassayed using leaf disks of several forest weeds.[85] Results indicated phytotoxicity of all these pathogens, especially *C. destructans*.

Microbial phytotoxin selectivity

All phytotoxins have an inherent range of specificity or selectivity on plant species, analogous to that of synthetic herbicides. However, some phytotoxins from fungi possess a unique type of phytotoxin specificity. Generally, fungal phytotoxins can be classified into two groups based on their specificity and biological activity. Host-specific phytotoxins elicit disease only in plants that are hosts of the pathogen that produces the toxin. In many cases toxin production and secretion by the pathogen are prerequisites for pathogenicity. Nonhost toxins exhibit phytotoxic effects on hosts and nonhost species. This latter group of allelochemicals may still show a wide degree of phytotoxicity when tested on various plants, ranging from no effect to mortality of nonhost plants. This can be due to differential uptake/translocation and metabolism, or to a lack of active site or receptor sites. It is necessary to discern the nature (host-specific or nonhost specific) of phytopathogens, to understand fully the virulence and pathogenicity of isolates of various fungal pathogens. The discovery, isolation, and purification of host-specific phytotoxins can be a major problem in developing a plant bioassay to detect such compounds, because if nonhost species are used solely as bioassay species, the toxin will go undetected.

As pointed out above, bioassay choice is extremely important since different sensitivities and different responses are observed that are dependent on the bioassay specimen. For example, comparison of four bioassay methods for detection of destruxin B, a host-specific toxin produced by *Alternaria brassicae*, indicated a lack of rapid electrolyte leakage and insensitivity of host protoplasts. However, leaf and pollen bioassays were sensitive to low toxin levels. Pollen germination was the most rapid, sensitive and quantitative bioassay for the toxin. Significant inhibition of *Brassica campestris* pollen germination and pollen tube growth occurred only 30 min after incubation with 2.5 μg mL^{-1}, and 7.5 μg mL^{-1} caused total inhibition.[3]

Plant bioassays for crop pathogens

In contrast to the small amount of research carried out on the development of rapid bioassay methods for pathogens with bioherbicidal potential, various bioassay procedures have been used for many years to evaluate crop plant susceptibility to pathogens. Examples include *C. lagenarium* in cucumber (*Cucumis sativus*),[7] *C. fragariae* in strawberry (*Fragaria* spp.)[105] and *Fusarium oxysporum* in muskmelon (*C. melo* L.).[68] Most of the standard methods for evaluating crop pathogens involve fungal conidial injection, pipette droplet techniques or spray application to foliage using greenhouse- or growth chamber-grown plants. All these techniques are time-consuming and require considerable labor and space. Freeman and Rodriguez[31] developed a dip method bioassay for *Colletotrichum* sp. on cucurbit crops that may be useful in other crop pathogen screening programs. Although the method is rapid, it measures only mortality, with no consideration of the direct effect on growth. A rolled-towel method was developed for evaluating the health of corn (*Zea mays* L.) and pea (*Pisum sativum* L.) seeds and seedlings exposed to soil containing seedling blight fungi,[66] but again, growth inhibition was not quantitatively determined. Chakraborty and Jones[12] developed a bioassay for measuring the pathogenicity of *C. gloeosporioides* on a forage legume. This system used a visual injury rating scale, rather than quantitative growth measurements.

Many species of the fungal genus *Colletotrichum* cause disease in various agricultural crops. Standard leaf inoculation procedures for evaluation of resistance to anthracnose require application of fungal conidia to foliage by spraying, pipetting, or injection.[7,12,20] All these procedures require a relatively long time, a high labor commitment, and considerable growth chamber or greenhouse space. A rapid and reliable bioassay for large-scale screening of the pathogenicity of *Colletotrichum magna* on cucurbits was developed.[31] This bioassay could also differentiate between susceptible and resistant cucurbits. The procedure consisted of a continuous dip method exposing whole plants or cuttings to conidia for 48 to 72 h, and then mortality was recorded. The method was proposed for the pathogenicity screening of other soil-borne fungal isolates and for evaluating cultivar resistance. *Colletotrichum acutatum* J.H. Simmonds, a soil-borne fungal pathogen of strawberry and other *Colletotrichum* spp. have been bioassayed via spray inoculation of plants with pathogen-infested soil.[22] Pathogenic variation of isolates of *Colletotrichum gloeosporioides* that cause anthracnose in the legume *Stylosanthes scraba*, was assessed using 3-day old vs. 6-week old seedlings. There was a high positive correlation of responses of the young seedlings vs. the old plants. Therefore, compared to the older plants, the 3-day old seedling assay was more rapid, used less inoculum and space, and could be carried out under more controlled conditions.

A more rapid (10 min) tomato seedling bioassay was developed to determine wilt-capacity and wilt-resistance of cell-free cultivar culture filtrates of races of *Verticillium dahliae* which causes wilt in tomato and potato (*Solanum tuberosum*).[73] Incubation conditions were manipulated to promote wilting response. This included induced stem elongation by growing a susceptible tomato cultivar used for the bioassay under low light intensity, concentration of pathogen cell-free culture filtrates, and increasing uptake of wilting factors by increasing transpiration [high light intensity, wind stream, low relative humidity, and high temperature (30°C)]. This points out the importance of modifying various conditions of a routine bioassay (or when a new bioassay is being developed), so that sensitivity can be maximized.

Table 16.3

Selected bioassays for allelopathic microbial interactions on plants.

Host plant	Microbe	Parameter Measured
Eggplant (*Solanum* sp.)	Bacterial and fungal pathogens	Screen for bacterial ring rot in potato[5]
Stylosanthes scabra (legume seedling)	*Colletotrichum gloeosporioides*	Resistance (visual)[13]
Citrus sp. (excised leaves; disks)	*Alternaria citri*	Infectivity; phytotoxin host range[33]
Brassica spp. (whole plant)	*Xanthomonas campestris*	Infectivity[101]
Various plants (whole; disks)	Various pathogens	Bioluminescence in transformed pathogen[100]
Triticum aestivum (wheat, excised leaves)	*Puccinia recondita*	Infectivity (visual)[28]
Wheat (hydroponic)	*Cephalosporium gramineum*	Leaf chlorophyll content[18]
Lotus corniculatus (hairy roots-transformed, *Agrobacterium rhizogenes*)	*A. rhizogenes*	Root tip elongation; Root proton excretion; Protoplast electrical potential[102]
Cucumis sativus (protoplasts)	*Phytophthora drechsleri*	Phytotoxin effects on protoplasts[109]
Turfgrasses	Soil bacteria	Suppression of *Phythium* blight[80]

Various other selected examples of bioassays for some other pathogen-plant interactions are presented in Table 16.3. Detached leaf segments have been widely used to evaluate pathogenicity, host resistance genetics, and sensitivity to fungicidal compounds produced by wheat and barley mildew pathogens, and more recently to test rust pathogens of cereals.[28] In this latter bioassay, disease response (visual scale) was measured in intact seedling leaves and leaf segments after incubation for 9 and 11 days, respectively. In these tests, several genetic lines of wheat gave inconsistent or poor expression in leaf segments compared to responses in intact leaves. Sensitivity to temperature and changes in biochemical or physiological changes were proposed as causal factors.

BIOASSAYS USING AQUATIC PLANTS

Most often, terrestrial plants have been examined in allelopathy research, but allelopathic phenomena can also occur in aquatic plants, and aquatic plants have been used as bioassay specimens. Aquatic plants, including macrophytes and various algal species, are important in sediment stabilization, water quality control, nutrient cycling, and oxygen production. These organisms are the primary producers in aquatic ecosystems. On the other hand, aquatic weeds cost hundreds of millions of dollars annually for control and damage in the U.S. alone. Chemical (herbicide) and allelochemical control of many aquatic plants has been investigated. The culture and testing of many aquatic plants are generally more difficult [duckweed (*Lemna* spp.) is an exception] than developing bioassays for terrestrial plants because of dilution effects of large culture volumes, large space requirements, and higher costs. This important topic has been studied and reviewed elsewhere [24-26,60,71] and is not discussed in detail here.

ALLELOPATHIC STIMULATORY ACTIONS AND SOME RELEVANT BIOASSAYS

Although Molisch[77] pointed out that allelopathic interactions could be either inhibitory or stimulatory to growth, only a few publications exist on stimulatory responses to allelochemicals. Important examples of stimulatory allelopathic interactions of plants on plants are the plant host effects (and other allelochemicals) on witchweed (*Striga asiatica*) seed germination. Witchweed is an obligate parasitic weed of many agronomically important legumes and cereals. The seeds of this parasite germinate only when exposed to chemical stimulants that are usually exuded from host plant roots.[79] Presently, there is little or no evidence that the detrimental effects of witchweed on host plants is caused by production of allelochemicals, an interaction that would be termed as an inhibitory allelopathic effect. Sufficient control of this serious weed was not achieved using synthetic herbicides; thus the use of non-herbicidal chemical stimulants to cause seed germination in the absence of host plants was attempted.[23,59] Several compounds have been isolated that stimulate witchweed seed germination.[125] Ethylene has been very effective in reducing viable *Striga* seed populations in soil.[27] Two other natural products, sorgoleone, isolated from sorghum (*Sorghum bicolor* L.) roots[15] and strigol, isolated from cotton (*Gossypium hirsutum* L.) root exudates,[16,17] can trigger germination of witchweed seed germination and may be useful to reduce the seed bank of this weed in infested soils. Bioassays using witchweed seeds have been useful to evaluate the chemicals cited above, as well as other compounds.

OTHER CONSIDERATIONS RELATED TO BIOASSAYS FOR ALLELOPATHY

Allelochemicals are released from plants via exudation, leaching or decomposition and decay of plant tissues. These compounds can enter or affect another plant directly by uptake of the affected plant or indirectly by effects of the allelochemical on soil microorganisms that are either plant growth promotors or that are pathogenic. This is well documented in the case of inhibition of nitrogen-fixing and nitrifying microorganisms by various plants during old field succession.[88] In nature, allelopathic compounds may act on (or be acted upon) many living organisms before an allelopathic action can be measured. The influence of secondary plant compounds on the associations of rhizosphere microorganisms and plant roots has been reviewed.[42] The interactions associated with soil microorganisms and plant roots are very complex, which makes actual proof of an allelopathic action in the field a formidable challenge. Proof of such phenomena may require the development and use of several innovative bioassays.

Another problem or consideration that may interfere with some aspects of allelopathy (especially in field experiments) has arisen over the past fifty years with the introduction and use of a multitude of xenobiotic compounds as insecticides, fungicides, plant growth regulators, harvest aids, and herbicides. A variety of such compounds have been and continue to be used on a world-wide scale. Some of these chemicals and/or their transformation products are persistent in soils and water. Other such chemicals are routinely applied at various times during a year. These xenobiotics may interfere or interact with naturally occurring allelochemics and thus alter or even mask certain natural allelochemical effects. Researchers will have a more difficult time to find natural areas that do not contain xenobiotic residues since these chemicals are used in agricultural areas, lawns, gardens, forests, roadsides, and parklands. Even some waters flowing into pristine natural settings contain low levels of various xenobiotics. It may be quite difficult to find natural ecosystems untainted by xenobiotics and their metabolic products.

IMMUNOLOGICAL METHODS AS POTENTIAL BIOASSAY TOOLS

Immunological techniques can be powerful molecular tools. Although immunological techniques have not been used to their fullest potential by plant and weed scientists, it is perceived that these protocols can provide fundamental insight to the areas of weed biology, mode and targets of action of herbicides, and in soil science.[103] Protocols such as immunofluorescence, immunogold electron microscopy, ELISA (enzyme-linked immunosorbant assay), immunoblotting and immunocytochemistry have provided valuable information on the detection of herbicides in soil, water and plants; the location, action and activity of various enzymes at the subcellular level; and data on modified binding sites in herbicide resistant vs. susceptible weeds.

These techniques are very sensitive and less costly and labor intensive than many other analytical methods. Such protocols may be utilized to answer fundamental questions in allelopathy, especially when unique allelochemicals are isolated from plants or microbes. This would include qualitative and quantitative subcellular distribution of proteins involved at binding sites or that are inhibited/activated by allelochemicals; subcellular distribution of an allelochemical in the plant/microbe that produces it; and distribution of an allelochemical and certain effects in plant tissues after treatment.

A recent patent contains a proposed method for evaluating plants in a pesticide screening system.[10] The method provides a system to test asexually propagated plants, whereby a segment of the mother plant (crop, transgenic plant, weed, etc.) is treated with the test pesticide, transferred to agar, soil, etc.; and regenerated into a whole plant followed by monitoring growth of the regenerated plant. Such test systems may have utility in testing allelochemicals from plants and microorganisms.

CONCLUDING REMARKS

It is evident from this overview and from examination of the allelopathy literature in general that many of the more sensitive and the more molecular bioassay techniques and ultrastructural methods as outlined here are under-utilized or not used at all by scientists studying allelopathy. Allelopathic interactions are complex and subtle, and bioassays that only measure germinating seeds after treatment with a crude extract from a suspected allelopathic plant do little to advance the scientific literature on allelopathy. Also few studies on allelopathy follow activity detection in crude extracts by purification and identification techniques.

We recommend the use of several bioassay systems during the isolation, purification, and characterization for a given allelochemical. If allelochemical compounds for potential herbicides are being sought, testing on a large variety of crop and weed species is desirable. Furthermore, as pointed out, different bioassays can give different or opposing results with a given compound due to differential uptake, translocation and/or metabolism, etc.

There are problems with using crude extracts or insensitive bioassays with crude extracts. The concentration of the allelopathic compound(s) is generally low in a crude extract. Thus, if bioassays are not sufficiently sensitive, potential allelopathic actions will go undetected. Also, nonactive compounds in crude extracts may physically or chemically bind or mask the action of an allelochemical. It is essential to use a sensitive assay and to have an adequate amount of active ingredient in any extract tested. However, this is only the first step in the extraction, isolation and identification of the compound(s). During purification and identification studies using TLC, UV/Vis and IR spectrophotometry, MS-GC, HPLC, etc., the bioassay should become more specific and sensitive. As purification proceeds, the bioassay response may decrease due to the separation of multiple compounds acting at the same or at different biochemical/physiological sites. During any study the possible additive, synergistic and antagonistic effects of multiple allelochemicals need to be considered, especially in crude extracts and multiple compound mixtures.

Stowe,[108] Inderjit and Dakshini,[58] and Foy[30] have suggested that laboratory bioassays may provide unrealistic observations that cannot be extrapolated to field observation. Their main concern is that bioassays cannot duplicate the dynamic interactions taking place in the field. Further, bioassays often ignore the role that the rhizosphere may play in allelopathic interactions. However, since allelopathy is a complex and dynamic process, it is necessary to study component parts in order to elucidate the overall mechanism. A properly chosen bioassay can provide researchers with a convenient exploratory tool that augments field observations. Even if the bioassay system is used without any regard to field observations, the bioassay may provide interesting data regarding chemical-plant interaction phenomena. Although such observations may be interesting, they do not necessarily increase our understanding of allelopathy in natural settings.

ACKNOWLEDGMENTS

We thank Richard Huckleberry and Velma Robertshaw for their assistance in typing and in literature searches involved in the preparation of this manuscript. We appreciate the expert assistance and untiring support of Rhonda Watson, librarian at Mississippi State University, Delta Research and Extension Center, Stoneville, MS.

REFERENCES

(1) Appleby, A. P. **1985**. Factors in examining fate of herbicides in soil with bioassays. *Weed Sci.* **33**, (Suppl. 2), 2-6

(2) Ashwood-Smith, M. J., Poulton, G. A., Ceska O., Liu, M. and Furniss, E. **1983**. An ultra-sensitive bioassay for the detection of furocoumarins and other photosensitizing molecules. *Photochem. Photobiol.* **38**, 113-118

(3) Bains, P. S. **1990**. Purification, chemical characterization, host-specificity, bioassay, mode of action, and herbicidal use of the toxin produced by *Alternaria brassicae*. *Dissertation Abst. Internat.* **50**, 2708-B

(4) Baud-Grasset, F., Baud-Grasset, S. and Safferman, S. I. **1993**. Evaluation of the bioremediation of a contaminated soil with phytotoxicity tests. *Chemosphere* **26**, 1365-1374

(5) Behn, H., Radtke, W. and Rudolph, K. **1992**. Effect of bacterial and fungal pathogens on the eggplant bioassay for detection of *Clavibacter michiganensis* sp. *Sepedonicus* (bacterial ring rot of potatoes). *J. Plant Dis. Protect.* **99**, 113-126

(6) Bennett, P. H. and deBeer, P. R. **1984**. A new rapid quantitative bioassay for the determination of biologically available bromacil in soils. *Pestic. Sci.* **15**, 425-430

(7) Biles, C. L., Ables, F. B., and Wilson, C. L. **1990**. The role of ethylene in anthracnose of cucumber, *Cucumis sativus*, caused by *Colletotrichum lagenarium*. *Phytopathology* **80**, 732-736

(8)   Blum, U. **1999**. Designing laboratory plant debris-soil bioassays: some reflections. In: Inderjit, Dakshini, K. M. M., Foy, C. L. (Eds.), *Principles and Practices in Plant Ecology — Allelochemical Interactions*. CRC Press, Boca Raton, FL, 17-23

(9)   Böger, P. and Sandmann, G. (Eds.), **1993**. *Target Assays for Modern Herbicides and Related Phytotoxic Compounds*. Lewis Publishers, Boca Raton, FL, 299 p.

(10)   Boutsalis, P. **1998**. Pesticide screening system. PCT Int. Appl. WO 98 55,860. (Patent application)

(11)   Boyette, C. D. **1991**. Control of hemp sesbania with a fungal pathogen, *Collectotrichum truncatum*. US Patent No. 5,034,328

(12)   Chakraborty, S. and Jones, P. N. **1993**. A rapid bioassay for the assessment of pathogenic variation in *Colletotrichum gloeosporioides* infecting *Stylosanthes scabra*. *Plant Dis.* **77**, 1016-1020

(13)   Chakraborty, S. **1990**. Expression of quantitative resistance to *Colletotrichum gloeosporioides* in *Stylosanthes scabra* at different inoculum concentrations and day-night temperatures. *Aust. J. Agric. Res.* **41**, 89-100

(14)   Chaleff, R. S. and Mauvais, C. J. **1984**. Acetolactate synthase is the site of action of two sulfonylurea herbicides in higher plants. *Science* **244**, 1443-1445

(15)   Chang, M., Netzly, D. H., Butler, L. G., and Lynn, D. G. **1986**. Chemical regulation of distance: Characterization of the first natural host germination stimulant for *Striga asiatica*. *J. Am. Chem. Soc.* **108**, 7858-7860

(16)   Cook, C. E., Shichard, L. P., Turner, B., Wall, M. E., and Egley, G. H. **1966**. Germination of witchweed (*Striga lutea* Lour.): isolation and properties of a potent stimulant. *Science* **154**, 1189-1190

(17)   Cook, C. E., Wichard, L. P., Wall, M. E., Egley, G. H., Coggon, P., Luhan, P. A., and McPhail, A. T. **1972**. Germination stimulants II. The structure of strigol — a potent seed germination stimulant for witchweed (*Striga lutea* Lour.). *J. Amer. Chem. Soc.* **94**, 6198-6199

(18)   Cowger, C. and Mundt, C. C. **1998**. A hydroponic seedling assay for resistance to *Cephalosporium* stripe of wheat. *Plant Dis.* **82**, 1126-1131

(19)   DaSilva, J. F., Fadayomi, R. O., and Warren, G. F. **1976**. Cotyledon disc bioassay for certain herbicides. *Weed Sci.* **24**, 250-252

(20)   Dean, R. and Kuć, J. **1986**. Induced systemic protection in cucumber: effects of inoculum density on symptom development caused by

Colletotrichum lagenarium in previously infected and uninfected plants. *Phytopathology* **76**, 186-189

(21) Duke, S. O. and Kenyon, W. H. **1993**. Peroxidizing activity determined by cellular leakage. In: Böger, P., Sandmann, G. (Eds.), *Target Assays for Modern Herbicides and Related Phytotoxic Compounds*. Lewis Publishers, Boca Raton, FL, 61-66

(22) Eastburn, D. M. and Gubler, W. D., **1990**. Strawberry anthracnose: detection and survival of *Colletotrichum acutatum* in soil. *Plant Dis.* **74**, 161-163. New York. pp. 33-50

(23) Egley, G. H. and Dale, J. E. **1970**. Ethylene, 2-chloroethyl phosphonic acid, and witchweed germination. *Weed Sci.* **18**, 586-589

(24) Elakovich, S. D. and Wooten, J. W. **1994**. Allelopathic, herbaceous, vascular hydrophytes. In: Inderjit, Dakshini, K.M.M., Einhellig, F.A., (Eds.), *Allelopathy: Organisms, Processes, and Applications*. ACS Symposium Series **582**, ACS Books, Washington, D.C., 58-73

(25) Elakovich, S. D. **1995**. Bioassay methods applied to allelopathic herbaceous vascular hydrophytes. *Proc. Plant Growth Regul. Soc. Amer.* pp 39-43

(26) Elakovich, S. D. **1999**. Bioassays applied to allelopathic herbaceous vascular hydrophytes. In: Inderjit, Dakshini, K. M. M., and Foy, C.L. (Eds.), *Principles and Practices in Plant Ecology — Allelochemical Interactions*. CRC Press, Boca Raton, FL, 45-56

(27) Eplee, R.E. **1975**. Ethylene: a witchweed seed germination stimulant. *Weed Sci.* **23**, 433-436

(28) Felsenstein, F. G., Park, R. F., and Zeller, F. J. **1998**. The use of detached seedling leaves of *Triticum aestivum* to study pathogenicity in *Puccinia recondita* f. sp. *tritici*. *J. Phytopathol.* **146**, 115-121

(29) Finey, D. J. **1971**. *Probit Analysis*. Cambridge University Press, New York. 333 p.

(30) Foy, C. L. **1999**. How to make bioassays for allelopathy more relevant to field conditions with particular reference to cropland weeds. In: Inderjit, Dakshini, K. M. M., and Foy, C. L. (Eds.), *Principles and Practices in Plant Ecology — Allelochemical Interactions*. CRC Press, Boca Raton, FL, 25-34

(31) Freeman, S. and Rodriguez, R. J.,**1992**. A rapid, reliable bioassay for pathogenicity of *Colletotrichum magna* on cucurbits and its use in screening for nonpathogenic mutants. *Plant Dis.* **76**, 901-905

(32) Galun, E. and Breiman, A. **1992**. Quantitative assays of phytotoxins using plants, protoplasts and isolated cells. In: Linskens, H. F. and Jackson, J. F. (Eds.), *Plant Toxin Analysis*. Springer-Verlag. 33-50

(33) Gardner, J. M., Kono, Y., and Chandler, J. L. **1986**. Bioassay and host-selectivity of *Alternaria citri* toxins affecting rough lemon and mandarins. *Physiol. Molec. Plant Pathol.* **29**, 293-304

(34) Gasson, M. J., **1980**. Indicator technique for antimetabolic toxin production by phytopathogenic species of *Pseudomonas*. *Appl. Environ. Microbiol.* **39**, 25-29

(35) Gressel, J., **1990**. Synergizing herbicides. *Rev. Weed Sci.* **5**, 49-82

(36) Gressel, J. **1993**. Synergizing pesticides to reduce use rates. *ACS Symposium Series* **524**, 48-61

(37) Grossmann, K. **1993**. Screening for growth retardants and herbicides (conductivity measurement) with heterotrophic plant cell suspension cultures. In: Böger, P. and Sandmann, G. (Eds.), *Target Assays for Modern Herbicides and Related Phytotoxic Compounds*. Lewis Publishers, Boca Raton, FL, 269-275

(38) Hatzios, K. K. and Penner, D. **1985**. Interactions of herbicides with other agrochemicals in higher plants. *Rev. Weed. Sci.* **1**, 1-63

(39) Heisey, R. M. **1990**. Evidence of allelopathy by the tree-of-heaven (*Ailanthus altissima*). *J. Chem. Ecol.* **16**, 2039-2055

(40) Hensley, R. **1981**. A method for identification of triazine resistant and susceptible biotypes of several weeds. *Weed Sci.* **29**, 70-73

(41) Hoagland, R. E. and Cutler, S. J. **2000**. Plant and microbial compounds as herbicides. In: Narwal, S. S., Hoagland, R. E., Dilday, R. H., and Reigosa, M. J. (Eds.), *Allelopathy in Ecological Agriculture and Forestry*. Kluwer Academic Publishers, Boston, 73-99

(42) Hoagland, R. E. and Williams, R. D. **1985**. The influence of secondary plant compounds on the associations of soil microorganisms and plant roots. In: Thompson, A.C., (Ed.), *The Chemistry of Allelopathy*. ACS Symposium Series **268**, ACS Books, Washington, DC., 301-325

(43) Hoagland, R. E., (Ed.) **1990**. *Microbes and Microbial Products as Herbicides*. ACS Symposium Series **439**, ACS Books, Washington, DC. 341 p.

(44) Hoagland, R. E. **1995**. Hydroponic seedling bioassay for the bioherbicides *Colletotrichum truncatum* and *Alternaria cassiae*. *Biocontrol Sci. Technol.* **5**, 251-259

(45) Hoagland, R. E. **1996**. Chemical interactions with bioherbicides to improve efficacy. *Weed Technol.* **10**, 651-674

(46) Hoagland, R. E. **1999**. Plant pathogens and microbial products as agents for biological weed control. In: Tewari, J. P., Lakhanpal, T. N., Singh, J., Gupta, R., and Chamola, B. P. (Eds.), *Advances in Microbial Biotechnology*. APH Publishing Corp., New Delhi, India, 213-255

(47) Hoagland, R. E. **1999**. Biochemical interactions of the microbial phytotoxin phosphinothricin and analogs with plants and microbes. In: Cutler, H. G. and Cutler, S. J. (Eds.), *Biologically Active Natural Products: Agrochemicals*. CRC Press, Boca Raton, FL, 107-125

(48) Hoagland, R. E. **2001**. Bioherbicides: phytotoxic natural products. In: Baker, D. R. and Umetsu, N. K., (Eds.), *Agrochemical Discovery*. ACS Symposium Series **774**, ACS Books, Washington, DC, 72-90

(49) Hoagland, R. E. **2001**. Microbial allelochemicals and pathogens as bioherbicidal agents. *Weed Technol.* **15**, 835-857

(50) Hoagland, R. E. **2001**. The genus *Streptomyces*: a rich source of novel phytotoxins. In: Prakash, I. (Ed.), *Ecology of Desert Environments*. Scientific Publishers, Jodhpur, India, 139-169

(51) Hoagland, R. E. **1990**. Microbes and microbial products as herbicides, an overview. In: Hoagland, R.E. (Ed.), *Microbes and Microbial Products as Herbicides*. ACS Symposium Series **439**, ACS Books, Washington, DC, 1-52

(52) Hoagland, R. E. **1999**. Allelopathic interactions of plants and pathogens. In: Macias, F. A., Galindo, J. C. G., Molinillo, J. M. G., and Cutler, H. G. (Eds.), *Recent Advances in Allelopathy. A Science for the Future*. Vol. 1. Servicio de Publicaciones-Universidad De Cádiz, Spain. 423-450

(53) Hoerlein, G. **1994**. Glufosinate (phosphinothricin) a natural amino acid with unexpected herbicidal properties. *Rev. Environ.Contam. Toxicol.* **138**, 73-145

(54) Holappa, L. D. and Blum, U. **1991**. Effects of exogenously applied ferulic acid, a potential allelopathic compound on leaf growth, water utilization, and endogenous abscisic acid levels of tomato, cucumber and bean. *J. Chem. Ecol.* **17**, 865-886

(55) Horowitz, M. **1976**. Application of bioassay techniques to herbicide investigations. *Weed Res.* **16**, 209-215

(56) Hoshmand, A. R. **1998**. *Statistical Methods for Environmental and Agricultural Sciences*. CRC Press, Boca Raton, FL, 439 p.

(57) Hyeon, S.-B., Nishida, M., Ohaskka, A., Kim, J.-M. and Suzuki, A., **1982**. A simple bioassay for chemicals active on the photosynthetic or respiratory systems of plants. *Agric. Biol. Chem.* **46**, 811-812

(58) Inderjit and Dakshini, K. M. M. **1999**. Bioassays for allelopathy: interactions of soil organic and inorganic constituents. In: Inderjit, Dakshini, K. M. M., and Foy, C. L. (Eds.), *Principles and Practices in Plant Ecology — Allelochemical Interactions*. CRC Press, Boca Raton, FL, 35-42

(59) Johnson, A. W., Roseberry, G., and Parker, C. **1976**. A novel approach to *Striga* and *Orobanche* control using synthetic germination stimulants. *Weed Res.* **16**, 223-227

(60) Joye, G. F. **1990**. Biological control of aquatic weeds with plant pathogens. In: Hoagland, R. E., (Ed.), *Microbes and Microbial Products as Herbicides*. ACS Symposium Series **439**, ACS Books, Washington, DC, 155-174

(61) Keen, N. T. **1975**. Specific elicitors of plant phytoalexin production: determinants of rice specificity in pathogens? *Science* **187**, 74-75

(62) Khan, M. I. **1980**. Gibberellic acid bioassay based on the inhibition of anthocyanin production in tomato seedlings. *Biol. Plant.* **22**, 401-403

(63) Kinsman, L. T., Pinfield, N. J., and Stobart, A. K. **1975**. A gibberellin bioassay based on betacyanin production in *Amaranthus caudatus* seedlings. *Planta* **127**, 149-152

(64) Koga, J., Ogawa, N., Yamauchi, T., Kikuchi, M., Ogasawara, N., and Shimura, M. **1997**. Functional moiety for the antifungal activity of phytocassane E, a diterpene phytoalexin from rice. *Phytochemistry* **44**, 249-253

(65) Koga, J., Oshima, K., Ogawa, N., Ogasawara, N., and Shimura, M. **1998**. A new bioassay for measuring elicitor activity in rice leaves. *Ann. Phytopathol. Soc. Jpn.* **64**, 97-101

(66) Kommedahl, T. and Burnes, P. M. **1989.** Cold test for corn and peas. In: *Manual of Laboratory Methods for Fungal Plant Disease*. University of Minnesota Press, St Paul, MN. pp. 1-3

(67) Kriedeman, P. E., Loveys, B. R., and van Dijk, H. M. **1980**. Photosynthetic inhibitors from *Capsicum annuum* L: Extraction and bioassay. *Aust. J. Plant Physiol.* **7**, 629-633

(68) Latin, R. X. and Snell, S. J. **1986**. Comparison of methods for inoculation of muskmelon with *Fusarium oxysporum* f. sp. *melonis*. *Plant Dis.* **70**, 297-300

(69) Lavy, T. L. and Santelmann, P. W. **1986**. Herbicide bioassay as a research tool. In: *Research Methods in Weed Science*. South. Weed Sci. Soc., Champaign, IL, 201-217

(70) Leather, G. R. and Einhellig, F. A. **1985**. Mechanism of allelopathic in bioassay. In: Thompson, A.C., (Ed.), *The Chemistry of Allelopathy*. ACS Symposium Series **268**, ACS Books, Washington, DC, 197-205

(71) Lewis, M. A. **1995**. Use of freshwater plants for phytotoxicity testing: a review. *Environ. Pollution* **87**, 319-336

(72) Lüsse, B., Schröder, D., and Soeder, C. J. **1986**. The bromothymol-blue test: A simple algal bioassay and its use for herbicide screening. *Arch. Hydrobiol. Suppl.* **73**, 147-152

(73) Madhosingh, C. **1996**. A rapid tomato seedling assay for *Verticillium* wilt. *Phytopathology* **144**, 131-134

(74) Mangnus, E. M., Stommen, P. L. A., and Zwanenburg, B. **1992**. A standardized bioassay for evaluation of potential germination stimulants for seeds of parasitic weeds. *J. Plant Growth Regul.* **11**, 91-98

(75) Mitchell, J. and Rook, A. **1979**. *Botanical Dermatology*, Greengrass, Vancouver, BC, Canada

(76) Mitchell, R. E. and Durbin, R. D. **1981**. Tagetitoxin, a toxin produced by *Pseudomonas syringae* pv. *tagetis*: purification and partial characterization. *Physiol. Plant Pathol.* **18**, 157-168

(77) Molisch, H. **1937**. *Der Einfluss einer Pflanze auf die andere-allelopathie*. G. Fischer, Jena, Germany

(78) Müller, C. H. **1969**. Allelopathy as a factor in ecological processes. *Vegetatio* **38**, 348-357

(79) Musselman, L. J. **1980**. The biology of *Striga*, *Orobanche*, and other root-parasitic weeds. *Annu. Rev. Phytopathol.* **18**, 463-489

(80) Nelson, E. B. and Craft, C. M. **1992**. A miniaturized and rapid bioassay for the selection of soil bacteria suppressive to *Pythium* blight of turf grasses. *Phytopathology* **82**, 206-210

(81) Parker, C. **1965**. A rapid bioassay method for the detection of herbicides which inhibit photosynthesis. *Weed Res.* **5**, 181-184

(82) Paxton, J. D., **1988**. Fungal elicitors of phytoalexins and their potential use in agriculture. In: Cutler, H. G., (Ed.), *Biologically Active Natural Products*. ACS Symposium Series **380**, ACS Books, Washington, DC, 109-119

(83) Pestemer, W. and Günther, P. **1995**. Growth inhibition of plants as a bioassay for herbicide analysis. *Chem. Plant Protect.* **11**, 119-231

(84) Poulton, G. A. and Ashwood-Smith, M. J. **1983**. Photosensitizing plant products. In: Stitch, H. (Ed.), *Carcinogens and Mutagens in the Environment*, Vol. II. Chemical Rubber Publishing Co., Cleveland, OH

(85) Prasad, R. and Dixon-Warren, H. **1992**. Bioherbicides for forestry: development of some procedures for bioassay of phytotoxins. *Plant Protect. Quart.* **7**, 154-156

(86) Ratkowsky, D. A. **1983**. Asymptotic regression model. In: Owen, D.B. (Ed.), *Non-linear Regression Modeling–a Unified Practical Approach*. Marcel Dekker, Inc., New York, NY, 93-104

(87) Ren, Y. Y. and West, C. A. **1992**. Elicitation of diterpene biosynthesis in rice (*Oryza sativa* L.) by chitin. *Plant Physiol.* **99**, 1169-1178

(88) Rice, E. L., **1974**. *Allelopathy*. Academic Press, New York, NY, 353 p.

(89) Rice, E. L., **1984**. *Allelopathy*. Academic Press, Orlando, FL, 422 p.

(90) Rice, E. L., **1995**. *Biological Control of Weeds and Plant Diseases: Advances in Applied Allelopathy*. University of Oklahoma Press, Norman, OK, 439 p.

(91) Ridley, S. M. and McNally, S. F. **1985**. Effects of phosphinothricin on the isoenzymes of glutamine synthetase isolated from plant species which exhibit varying degrees of susceptibility to the herbicide. *Plant Sci.* **39**, 31-36

(92) Saka, H. and Chisaka, H. **1985**. Photosynthesis measurement by oxygen electrode as a simple bioassay method. *JARQ* **18**, 252-259

(93) Saltzman, S. and Heuer, B. **1985**. A rapid bioassay for the detection of photosynthesis inhibitors in water. *Weed Sci.* **24**, 250-252

(94) Sandmann, G. **1993**. Spectral determination of carotenoid precursors in *Scenedesmus* cells treated with bleaching herbicides. In: Böger, P., Sandmann, G. (Eds.), *Target Assays for Modern Herbicides and Related Phytotoxic Compounds*. Lewis Publishers, Boca Raton, FL, 3-8

(95) Santelmann, P. W. **1977**. Herbicide bioassay. In: *Research Methods in Weed Science*. South. Weed Sci. Soc. Champaign, IL, 82-87

(96) Sato, R., Nagano, E., Oshio, H., and Kamoshita, K. **1988**. Activities of the *N*-phenylimide S-23142 in carotenoid-deficient seedlings of rice and cucumber. *Pestic. Biochem. Physiol.* **24**, 213-218

(97) Selim, S. A., O'Neal, S. W., Ross, M. A., and Lembi, C. A. **1989**. Bioassay of photosynthetic inhibitors in water and aqueous soil extracts with Eurasian watermilfoil (*Myriophyllum spicatum*). *Weed Sci.* **37**, 810-814

(98) Shaner, D. L., Anderson, P. C., and Stidham, M. A. **1984**. Imidazolinones (potent inhibitors of acetohydroxyacid synthase). *Plant Physiol.* **76**, 545-546

(99) Sharp, J. K., Valent, B., and Albersheim, P. **1984**. Purification and partial characterization of a β-glucan fragment that elicits phytoalexin accumulation in soybean. *J. Biol. Chem.* **259**, 11312-11320

(100) Shaw, J. J. and Kado, C. I. **1986**. Development of a *Vibrio* bioluminescence gene-set to monitor phytopathogenic bacteria during the ongoing disease process in a non-disruptive manner. *Bio/Technol.* **4**, 560-564

(101) Shaw, J. J. and Kado, C. I. **1988**. Whole plant wound inoculation for consistent reproduction of black rot of crucifers. *Phytopathology* **78**, 981-986

(102) Shen, W. H., Petit, A., Guern, J., and Tempe, J. **1988**. Hairy roots are more sensitive to auxin than normal roots. *Proc. Natl. Acad. Sci. USA* **85**, 3417-3421

(103) Sherman, T. D., and Vaughn, K. C. **1991**. Immunology in weed science. *Weed Sci.* **39**, 514-520

(104) Siehl, D. L., Subramanian, V., and Toschi, A. G. **1998**. Method for the detection of herbicides. US 5,780,254. (Patent application)

(105) Smith, B. J., Black, L. L., and Galletta, G. J. **1990**. Resistance to *Colletotrichum fragariae* in strawberry affected by seedling age and inoculation method. *Plant Dis.* **74**, 1016-1021

(106) Staskawicz, B. J. and Panopoulos, N. J. **1979**. A rapid and sensitive microbiological assay for phaseolotoxin. *Phytopathology* **69**, 663-666

(107) Steel, R. G. D. and Torrie, J. H. **1960**. *Principles and Procedures of statistics*. McGraw-Hill, NY, 481 p.

(108) Stowe, L. G., **1979**. Allelopathy and its influence on the distribution of plants in an Illinois old-field. *J. Ecol.* **67**, 1065-1085

(109) Strange, R. N., Pippard, D. J. and Strobel, G. A. **1982**. A protoplast assay for phytotoxic metabolites produced by *Phytophthora drechsleri* in culture. *Physiol. Plant Pathol.* **20**, 359-364

(110) Streibig, J. C. and Kudsk, P., (Eds.), **1993**. *Herbicide Bioassays*. CRC Press, Boca Raton, FL, 270 p

(111) Subramanian, V. and Toschi, A. G. **1998**. Screening method for detection of herbicides by inhibition of plant purine biosynthetic pathway enzyme. US 5,780,253. (Patent application)

(112) Tanaka, T., Abbas, H. K., and Duke, S. O. **1993**. Structure-dependent phytotoxicity of fumonisins and related compounds in a duckweed bioassay. *Phytochemistry* **33**, 779-785

(113) Tchan, Y. T., Roseby, J. E., and Funnell, G. R. **1975**. A new rapid specific bioassay method for photosynthesis inhibiting herbicides. *Soil Sci. Biochem.* **7**, 39-44

(114) TeBeest, D. O. (Ed.), **1991**. *Microbial Control of Weeds*. Chapman and Hall, New York. 284 p.

(115) Truelove, B., Davis, D. E. and Jones, L. R. **1974**. A new method for detecting photosynthesis inhibitors. *Weed Sci.* **22**, 15-17

(116) Turner, J. G. and Taha, R. R. **1984**. Contribution of tabtoxin to the pathogenicity of *Pseudomonas syringae* pv. *tabaci*. *Physiol. Plant Pathol.* **25**, 55-69

(117) Walker, H. L. and Riley, J. A. **1982**. Evaluation of *Alternaria cassiae* for the biocontrol of sicklepod (*Cassia obtusifolia*). *Weed Sci.* **30**, 651-654

(118) Walker, H. L. **1982**. A seedling blight of sicklepod caused by *Alternaria cassiae*. *Plant Dis.* **66**, 426-428

(119) Waller, G. R. (Ed.) **1987**. *Allelochemicals: Role in Agriculture and Forestry*. ACS Symposium Series **330**, ACS Books, Washington, DC, 606 p

(120) Waters, E. R. and Blum, U. **1987**. Effects of single and multiple exposure of ferulic acid on the vegetative and reproductive growth of *Phaseolus vulgaris* BBL-290. *Amer. J. Bot.* **74**, 1635-1645

(121) Wei, Y. D. and Mortensen, C. N. **1992**. Bioassays of toxins as a diagnostic method for *Pseudomonas syringae* pathovars. *J. Phytopathol.* **134**, 110-116

(122) Went, F. and Thimann, K. V. **1937**. *Phytohormones*. MacMillan, Inc., New York. 294 p.

(123) Williams, R. D. and Hoagland, R. E. **1982**. The effects of naturally occurring phenolic compounds on seed germination. *Weed Sci.* **30**, 206-212

(124) Williams, R. D. and Hoagland, R. E. **2000**. The effect of mimosine on seed germination and seedling growth. *Weed Sci. Soc. Amer. Abstr.* **40**, 25

(125) Worsham, A. D. **1987**. Germination of witchweed seeds. In: Musselman, L. (Ed.), *Parasitic Weeds in Agriculture*, Vol. I. CRC Press, Boca Raton, FL, 45-61

(126) Wu, H., Pratley, J., Lemerle, D., Haig. T., and An, M. **2001**. Screening methods for the evaluation of crop allelopathic potential. *Bot. Rev.* **67**, 403-415

(127) Yamada, A., Shibuya, N., Kodama, O. and Akatsuka, T. **1993**. Induction of phytoalexin formation in suspension-cultured rice cells by N-acetylchito-oligosaccharides. *Biosci. Biotech. Biochem.* **57**, 405-409

(128) Yanase, D., Andoh, A., and Yasudomi, N. **1990**. A new simple bioassay to evaluate photosynthesis electron-transport inhibition utilizing paraquat phytotoxicity. *Pestic. Biochem. Physiol.* **32**, 92-98

(129) Yopp, J. H., Aung, L. H., and Stefens, G. L. **1986**. Bioassays and other special techniques for plant hormones and plant growth regulators. *Plant Growth Regulat. Soc. Amer.* 208 p.

Index

A

AAL-toxin, 325
abcisic acid, 223, 228, 260
    biosynthesis, 223
    levels, 58
absolute configuration, 108, 137
absolute stereochemistry, 106, 108, 109
Abutilon theophrasti, 224
Acacia sedillense, 271-273, 279
Acanthaceae, 77-82
Acanthus ebracteatus, 80
Acanthus mollis, 80, 81
ACCase inhibition, 22
Acetabularia mediterranea, 259
acetic acid, 62, 248, 249
acetolactate synthase, 327
acetone, 26, 143, 309, 320
15-acetoxyscirpendiol, 266
acetylcholine, receptor, 167, 195
acid phosphatase, 224
acifluorfen, 17, 208
Acremonium, 265
actinomycetes, 61, 67-68, 71, 330
ACH receptor, 167, 177
additive effect, 321, 340
adenylate cyclase, 177
adenylosuccinate lyase, 326
adenylosuccinate synthase, 326
AD-mix, 112
Agallinis purpurea, 127
agar, 151
Agrobacterium rhizogenes, 336
Agropyron repens, 81, 90
Agrostemma githago, 90, 92
Agrostis tenuis, 26
ailanthone, 210
albine, 193

albizziin, 325
Alectra vogelii, 27
alectrol, 128, 130
alert pheromone, 176
alga, 13-15, 35-36, 43, 47-48, 226, 230, 259, 290-292, 303, 306-310, 323, 325, 327, 337
algal growth, 292
algaltoxikit FTM, 52
alkaloid, 163-177
    parsimony, 163, 164, 175
    ant-derived, 172
    α-pyridone, 187
    bipyrrolic, 170
    diterpenes, 173
    guanidium, 169
    lupine, 166
    monoterpene, 173
    protective role, 171
    pyrrolizidine, 167, 171-175
    quinolizidine, 167, 183-193
        biosynthesis of, 187
        mode of action of, 192
    sequestration, 170
    stereoidal, 165-166, 169-170
    toxic, 165
    tropane, 169
    ergot, 258
alkylpyrazine, 176
alpha1 receptor assay, 191
alpha2 receptor assay, 191
ALS, 327
Alternaria alternata, 325
    A. brassicae, 334, 331
    A. cassiae, 332
    A. citri, 336
Alteromonas, 170
allelochemical interaction, 188
Allium cepa, 26, 117, 120, 212

A. sativum, 262
allomone, 170
amanitine, 177
amaranthin, 319
Amaranthus albus, 91-93
    A. caudatus, 242, 319
    A. hybridus, 328
    A. palmeri, 241
δ-aminolevulinic acid, 210
2-aminophenol, 85, 89
ammodendrine, 192
amoebicidal, 195
amphibian, 172
Amphicallia bellatrix, 173
ampholyte, 282
Amyema sanguineum, 127
α-amylase, 224, 277, 280
    inhibitor, 272, 277, 280
    synthesis, 210
amyloplast, 300
anagyrine, 187, 192-193
analytical techniques, 149
anaphylactic reactions, 166
andrograpanine, 39
Andrographys paniculata, 39, 54-55
angelicin, 324
angustifoline, 193
p-anisic acid, 225
Ankistrodesmus braunii, 289-294, 302-307, 310
anostracans, 35, 45
antagonism, 321, 340
antagonist, 229
anthocyanin, 319
anthracnose, 335
anthraquinone, 13, 15, 16, 21-25, 31
antialgal, 43, 48
antiarrhytmic, 194
antibacterial, 16, 48, 176, 192
antibiotic, 315, 330
antifungal, 176
antihistamine, 16
antimalarial, 205, 207
antimicrobial, 265

antineoplasic, 259
antioxidant, 21
antiviral, 16
apex, 290, 295
Aphelandra tetragona, 81, 82
aphids, 168, 194
Aphis jacobaeae 168
apicoplastid, 207
Apocynaceae, 174
aposematic, 171
    compounds, 169
appresorium, 255
aquatic invertebrates, 45
    macrophytes, 35
    organisms, 35
    plants, 35, 315, 337
aqueous extract, 273, 275, 277
Arabidopsis seeds, 326
    A. thaliana, 28, 287
arbutin, 225
Arceutobium abietinum, 127
    A. americanum, 127
    A. pusillum, 127
    A. verticilliflorum, 127
arctiid moth, 175
arginine, 330
    biosynthesis, 329
Artemia salina, 35, 45, 48, 53, 272
Artemisia annua, 205
artemisinin, 201, 205-207
artifact, 323
asparagine, 205
asparagine synthetase, 201, 205
    inhibition, 105, 201, 205
aspartate amino transferase, 205
association, 171
Asteraceae, 174
asymmetric dihydroxilation, 112
    synthesis, 108, 136
ATP binding site, 21
    hydrolysis, 230
    production, 226, 230
ATPase activity, 242, 244

atropine, 191
autospores, 302
auxin, 248
    activity, 224, 264
Avena fatua, 22, 87, 93, 242
    seedlings, 242
Avena sativa, 87, 93, 242, 317, 325
    seedlings, 242
avicine, 177
AZOB, 240

B

bacteria, 64-67
bactericide, 265
bacto-agar medium, 309
barbatic acid, 15
barley, 261
    embryos, 210
    seedlings, 28
batrachotoxin, 164-165, 176
β-carboxy-*cis,cis*-muconic acid, 62
bean, 255, 263, 265, 275, 279-280
    roots, 271
beetle, 173, 194
Bemisia tabaci, 14, 25
bentgrass, 26
benzaldehyde, 225
benzo-1,4-quinone, 142
benzoic acid, 219-220, 225-226
benzoic acids, 217-227
benzoquinone, 142
benzoxazinoids, 78-83, 150-155
benzoxazinones, 84, 90-92, 241, 242
    biosynthesis, 79, 83
benzoxazolin-2(3*h*)-ones, 77, 239-240
benzoxazolinones, 77, 83-85, 92, 239-240
berbamine, 177
berberine, 166, 176-177
betulin, 210
bialaphos, 104, 205, 330
binding assay, 191
bioactiphore, 129

bioassay, 315
    selection, 319
    algae, 43, 310
    aquatic invertebrates, 45
    aquatic plants, 337
    crustacean, 53
    DNA-polymerase, 189
    dose-response, 26
    electrolyte leakage, 326
    enzyme target, 327
    germination, 323
    germination, 132, 142
    herbicide, 318
    hydroponic seedling, 331
    immunological methods, 339
    oat coleoptile, 317, 324
    patogenicity, 328
    Petri dish, 115-117, 120, 242
    photosynthesis inhibition, 327
    pigment analysis, 325
    plant growth, 324
    plant growth regulator, 319
    Protox, 28
    radicle growth, 281
    seed germination, 323
    seedling, 336
    standard, 321
    stimulatory actions, 337
    wheat coleoptiles, 117, 253, 258, 265
    whitefly, 17, 25, 29
bioherbicide, 329
biological control, 329
bioluminiscence, 336
bioremediation,. 325
biosynthesis, 187
γ-bisabolene, 108
bleaching, 13, 17, 20, 21
Blepharis edulis, 82
blight bungi, 335
blood system, 177
BOA, 78, 79, 82-86, 239-248
    treatment, 89
    -acetyl, 83

detoxification of, 86-87, 90-91
fate of, 89
BOA-6-O-glucoside, 86-92
content, 86
BOA-6-OH, 78, 87-92
BOA-N-glucoside, 86-88
Bold Basal Medium, 310
Boraginaceae, 174
bovine serum albumin, 190
β-oxoadipic acid, 62
β-phenylacetic acid, 241
BQ, 142
Brachionus calyciflorus, 35, 45-47, 52
Bradford method, 282
Brassica spp., 326
B. kaber, 155
brassinolide, 133
brassinosteroid, 132, 133
broadbean, 127, 139
Brongniartiaceae, 185
Broom, 194
Broomrape, 125, 135, 139
Brucine, 164
Bryozoo, 170
buffer, grinding, 28
resuspension, 28
Bufo marinus, 172
Bufonidae, 169
bulk soil, 67
2-butenoic acid, 4-carboxy-2-methyl, 130
γ-butyrolactone, 135
butterfly, 173

C

Ca^{2+} homeostasis, 302
uptake, 210
cadaverine, 187
caffeic acid, 59, 62, 218, 223-225, 294, 307
caffeine, 166, 176-177
California chaparral, 219
calyptra, 296, 300
calystegine A-3, 173

Callicarpa acuminata, 271-272, 277, 280
Callimorphine, 175
camphor, 203
Candida albicans, 265
C. antarctica lipase, 112
Capsaicin, 177
Capsella bursa pastoris, 90, 92
Capsicum annuum, 328
carbohydrate polymer, 255
carbon allocation, 14, 58
fixation, 226
flow, 217, 227
isotope ratio, 223
Carduus nutans, 89, 92
carotenoid biosynthesis, 13, 16, 19, 22
content, 19, 22, 26
carrot, 127
castasterone, 133
catalase, 223-224
catechin, 220, 230
catechol, 290, 295, 302-303, 306-307
Caulerpa prolifera, 259
Cecil A_p soil, 63-66
Cecil B_t soil, 62
celery, 127
cell, 326
cycle, 244, 247-248
analysis, 248
division, 295
expansion, 290
growth, 230
membrane, 223, 232
potential, 326
wall, 208
peroxidases, 142
cellular leakage, 207
organelles, 302, 306, 326
Centaurea cyanus, 90, 92
Cephalosporium gramineum, 336
cercosporin, 326
cereals, 79
chaetoglobosin k, 253, 258, 262
chaetoglobosins, 258

channel proteins, 223
chaparrinone, 210
Chara corralina, 260-261
chemical defense, 164
    ecology, 220
Chenopodium album, 70, 90-92, 241, 328
chickpea, 127
chitin, 329
Chlorella spp., 328
    C. fusca, 48
    C. pyrenoidosa, 229
chlorogenic acid, 218, 220, 223-225
chlorophyll, 17
    content, 17, 21, 226, 325, 336
    development, 325
    fluorescence, 226
    loss, 18
    production, 316
    reduction, 225
    synthesis, 209-210
chloroplast, 15, 187, 226, 229-232, 242, 302, 304-306, 327-328
    functions, 218
chromatography, 292
chromones, 15
cinchonine, 176
1,4-cineole, 104-105, 201-206
    cis-2-hydroxy-, 205
1,8-cineole, 202, 203
cinmethylin, 104, 201, 205-206
cinnabar moth, 173
cinnamic acid, 219-220, 225-227
cinnamic acids, 218-227
citokynin, 132
citrulline, 330
Citrullus vulgaris, 328
Citrus spp., 326, 336
CK2 enzyme, 21
Cladoceran, 35, 45, 52
Claviceps purpurea, 167
C-metaphase, 257, 260
C-NMR, 38, 39, 42, 184
CNS-depressor, 195
coca, 169

cocaine, 167, 169, 176
cockerel, 256
codon, 266
Coix lachryma jobi, 80, 82
colchicine, 176-177, 258, 260
coleoptile growth, 316, 324
Colletotricum fragariae, 335
    C. acutatum, 335
    C. demantium, 333
    C. gloeosporioides, 335-336
    C. magna, 335
    C. truncatum, 331-332
communication, 169-170
complementation studies, 201, 211
conditioning period, 133, 143
conductance, 223, 226, 228
conductivity, 211, 326-327
coniine, 176-177
connexin, 263
Consolida orientalis, 80, 81, 91-93
    C. regalis, 89, 92
Convolvulaceae, 127
coral reef fish, 170
coralyne, 177
coremata, 174
Coriandrum sativum, 89, 92
corn, 335
corn earworm, 171
cornexistin, 205
cortex, 296
cortical microtubules, 297
cotton, 128, 130, 338
cotyledon, 187, 325
cotylenin, 133
p-coumaric acid, 8-59, 63-69, 151, 154-156, 220, 223-226, 307, 322
    transformation of, 63
    utilization, 65
coumarin, 229
    4,7-dimethyl, 114
coumarins, 106, 132 217-220, 223, 227-232, 324
cowpea, 127, 130
m-CPBA, epoxidation, 114

creatonotine, 175
Creatonotus transiens, 174
cress, 259, 264
    inhibition, 117
cristae, 299-300
crosemperine, 173
cross resistance, 91
Crossandra pungens, 81
cross-linking, 223
Crotalarieae, 185
crotocal, 266
crotocin, 266
Crotolaria semperflorans, 173
crustacean, 35, 45
    toxicity test, 53
cucumber, 335
    cotyledons, 18, 27, 327
    seedlings, 27, 67, 70-74
Cucumis sativus, 27, 328, 335-336, 341
    C. melo, 335
Cucurbita pepo, 328
culture filtrate, 326
curcuphenol, 108
curcuquinone, 108
Cuscuta campestris, 127
Cuscutaceae, 127
cutin, 255
cutinase, 255
cyanobacteria, 221
cyclopentenone, 207-208
Cylindrocapsa, 303
Cylindrocarpon spp., 265
    C. destructans, 333
cysteine, 207
cyt *b6f*, 24
cytisine, 193, 195
Cytisus scoparius, 195
cytochalasin B, 253, 259-260
cytochalasin H, 253, 263
    deacetyl-, 258
    deacetylepoxy-, 258
    2,7,18-triacetoxy, 258
cytochalasins, 253, 258-264

cytochrome, 226
    P-450, 175
cytokinesis, 299, 303-304, 306
cytological responses, 291, 295, 301, 306
cytoplasm, 297
cytoplasmic cleavage, 258
    streaming, 253, 259-261
    vacuoles, 261
cytosine, 187
cytotoxicity, 218

D

danaidone, 174
Daniella oliveri, 36
Daphnia magna, 35, 45-48, 52-55
daphtoxkit FTM, 52
data analysis, 53
Daucus carota, 90, 92
deacetoxyscirpenol, 267
decarboxylation, 224
decomposition, 338
defense, 163, 166-171, 184, 219, 315, 329
    compounds, 195
defensive role, 164
dehydrozaluzanin C, 201, 207
demethylsorgolactone, 136
Dendropthoe curvata, 127
Dendrostibella, 265
densitometry, 271
    analysis, 281
deoxynivalenol, 267
depolarization, 222, 302
depsides, 14
depsidone, 14
depsone, 14
derivatization, 158
destruxin B, 334
desymmetrization, 112, 114
detergent activity, 210
deterrence, 169
deterrent, 15, 25, 163-176, 192, 195
detoxification, 45, 77-78, 86-92, 250, 280, 285, 325

capacity, 86, 90-92
DHBOA, 151-152
Diabrotica virgifera, 83, 163
diacetoxyscirpenol, 264, 266
dibenzofuran, 14-16
DIBOA, 78, 81, 84, 94-96, 151-152, 156, 240-242
DIBOA-Glc, 78-80, 82, 84
    7-chloro, 80
    degradation, 79, 82
dictyosome, 297
dieldrin, 262-263
digestive processes, 177
Digitaria sanguinalis, 87, 91-93, 241-242
digitonin, 210
dihydroparthenolide, 137
3,4-dihydroxyphenylacetic acid, 307
3,4-dihydroxyphenylglycol, 307
DIM_2BOA-Glc, 80-82
    degradation, 82
DIMBOA, 81-86, 149, 151, 156, 240, 242
DIMBOA-Glc, biosynthesis, 80, 82-84
    degradation, 82
2,5-dimethyl-3-isopentylpyrazine, 176
2,6-dimethoxy-1,4-benzoquinone, 141
dimethyl sulfoxide, 26, 52, 118, 120, 260, 320, 325
Diplotaxis tenuifolia, 91-92
disruption, 255
diterpene, 15, 35-39, 43-49, 106, 173, 201, 208-209
DMBOA, 4-chloro, 82
DMBQ, 142
DMSO, 26, 52, 118, 120, 260, 320, 325
DNA, 192
    interaction with, 188
    processing enzymes, 177
    topoisomerase, 210
DNA-polymerase, 177, 189, 229
    inhibition assay, 189
dopamine, 167
    receptor, 192
dormancy, 128
dormant bacteria, 64
dose-response, 26, 43, 156, 211

bioassay, 26
duckweed, 205, 325-327, 337
duvatriene-diol, 208

E

Echinochloa crus-galli, 208, 241
    E. indica, 241
Edman sequencing, 272
EDTA, 189-190
eggplant, 336
electrochemical potential, 222, 336
electrolyte leakage, 18, 20, 27, 207, 211, 267, 316, 326-328, 334
electron chain, 177
    microscopy, 207, 290, 297, 303, 307
    spray ionisation, 93
    transport, 15, 226, 229, 230-242, 242, 272
elicitor, 329
elongation inhibitors, 266
Eloria noysei, 169
ellipticine, 177
emetine, 177
emodin, 13, 15, 20-25
endoplasmatic reticulum, 298
    vacuoles, 261
endosulfan, 262-263
endosymbiont, 170
energy metabolism, 218, 230
    system, 225
enteric fluid, 171
ent-labdanes, 35-39, 47
    furano, 39
    γ-lactones, 39
    glicosides, 39
environmental conditions, 231, 280, 330
enzymatic action, 323
    cleavage, 78
    activity, 326
enzyme effects, 217, 224
ephedrine, 166
6-*epi*-cucurbic acid, 132
6-*epi*-9,10-dihydro-cucurbate, methyl, 132

epoxychalasin H, 258
12,13-epoxytrichothecene, 264
ergosterol, 15
ergotamine, 176, 258
ergothioneine, 167
erythrocytes, 190
esculetin, 227
Eschirichia coli, 28, 324, 330
ester hydrolisis, 174
ethane production, 316
ethylene, 130, 132, 338
    synthesis, 225, 228
etioplast preparations, 28
eucalyptus, 127
Euchresteae, 185
eudesmanolide, 137, 139
Euga rubipes, 169
eugenitin, 15
excised leaves, 336
exudate, 151, 154, 155
    collection of, 157
exudation, 338
exuded cytoplasm, 261

F

Fabaceae, 174, 221
fatty acid biosynthesis, 21
ferulic acid, 58-59, 62, 67-69, 151, 154-155, 158-159, 220-227, 231, 236, 294, 304, 307, 345
    transformation of, 62
fire ant, 172, 176
flavin, 242
flavonoids, 106, 217-218, 220, 230-232
    action, 218
flow cytometry, 244, 247-248
FMBOA, 84
food consumption, 7
formaldehyde, 308
Fragaria spp., 335
Fries rearrangement, 114
frog, 169
fulvic acid, 62

fumonisins, 325
fungal conidia, 335
    elicitor, 329
    metabolite, 133, 253
    pathogen, 329
    respiration, 226
fungi, 61-62, 67-74, 83, 167, 171, 184, 208, 226, 265, 272, 323, 329, 334-335
furanocoumarin, 324
furano-diterpene, 39
fusarenone, 267
Fusarium, 265
Fusarium equiseti, 264
    F. moniliforme, 253, 256, 325
    F. oxysporum, 333, 335
    F. solani, 255
fusicoccin, 133, 210

G

GABA receptor, 192
Galbraith buffer, 248
Galinsoga ciliata, 89, 90, 92
gallic acid, 220, 223-225
gallotannin, 229
gap junction, 262
GC/MS/MS, 149-150, 152
    analysis, 156, 158
gel electrophoresis, 281
    scan densitometry, 271
gelfiltration, 189
genes, 89, 105, 185, 261
geniculol, 48
Geniculosporium, 48
Genisteae, 185, 187
genistoid alliance, 185, 187
Gentiobioside carbamate, 78, 87-89
gentisic acid, 225-226
germ tube, 255
germination, 13-15, 21, 25, 89, 116, 119-120, 125-143, 175, 187, 192, 200, 209, 220, 237, 239-241, 249, 289, 291-294, 308-310, 320-325, 334, 338, 343

Index

data, 321
inductor, 128-130, 134-143
bioassay, 132, 142
gibberelic acid, 224, 229, 260
gibberelin, 128
    GA3, 132, 260, 319
    Synthesis, 132
Gibbonsia elegans, 170
gland secretion, 165
glaucarubolone, 210
D-glucitol, 2,5-anhydro-, 205
gluconeogenetic activity, 301
glucose carbamate, 78, 87, 89, 91
    mineralization, 63
β-glucosidases, 80
glucoside carbamate, 87-92
glucosylation, 45, 80, 87-88
glucosyltransferase, 87
glufosinate, 327, 330, 345
glutamine, 330
    Synthetase, 104, 327, 329
glutaraldehyde, 308
glutathione, 138, 207-208
    reduced, 207
    S-transferase, 280
        inhibitor, 272
glycine, 207, 267
Glycine max, 223, 328
goby fish, 169
Gossypium hirsutum, 128, 338
GR-24, 136, 139
GR-7, 136
gramine, 166
Gramineae, 240
green alga, 259, 291-292
green pea, 127
greenhouse, 257, 316, 330, 335
growth abnormalities, 203
    chamber, 26, 52, 310, 316, 330, 335
    inhibition, 207, 263
    inhibitor, 203
guaianolide, 139-140
gyrophoric acid, 15

H

Haliclona ?fascigera, 103, 106, 108
Hapalochlaena maculosa, 165, 169
harmaline, 176
haustorium, 126, 255
    induction, 131, 142
    inductors, 141
HBOA, 151, 152
HeLa cells, 259
heliannane, 103-120
    bioactivity, 115-117
    biosynthesis, 109
    total synthesis, 109-115
heliannuol A, asymmetric synthesis, 111-112
    total synthesis, 110-114
heliannuol D, asymmetric synthesis, 111-112
    biomimetic synthesis, 113-114
    total synthesis, 110-114
heliannuol E, synthesis, 112-114
heliannuols, 106-108
Helianthus annuus, 89, 92, 103
helibisabonol A, 109, 115-117
Helicoverpa zea, 171
Heliotropium europaeum, 91
Helminthosporium maydis, 326
hemiparasite, 126
hemp sesbania, 325, 331
Hendersonula toruloidea, 62
hepatotoxic, 174
herbicidal target site, 202, 327
herbicide-pathogen, 333
herbicide-resistant, 221
hervibores, 167, 184
Hibiscus trionum, 91, 92
histidine, 207
histological responses, 295, 301
HMBOA, 151, 152, 240
H-NMR, 37-40, 51-52, 93, 184, 256, 307
Hoagland's solution, 64
Holocanthone, 210
holoparasite, 126
homobatrachotoxin, 172

- 361 -

homogenization buffer, 282
Hordeum vulgare, 81, 117, 120, 261, 269
hormone regulation, 232
host, 125, 337
    recognition, 135
HPLC, 27, 29, 49-50, 64, 85, 89, 93, 120, 154, 292, 318, 340
    analysis, 29, 93, 292
    RP-18, 49-50
HPPD, 19-20, 28
    assay, 28
    inhibitors, 19
HT-2 toxin, 266
humic acid, 62
humidity, 330
humus phenolics, 219
hydantocidin, 205
hydraulic conductivity, 58, 218, 223
hydrocinnamic acid, 223-224
hydrogen peroxide, 142
hydroponic, 324, 336
hydroquinone, 130-131, 223
    2-methyl, 114
hydroxamic acid, 239-240
hydroxamic acids, 77, 78, 155, 239-242, 248
p-hydroxybenzaldehyde, 226
p-hydroxybenzoic acid, 58-59, 63-67, 151, 154-156, 159, 220, 223-226, 237, 307
β-hydroxybutyric acid, 241
hydroxydanaidal, 174
3-hydroxycyclobut-3-ene-1,2-dione, 256
16α-hydroxykaurane, 15
13-hydroxylupanine, 193, 195
p-hydroxyphenylacetic acid, 307
    3,5-dimethoxy, 307
4-hydroxyphenylpyruvate dioxigenase, 13, 16, 29
hydroxytyrosol, 307
Hydroxyurea, 244, 248
Hylidae, 169
Hyphae, 255
hypocothyl growth, 260, 294-295, 324
hypotensive, 195

I

IAA, 224, 317, 319
    oxidase, 224
    oxidation, 228
    induced growth, 224
IAS, 2
immunoblotting, 339
imidazolinone, 327
immunogold electron microscopy, 339
immunocytochemistry, 339
immunofluorescence, 339
immunological methods, 315, 339
indole-3-acetic acid, 224, 258, 260, 317
indolylcytochalasin, 253
infectivity, 336
initiation inhibitors, 266
inositol, 132
insect, 171, 173
    deterrent, 170
    toxicity, 192
insecticidal, 167
integrated pest management, 4, 25
International Allelopathy Society, 2
invertase, 224
ion balance, 218, 232
    channels, 177
    flux, 223
    pumps, 177
    release, 242
    retention, 223
    -trap, 150
    uptake, 222, 272
isoleucine, 327
isoquinoline, 173
isotox®, 25, 29
isovanillin, 225

J

Jania, 170
jasmonate, methyl, 132
jasmonates, 132

jasmonic acid, 132
juglone, 142
Julia coupling, 110

K

K⁺ absortion, 222
    channel, 183, 192, 195
kaempferol, 230
kelpfish, 170
[3H]-ketanserine, 191
α-ketoglutarate, 257
kidney bean, 277
    function, 177
Kochia scoparia, 224

L

labdane diterpenes, 35, 36, 45, 48
lactone, 39, 45, 406, 125, 128-129, 132, 135-140, 201, 207, 210
Lactuca sativa, 227, 241-242, 248, 325
ladybird beetle, 172
Lantana camara, 271-272, 275, 279
latent phase, 126
LC/MS/MS, 128-129, 150
leaching, 338
leaf disc, 326, 327
    bioassay, 25, 29, 328, 334
    expansion, 58-59, 70-72
    miners, 194
leakage, 17-18, 20-21, 27, 207, 211, 267, 316, 326-328, 334
lecanoric acid, 15
lecanorin, 15
lectin, 277
Legousia speculum veneris, 90, 92
legumes, 185
Lemna minor, 207, 225, 229, 327-328, 337
Lemna parasicotata, 325-326
lentil, 127
Lepidium sativum, 242, 259
Lepidoptera, 173

lettuce, 18, 19, 22, 26, 116, 117, 199-120, 127, 206-207, 241-249, 260, 264, 320, 325
    cotyledons, 18, 19
    germination, 116
    inhibition, 117
    seedlings, 227
leucine, 327
lichen, 13-16, 20, 23, 25-26, 29
    yellow-green, 16
lichexanthone, 15
light, 330
    cycle, 26
    microscopy, 290, 295, 302, 307
lignification, 219
Lilium longiflorum, 259
lily, 259
Liparieae, 185
lipid, 291, 297, 299
    peroxidation, 20, 223, 244
    production, 227
    solubility, 222
lithiation, 110
liver function, 177
lobeline, 177
log-linear, 318
logran, 115
Lolium multiflorum, 242
    L. perenne, 87, 93, 258
    L. rigidum, 149, 151, 154
Loranthaceae, 127
L-ornitine carbayltransferase, 329
Lotus corniculatus, 336
Lowry method, 190
LSD, 258
lupanine, 177, 187, 192-195
lupin mutants, 194
lupinine, 193
lupins, 183
Lupinus spp., 167
    L. albus, 195
Lycopersicon esculentum, 241, 261, 267, 319
lycopsamine, 173

M

maize, 17, 78-84, 127, 224, 326
    protoplasts, 326
    seedlings, 224
maleic hydrazide, 257
maltase, 224
mammalian toxicity, 169, 176
Manduca sexta, 167, 169
marine alga, 259
    organisms, 103, 108, 169, 184
mass spectrometry, 272
Matricaria chamomilla, 90, 92
matrine, 195
maytansine, 177
MBOA, 82, 83, 85, 240
    5-chloro, 82
Cl-MBOA, 240
6-MBOA, 240
MBQ, 142
melampolide, 139
melinone, 174
melting point determination, 188
membrane, 302
    disruption, 177, 208, 326
    effects, 217, 222, 316
    function, 231
    hormone-binding sites, 223
    integrity, 27, 192, 201, 210-211, 223
    lipids, 202, 210, 230
    permeability, 190, 192, 222, 242, 272, 326
    proteins, 201, 210
    transporters, 223
Menispermum dauricum, 129-130
meristem, 296
MES buffer, 309
mesophyll cells, 326
metabolic activation, 205
metabolism, 170
metabolization, 86, 90
methanol, 26, 27, 29, 93, 158, 282, 309
methionine, 132, 267
methoxy-1,4-benzoquinone, 142

methylcytisine, 193
methylen-γ-lactone, 207
methylgreen, 189
    assay, 189
Mg^{2+} absorption, 222
    chelatase, 225
mice, 256, 265
microalgae, 36, 289, 309
    culture, 308
microbes, 315
microbial activity, 57, 59, 60, 62, 65, 317
    compounds, 315
    enzyme activity, 61
    metabolism, 324
    metabolites, 254
    populations, 58-60, 62, 65-67, 69-70, 72
    transformation, 57, 59, 62
    rate of, 59
microbiotest, 52
microfilaments, 260
microfiltration, 307
microorganism, 315
microscopic techniques, 289
microsequencing, 280, 282
microtubule, 177, 258, 303
    disruption, 258
Michael acceptor, 137
    addition, 110, 132, 135-140
millet, 127, 325
mimosine, 325
mineral uptake, 222
mitochondria, 17, 202, 209, 214, 218, 226, 229-232, 242, 290, 297-303, 306, 311
mitochondrial function, 218, 226
    membrane, 230
    respiration, 202
    swelling, 300-302
mitosis, 201, 203, 210, 227, 257, 262, 303
mitotic index, 212, 244, 247-248
mode of action, 17, 21, 24, 77, 84, 110, 164, 192, 194-195, 201-202, 205-209, 217, 221-223, 228, 230, 232, 239, 242, 253-255, 258-260, 263, 266, 272, 280,

mode of action, 316, 318, 320, 327
modified proteins, 279
molecular assays, 316, 326
    modeling, 140
    target site, 202, 210
monilliformin, 256
Monomorium, 165
monoterpene, 201, 202
morphine, 172, 176-177
morphological responses, 291
Mosher's method, 39, 123
moth, 173
    larvae, 169
MS, 93
MS/MS/MS, 156
mulching, 7
multiflorine, 193
multinucleate cells, 303-304
multiple resistance, 86
mung bean, 207
muscarinic assay, 191
    acetylcholine receptor, 183, 192-193
muskmelon, 267, 335
mutagenic, 192
mycelium, 255, 329
myricetin, 220, 230
Myriophyllum spicatum, 328
Myrothecium, 265

N

Na^+ channel, 183, 192, 195
N-acetylchitosaccharide, 329
nanofiltration, 307
Naphtoquinone, 142
    5,7-dihydroxy, 141
naringenin, 230
necrosis, 176, 242, 264
nematodes, 255
neoandrographolide, 39
neurotransmitter, 167
N-glucosylation, 87
Nicotiana spp., 167

N. tabaccum, 229, 253, 257, 263, 265
nicotine, 166-169, 176-177
[3H]-nicotine, 191
nicotinic assay, 191
    acetylcholine receptor, 183, 191-193
nijmegen 1, 134
Nitelle translucens, 260-261
nitrate absortion, 222
    reduction, 227
nivalenol, 266-267
    4-acetoxy, 266
    3-acetyldeoxy, 267
NMDA receptor, 192
nodulating bacteria, 229
Nomuraea rileyi, 171
noradrenaline, 167
nor-diterpene, 39
n-oxides, 175
nuclear extrusion, 258
nucleus, 297
nudibranch, 170
nutrient stress, 323
    uptake, 58, 141

O

oat, 21, 86, 240, 324-325
    resistant, 21
o-coumaric acid, 155
octopus, 169
Odontomachus spp., 176
Olea europaea, 291
olefin metathesis, 110
oligosaccharide, 329
olive, 291
    oil, 291
    wastewater, 289-292
onion, 26, 117, 119-120, 127, 203-204, 207, 212, 262
    root tips, 203, 207, 262
organelles, 326
Orobanchaceae, 127
Orobanche spp., 128, 134

Orobanche aegyptiaca, 127
    O. cernua, 127, 139
    O. crenata, 127, 135, 139
    O. cumana, 127
    O. minor, 127-128, 132-133
    O. ramosa, 127, 139
    Inductors, 132
orobanchol, 128, 130
ortho metalation, 110
Oryza sativa, 219
Oscillatoria perornata, 221
osmotic potential, 231, 323
    pressure, 281
oxidation, 142, 174, 230, 257, 323
oxidative phosphorylation, 17
    Stress, 207
17-oxolupanine, 195
12-oxopotamogetonin, 39
17-oxosparteine, 193
oxygen electrode, 327-328
    evolution, 230, 316, 327-328
    singlet, 20
    uptake, 207, 226, 229-230
oxypeucedanin, 324

P

palladium, 110
Panicum milliaceum, 241, 325
pannaric acid, 15
Papaver rhoeas, 90, 92
papaverine, 177
Papilionoideae, 173, 185, 187
Paraformaldehyde, 308
parasite, 125
parasitic weeds, 125, 321, 337
pathogen virulence, 332
pathogenicity, 315
pea, 264, 335
peach, 326
pectin, 142, 291
pectinase, 255
peonidin, 141

peptidase, 330
peptidyl transferase, 253, 267
permeability, 232
    changes, 218
    transition pore, 302
peroxidase, 224
    activity, 223
    secretion, 207
    synthesis, 207
peroxidation, 326
pest resistance, 3, 85
pesticides, 3-4, 25, 231, 202, 253
pharmaceutical, 210
pharmacological properties, 194
pharmacophagy, 174
phaseolotoxin, 329
Phaseolus vulgaris, 255, 257, 263, 265, 277
phenolic acid, 57-62, 149-158, 161, 217-232
    as carbon source, 65-68
    depletion, 61
    compounds, 217
    content, 154
    fate of, 57
phenoxazinones, 85, 89, 96
phentolamine, 191
phenylalanine, 219
phenylpropanes, 219
phenylpropanoid esters, 142
    pathway, 219
pheromone, 174
    sex, 174
Phleum pratense, 242
Phloem, 141
    feeding insects, 14, 16, 25
Phomopsis sp., 263
Phoradendron bolleanum, 127
    P. juniperinum, 127
phosphate absorption, 222
    citrate buffer, 118
phosphinothricin, 104, 327, 330
photobleaching, 17, 325
photoperiod, 310
photophosphorylation, 230, 242

photosensitization, 324
photosynthesis, 24, 58, 126, 218, 222, 225-226, 229-230, 232, 272, 316, 320, 327-328
    rate of, 218, 229, 232
Phyllobates, 165
physiological action, 222
physodic acid, 15
Physostigmine, 177
phytoalexin, 329
phytoene desaturase, 19
Phytohormone, 132
    activity, 222
    interaction, 217, 224
Phytophthora citrophthora, 326
    P. drechsleri, 336
    P. infestans, 329
phytotoxicity, 315
picrolichenic acid, 15
pine, 127
piperidine, 2,6-dialkyl, 172, 176
piperine, 166
pipes buffer, 308
Pipidae, 169
Pistia stratiotes, 36
Pisum sativum, 335
Pitohui dichrous, 172
plant breeders, 194
    growth regulator, 317, 319
    recognition, 125
Plantago major, 90-91, 93
plant-microbe interaction, 329
plant-water, balance, 222
    relationship, 218
plasma membrane, 206-207, 211, 222-223, 229, 244
plasmid, 189
Plasmodium spp., 205, 207
plastid, 207, 297, 299
pneumotoxic, 174
Poaceae, 77-81, 86-87, 91-93
Podalyrieae, 185
podolactones, 209-210
polyalcohols, 291
polyethylene glycol, 231

Polygonum aviculare, 89-90, 93
polyphenol oxidase, 224
polyphenols, 291, 307
polyporic acid, 15
polysaccharide, 329
pollen bioassay, 334
    germination, 334
    tube growth, 334
porcin, pancreatic lipase, 114
    brains, 190
portsmouth soil, 59, 66-70
Portulacca oleracea, 91-93
Potamogeton natans, 35-36, 39, 48, 50, 53
potato-X, virus, 192
Potomagetonaceae, 35-36, 44-46
[3H]-prazosine, 191
precoccinelline, 172
pre-germination, 157
procambial area, 295
proherbicide, 205
protease, 224
protection, 184
protein, 273-280, 299
    bodies, 297, 299
    content, 282
    expression, 272
    extraction, 281-283
    function, 232
    microsequencing, 282
    patterns, 271-272
    synthesis, 132, 177, 192, 227, 229, 253, 266
        inhibition assay, 190
proteomic techniques, 271
protocatechuic acid, 58-59, 62, 225, 307
proton efflux, 210
    pumps, 223
protoplast, 326, 334, 336
protoporphyrinogen oxidase, 17-18
Protox, 17, 20, 28
    assay, 28
Pseudoguaianolide, 137, 139
Pseudomonas spp., 170
    P. syringae, 329-330

Psiadia altissima, 36
PSII, 13, 24, 226, 229-230
    Inhibition, 24
    Inhibitor, 131
psoralen, 324
psudane, 177
Pterophoridae, 173
Puccina recondita, 336
puffer fish, 169
pulvinic acid, 15
pumpkin cotyledons, 327
    leaves, 29
purine, 326
pyrenoid, 302
pyrrolidines, 2,5-dialkyl, 165, 172
    2,6-dialkyl, 166
pyrrolizidine, 174
pyruvate, 257

Q

quantification, 158
quassinoid, 201, 210
Quercetin, 220, 230
quiescent states, 60, 62, 64, 65
quinine, 164, 175, 177
quinone, 125, 130, 141
[3h]-quinuclidinyl benzylate, 191

R

radical, 324
radicle growth, 271, 275-280, 294
radioactivity, 191
radish, 259, 289-292, 306
ragwort, 173
Ranidae, 169
Ranunculaceae, 77-78, 80, 90-93
Raphanus sativus, 259, 291, 307
receptor binding studies, 190
red alga, 170
red clover, 129-130
redfir, 127

replication, 320-321
reproducibility, 319
resistance, 3, 20, 25, 77, 78, 83, 85-86, 91, 105, 219, 222-223, 226, 229, 240, 249, 326, 335-336
respiration, 207
    rate, 218, 232
respiratory chain, 272
respirometer, 225
response curve, 318
    time, 320
retronecine, 175
reversal studies, 205
reverse osmosis, 307
    transcriptase, inhibition assay, 189
reversion, 207, 224, 261
reynosin, 137, 139
rhexifoline, 173
Rhizobia, 185
rhizosphere, 57-60, 66-67, 69-72, 336, 340
    bacteria, 69-70, 75
    microbial population, 57-59, 69-72
    microorganisms, 57, 59-60, 70, 338
rhodocladonic acid, 23-25
    analogues, 23-25
ribosome, 60s, 253, 266
rice, 219
m-RNA, 266
RNA polymerase, 177
    processing enzymes, 177
Roboastra tigris, 170
Roccellaceae, 23
root apex, 296, 300
    exudates, 59, 128-129, 149, 153-155, 157, 250, 358
    exudation, 77, 80, 85
    hairs, 261
    malformation, 25
    meristems, 244
    nodules, 185
    tip elongation, 336
    tip homogenization, 281
    tips, 212, 262
roridin E, 267
Rotifer, 35, 45, 47-48, 52-53

rotoxkit FTM, 52
Ruppia maritima, 35, 36, 38
rust fungus, 255
rutin, 230-231
rye, 78-80, 83, 85, 91, 154, 156-157, 167, 240-241, 249, 251
    seedlings, 240-241
ryegrass, 154

S

Saccharum officinale, 81
salamander, 165, 169
Salicylic acid, 219, 223, 225-226, 231
samandarine, 165, 176
santamarin, 137, 139
saponin, 201, 210
sar, 103-104, 110, 129-130, 134, 136
saxitoxin, 177
Scenedesmus spp, 328
    S. acutus, 325
scent organ, 174
Scirpentriol, 264, 266
Scoparia dulcis, 82
Scopoletin, 132, 155, 220, 227-229
Schiff's reagent, 249
Schrophulariaceae, 77-78, 82, 127, 141
sea slugs, 170
Secale cereale, 80-82, 90, 249
seed bioassay, 309
    conditioning, 126
    dispersion, 126
    germination, 289-292, 316, 321, 337
    imbibition, 295
seedling, 260, 319
    growth, 289, 291
Selectivity, 319, 334
Selenastrum capricornutum, 35, 43, 47, 52, 229-230, 328
Senecio jacobaea, 168, 173
    S. vulgaris, 328
senecionine, 173
senescence, 132
Senna obtusifolia, 331

sensitivity, 319
Sephadex G, 189
    LH-20, 49-50
septa, 290, 302, 306
sequential cropping, 219
serotonin, 167
$serotonin_2$, receptor assay, 191
Sesbania exaltata, 331
sesquiterpene, 103, 106, 108, 125, 128, 132, 137-138, 201, 205, 207
    lactone, 106, 125, 132, 137-138, 201, 216
    endoperoxide, 205
Sessibubula translucens, 170
shoot growth, 331
sicklepod, 325, 331
signal compound, 184
simple phenols, 218
sinapic acid, 307
sinking disc technique, 328
sirodesmin, 326
sodium ascorbate, 28
soil bacteria, 336
    microbial population, 60, 62, 65-66
    microorganisms, 59, 61, 63-64, 70, 317, 338, 344
    respiration, 61-62
    transportation, 151
solenopsine, 177
Solenopsis spp., 165, 166, 172
solvent, 323
sophoreae, 185
sordinone, 15
sorghum, 127-131, 219, 338
Sorghum bicolor, 128, 130, 338
sorgolactone, 125, 128, 130-131, 134
sorgoleone, 125, 130, 338
soybean seedlings, 223, 228, 230
sparteine, 173, 177, 187, 192-195
spasmolytic, 16
Spinacia oleracea, 328
spindle microtubule, 258
sponge, 106, 108, 121, 124
sporangia, 302
spore, 255

concentration, 332
starch, 297-300
statistical analysis, 320
    treatment, 120
stem collapse, 331
sterilization, 157
steroid, 15, 132, 133, 165
stock solution, 26, 143, 309
stomatal closure, 228, 229, 232
    conductance, 225
    function, 218, 222
stomates, 255
storage, 187
strawberry, 335
Streptomyces spp., 104, 330
stress, 221, 231, 271, 289, 302, 323
    proteins, 280
Striga asiatica, 127, 128, 137, 337
    S. gesnerioides, 127
    S. hermonthica, 127-128, 132-135
strigol, 128, 130, 338
strigolactone, 128-129, 134, 140
structural alterations, 290, 302
Structure-Activity Relationship, 24, 103-104, 129, 134, 220, 318
strychnine, 166, 176
Student's t test, 45
Stylosanthes scraba, 335, 336
succesional sequence, 219
succinic acid, 62
sucrose, 118, 190
sugars, 291
sulcotrione, 17, 19-20
sulfhydryl groups, 223
sulfhydryl-containing, 207
Sulfonylurea, 327
sunflower, 17, 66, 69-71, 103, 106, 115, 127, 139-140
sustainable agriculture, 3, 4
sweet lupins, 194
swollen mitochondria, 299
    root tips, 203
SXSg, 130-131
symbiosis, 13-14, 167

symbiotic, 185
    microorganism, 184
synephrine, 173
synergism, 151, 231, 321, 340
syringaldehyde, 225
syringic acid, 58-59, 63, 142, 151, 154-155, 159, 225
systemic acquired resistance, 219

T

T-2 toxin, 266-267
tabtoxin, 329
taiga soil, 63
Tambja abere, 170
    T. eliora, 170
tambjamine, 170
tannic acid, 229, 308
tannins, 217, 229, 291
Tapinantus buchneri, 127
target, 176, 192, 201
tassel flower, 319
taxol, 177
TEM, 290
temperature stress, 323, 330
template, 317
termination inhibitors, 266
terpenoid, 201
terphenylquinone, 15
tetrafluorobenzo-1,4-quinone, 142
tetrahydrorombifoline, 193
tetrandrine, 177
tetrodotoxin, 164-165, 169
TFBQ, 142
Thamnocephalus platyurus, 35, 45-47, 52
thamnotoxkit ftm, 52
theobromine, 177
theophylline, 177
thermopsideae, 185
thief ant, 172
thylakoid, corn, 56
    spinach, 23
tiger moth, 173-174
tissue slices, 326

Index

TLC, 36, 49-51, 340
tobacco, 127, 139, 208, 229, 253, 257, 263, 265
    cells, 326
    seedlings, 263
    hornworm, 169
        larvae, 167
tolerant, 168
tomatine, 171
tomato, 127, 261, 267, 275, 277-280, 319, 336
    roots, 271, 280
toxicity, 35-36, 4348, 52,-53, 165, 169, 171-177, 220, 232, 262, 267, 292, 295
    test, 43, 52
trans,trans-germacranolide, 137, 139-140
transesterification, 112
transgenic crops, 105
    microorganisms, 105
translocation, 187
transpiration, 17, 58, 228, 336
    inhibition 17, 228
transport process, 177
trichdermol, 266
trichloroacetic acid, 190
Trichoderma spp., 265
    T. viride, 265
trichodermin, 265-266
trichodesmine, 173
Trichomes, 205
trichothecenes, 253, 264-266
trichothecin, 266
Trichothecium, 265
trichothecolone, 266
trichoverrin B, 265
Trifolium pratense, 128
Triketone, 16-17, 19
Triterpene, 15, 106, 201, 210
Triticum aestivum, 80-82, 87, 90, 93, 117, 149-151, 155, 257, 336
T-toxin, 326
Tubulosine, 177
Tukey's test, 310
turfgrass, 336

turgor pressure, 223
two-dimensional electrophoresis, 271
13-tigloyl-oxylupanine, 192
Tyria jacobaeae, 168, 173
tyrosine protein kinase, 21
tyrosol, 307

U

ubiquinone, 242
ultrafiltration, 307
ultrastructural alterations, 290, 302
umbelliferone, 220, 223, 227-228, 231
uncoupler, 230
unpalatability, 169
Uromyces appendiculatus, 255
Urtica urens, 89-90, 93
usnic acid, 13-20, 25
uterus contractor, 195

V

vacuole, 187, 297, 299
valine, 327
vanillic acid, 58-59, 62, 63, 67-68, 72, 151, 154, 155, 159, 220, 225-226, 307
vanillin, 223, 225
vector, 255
venom, 165, 169, 176
venom gland, 165
Verbenaceae, 271
verrucarin A, 265-266
verrucarol, 266
vertebrate toxicity, 192
Vibrio, 170
Vicia faba, 86, 88, 92, 242
    BOA sensitive, 88
Vigna unguiculata, 128
vinblastine, 177
vincrastine, 177
Viscaceae, 127
vulpinic acid, 14, 25

W

water tongue, 36, 39
    balance, 223
    content, 232
    potential, 223, 228
    relations, 217, 223, 230
    stress, 228, 265, 323
    uptake, 229
wax, 21
western corn rootworm, 83, 95
    juniper, 127
wheat, 3, 66-69, 78-80, 83, 85, 91, 103, 117, 120, 149-160, 240, 249, 263-264, 302, 336-337
    accesions, 154, 157
    coleoptile, 117, 253, 257-258, 262-263, 265
    cultivars, 155-156
    exudates, 158
    seedlings, 151
white spruce, 127
whiteflies, 17,25
wilt-capacity, 336
wilting, 265, 336
wilt-resistance, 336
witchweed, 125, 135, 137, 337-338
wittig olefination, 110
wounding, 194

X

XAD-4 resin, 151
Xanthomonas campestris, 326, 336
xanthones, 15
xanthotoxin, 5-nitro, 324
xenobiotics, 338
xenognosin, 141
X-ray crystallography, 256
xylem parenchyma, 298

Y

[3H]-yohimbine, 191

Z

Zea mays, 80-82, 86-87, 93, 224, 241, 256-257, 263, 265, 335
zeorin, 15